Structural Adhesives

Scrivener Publishing
100 Cummings Center, Suite 541J
Beverly, MA 01915-6106

Adhesion and Adhesives: Fundamental and Applied Aspects

The topics to be covered include, but not limited to, basic and theoretical aspects of adhesion; modeling of adhesion phenomena; mechanisms of adhesion; surface and interfacial analysis and characterization; unraveling of events at interfaces; characterization of interphases; adhesion of thin films and coatings; adhesion aspects in reinforced composites; formation, characterization and durability of adhesive joints; surface preparation methods; polymer surface modification; biological adhesion; particle adhesion; adhesion of metallized plastics; adhesion of diamond-like films; adhesion promoters; contact angle, wettability and adhesion; superhydrophobicity and superhydrophilicity. With regards to adhesives, the Series will include, but not limited to, green adhesives; novel and high-performance adhesives; and medical adhesive applications.

Series Editor: Dr. K.L. Mittal
P.O. Box 1280, Hopewell Junction, NY 12533, USA
Email: raareviews@gmail.com

Publishers at Scrivener
Martin Scrivener (martin@scrivenerpublishing.com)
Phillip Carmical (pcarmical@scrivenerpublishing.com)

Structural Adhesives

Properties, Characterization and Applications

Edited by

K.L. Mittal

and

S.K. Panigrahi

Scrivener Publishing

WILEY

This edition first published 2023 by John Wiley & Sons, Inc., 111 River Street, Hoboken, NJ 07030, USA and Scrivener Publishing LLC, 100 Cummings Center, Suite 541J, Beverly, MA 01915, USA

For more information about Scrivener publications please visit www.scrivenerpublishing.com.

Wiley Global Headquarters
111 River Street, Hoboken, NJ 07030, USA

For details of our global editorial offices, customer services, and more information about Wiley products visit us at www.wiley.com.

Library of Congress Cataloging-in-Publication Data

ISBN 978-1-394-17472-0

Cover image: Pixabay.com
Cover design by Russell Richardson

Set in size of 11pt and Minion Pro by Manila Typesetting Company, Makati, Philippines

Printed in the USA

10 9 8 7 6 5 4 3 2 1

Contents

Part 2: Applications 291

Preface

Adhesive bonding is utilized for mundane (e.g., gluing of toys) to highly-sophisticated applications. Structural adhesives constitute a special class of adhesives, as these must possess certain properties/characteristics (e.g., low shrinkage, low creep and good strength retention under system load, long pot-life, and good wetting property) to fulfill their intended function and performance. A structural adhesive can simply be described as a high-strength adhesive material which is isotropic in nature and bonds two or more parts together in a load-bearing structure. A structural adhesive material must be capable of transmitting the stress/load without loss of structural integrity within design limits. There are many types of established structural adhesives including epoxy, urethane, acrylic, silicone, etc.

A glance at the literature will evince that besides conventional structural adhesives, there is much interest and high tempo of research in harnessing nanotechnology (e.g., use of nanoparticles) in ameliorating the existing structural adhesives or devising new improved variants.

This book contains nine chapters and is divided into two parts: Part 1: Preparation, Properties, and Characterization; Part 2: Applications. The topics covered include: structural epoxy adhesives; biological reinforcement of epoxies as structural adhesives; marble dust reinforced epoxy structural adhesive composites; characterization of various structural adhesive materials; effects of shear and peel stress distributions on the behavior of structural adhesives in tubular composite joints; inelastic response of structural aerospace adhesives; structural reactive acrylic adhesives: their preparation, characterization, properties, and applications; application of structural adhesives in composite connections; and naval applications of structural adhesives.

The book reflects the cumulative wisdom of a number of internationally-renowned researchers actively engaged in the arena of structural adhesives. The book is profusely referenced and copiously illustrated. This book should be of much use and interest to adhesionists, materials scientists, adhesive technologists, polymer scientists, and those working in the construction, railway, automotive, aviation, bridge, and shipbuilding industries.

As more advanced and improved structural adhesives become available, new vistas will emerge for the use of structural adhesives. In this vein, it is expected that the utilization of various nanomaterials as reinforcement of structural adhesives will play an important role and will prove very beneficial. Also, we anticipate development of environmentally-friendly and "green" structural adhesives.

Now it gives us great pleasure to acknowledge all those who played essential roles in giving this book a body form. Naturally, first and foremost, our profound thanks go to the authors for their keen interest, sustained enthusiasm, unwavering cooperation, and sharing their valuable research experience in the form of written accounts (which essentially provided the grist for this book), without which this book could not be materialized. Also, the steadfast interest and whole-hearted support of Martin Scrivener (publisher) in this book endeavor is highly appreciated.

K.L. Mittal
Hopewell Junction, NY, USA
email: raareviews@gmail.com

S.K. Panigrahi
Defence Institute of Advanced Technology, Pune, India

January 2023

Part 1

PREPARATION, PROPERTIES AND CHARACTERIZATION

1

Structural Epoxy Adhesives

Chunfu Chen

Henkel Technology Center – Asia Pacific, Henkel Japan Ltd., Isogo-ku, Yokohama, Kanagawa, Japan

Abstract

Epoxy adhesives form very strong and durable bonds with most materials and are widely used in various structural bonding applications. Epoxy adhesives are formulated in the form of a compound containing epoxy resin and curing agent, catalyst with modifiers and additives. Typical epoxy resins include glycidyl ether epoxy resins such as Bisphenol A, Bisphenol F and novolac type, and glycidyl amine epoxy resin, glycidyl ester epoxy resin as well as cycloaliphatic epoxy resin. Epoxy adhesives are supplied in both one-component and two-component packages depending on curing agent used and curing method applied. Two-component epoxy adhesives are prepared by packaging epoxy composition and curing agent composition separately. Almost all room temperature cure epoxy adhesives are supplied in two-component packages. One-component epoxy adhesives are prepared and supplied by mixing all formulated components in advance including epoxy resin and curing agent. One-component epoxy adhesives usually need cure at elevated temperature and storage at low temperature conditions. Typical room temperature cure epoxy adhesives, thermal cure epoxy adhesives and UV cure epoxy adhesives are described. Toughened high performance epoxy adhesives, low temperature cure one-component epoxy adhesives, instant bonding epoxy adhesives and sustainable epoxy adhesives are recent progressions and new trend in epoxy adhesive technology developments.

Keywords: Epoxy adhesive, structural bonding, two-component, one-component, toughened, instant bonding, sustainable epoxy

E-mail: chunfu.chen@henkel.com

K.L. Mittal and S.K. Panigrahi (eds.) Structural Adhesives: Properties, Characterization and Applications, (3–30) © 2023 Scrivener Publishing LLC

1.1 Introduction

Epoxy resins are one of polymer materials containing at least one carbon-oxygen-carbon three ring known as the epoxy group, epoxide or oxirane. Epoxy resin was initially discovered in late 1890's. It was first synthesized by N. Prileschajew in 1909 via oxidation of olefin with benzoic acid peroxide. The first epoxy adhesive was invented in 1936 by Dr. Pierre Castan for dental application via curing bisphenol A epoxy resin with phthalic anhydride. Commercialization of epoxy adhesives was started around late 1940's in Europe and USA [1, 2]. Various epoxy adhesives have been developed and commercialized since then. Epoxy adhesives are widely used as typical reactive adhesives for various structural bonding applications ranging from general industry, construction, electronics assembly, automobile production to aerospace and defense markets [3–20]. Major global suppliers for epoxy adhesives are Henkel AG & Co. KGaA, H.B. Fuller Company, 3M, Huntsman Corporation, Sika Corporation, Arkema Corporation, Cemendine Co., Ltd., Three-Bond Co., Ltd., Huitian Adhesive, etc.

Epoxy adhesives show good adhesion to a wide range of materials such as metals, glass, concrete, ceramics, wood and many plastics. Curing shrinkage is very low. Cured epoxy resin possesses strong and rigid cross-linked chemical structure especially well-suited for structural bonding applications. By combination of various epoxy resins with different curing agents, a number of epoxy adhesives have been commercialized for various applications. On the other hand, room temperature and thermal cure epoxy adhesives need relatively long cure time. Most cured epoxy adhesives are rigid and thus are not suitable for bonding flexible substrates. In selection and use of epoxy adhesives, their pot-life, cure condition, cure method, physical properties of un-cured and cured resin as well as adhesion and reliability performance should be taken into consideration.

1.2 Epoxy Adhesive Chemistry

1.2.1 Epoxy Resins

Epoxy resins are usually synthesized from the reaction of active hydrogen in phenols, alcohols, amines or acids with epichlorohydrin at certain well-controlled conditions. Epoxy resins can also be prepared by oxidation of olefin with peroxide as in the case of cycloaliphatic epoxy resins. Typical epoxy resins are glycidyl ether resins such as Bisphenol A, Bisphenol F and novolac type, glycidyl amine epoxy resin, glycidyl ester epoxy resin and cycloaliphatic epoxy resin.

Bisphenol A glycidyl ether, often called DGEBA, was the first commercialized epoxy resin. It is still the most standard and widely used epoxy resin, constituting the majority, estimated over 75% in sales volume, of all epoxy resins used today. Bisphenol A epoxy resin is normally synthesized by the reaction of bisphenol A and epichlorohydrin at 70 – 80°C in alkaline condition as illustrated in Figure 1.1 [21]. Various grades of bisphenol A epoxy resins in either liquid or solid state with different viscosity or melting point, EEW (abbreviated for epoxy equivalent weight) and purity have been commercialized and supplied by all major epoxy resin manufacturers. Chemical structure and key features of functional groups of bisphenol A epoxy resin are illustrated in Figure 1.2 [22], indicating potential balanced properties of good reactivity, good chemical and thermal resistance as well as high adhesion.

Bisphenol F glycidyl ether, often called Bisphenol F epoxy resin, is synthesized by the reaction of bisphenol F and epichlorohydrin. Figure 1.3 shows its typical chemical structure. Bisphenol F epoxy resin has lower viscosity, better solvent and chemical resistance as compared to standard bisphenol A epoxy resin.

Novolac glycidyl ether, often called novalac epoxy resin, is synthesized by the reaction of novalac phenol and epichlorohydrin. Figure 1.4 shows its typical chemical structure. Novolac epoxy resin has multi-functional

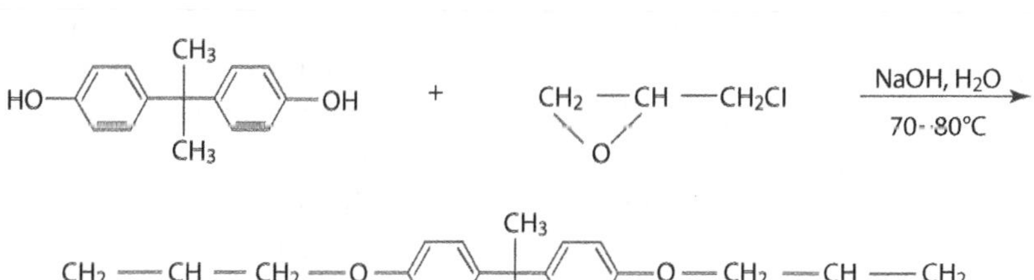

Figure 1.1 Synthesis of DGEBA from bisphenol A and epichlorohydrin.

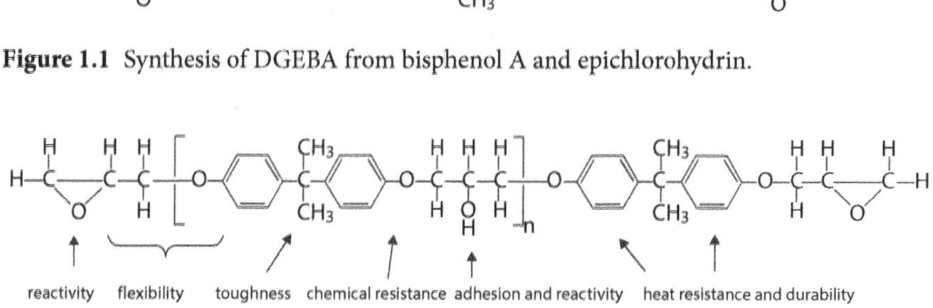

Figure 1.2 Chemical structure and key features of DGEBA.

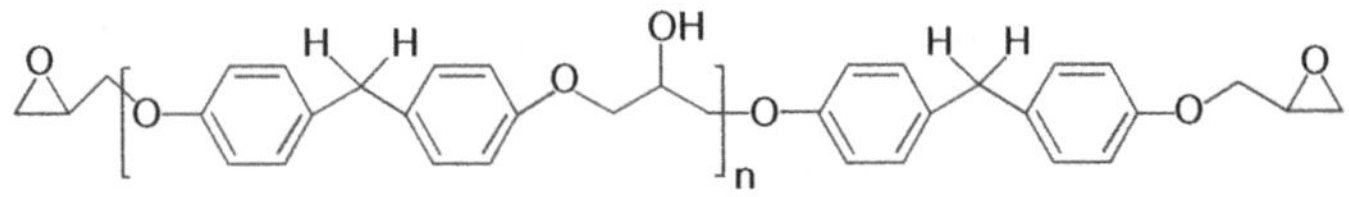

Figure 1.3 Chemical structure of bisphenol F epoxy resin.

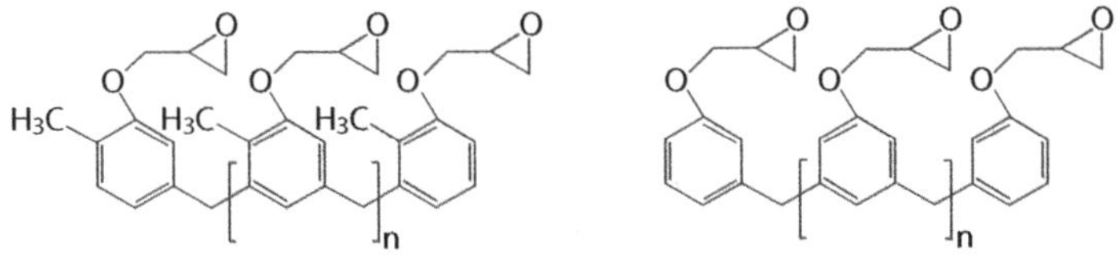

Figure 1.4 Chemical structures of cresol (left) and phenol (right) novolac epoxy resins.

epoxy groups and possesses especially high thermal resistance. Novolac epoxy resin is usually used in combination with bisphenol A or bisphenol F epoxy resin due to its high viscosity.

Glycidyl ether epoxy resins, synthesized from alcohols and epichlorohydrin, usually have much lower viscosity and are used mainly as reactive diluent to lower viscosity for better handling property adjustment.

Glycidyl amine epoxy resin is synthesized by the reaction of amino compound with epichlorohydrin. Figure 1.5 shows chemical structures of tri-functional and tetra-functional glycidyl amine epoxy resins. Glycidyl amine epoxy resins with less than tri-functional epoxy groups have low viscosity and are often used as epoxy diluent especially for applications requiring high thermal resistance. Tetra-functional glycidyl amine epoxy resin possesses extremely high glass transition temperature and excellent mechanical properties after full cure, suitable for use as base resin in applications requiring high thermal resistance and high performance.

Glycidyl ester epoxy resin is prepared by the reaction of carboxylic acid with epichlorohydrin. Figure 1.6 (left) illustrates chemical structure of typical glycidyl ester epoxy resin. It is often cured by anhydride, offering good

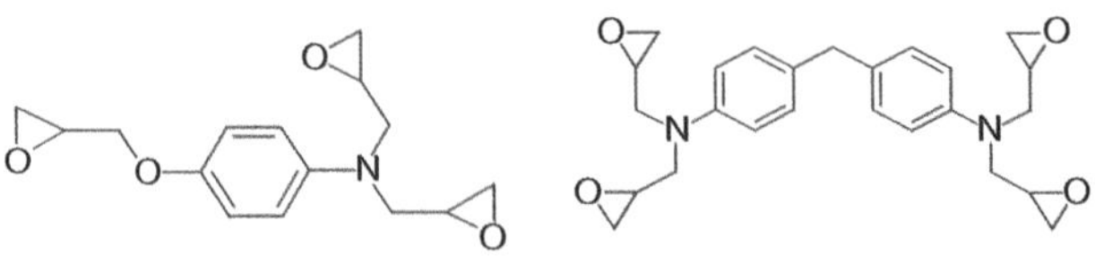

Figure 1.5 Chemical structures of tri- (left) and tetra-functional (right) glycidyl amine epoxy resins.

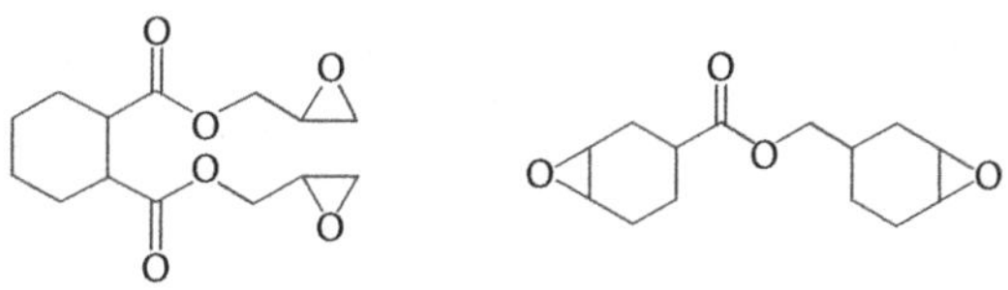

Figure 1.6 Chemical structures of glycidyl ester (left) and cycloaliphatic (right) epoxy resins.

insulation, high thermal resistance and good UV resistant performance. Cycloaliphatic epoxy resin is usually prepared by the oxidation of olefin with peroxide. Figure 1.6 (right) illustrates chemical structure of typical cycloaliphatic epoxy resin. Cycloaliphatic epoxy resin is principally cured with anhydrides or cationic initiators, offering good weathering and thermal resistant performance.

1.2.2 Curing Agents and Catalysts

Epoxide group is chemically very active. Epoxy resin can react, almost equivalently, with active hydrogen in polyamines, mercaptan compounds, phenols and anhydrates via polyaddition mechanism at certain conditions to become cross-linked strong thermoset polymers. Epoxy resin can polymerize homogeneously via anionic polymerization mechanism by initiating it with Lewis bases such as tertiary amines or imidazole compounds. It can also polymerize via cationic polymerization by initiating it with Lewis acids such as onium salts, iodonium salts. Table 1.1 lists typical curing agents, catalysts and initiators used for epoxy adhesives.

Table 1.1 Typical epoxy resin curing agents, catalysts and initiators.

Polymerization mechanism	Curing agent, catalyst
Polyaddition	Polyamines Modified polyamines Mercaptans Phenols Anhydrides
Anionic	Tertiary amines Imidazole compounds
Cationic	Onium salts Iodonium salts

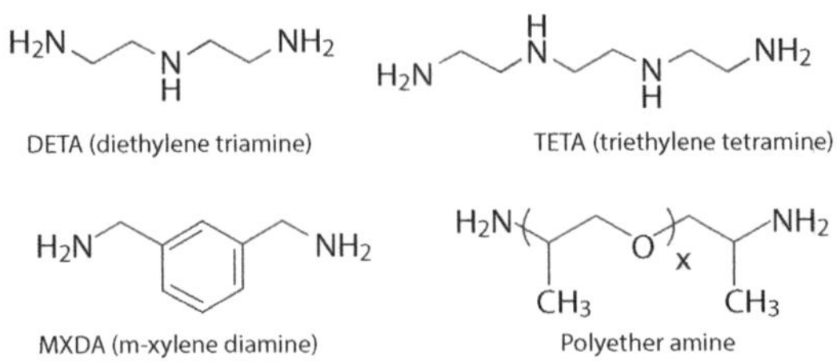

Figure 1.7 Chemical structures of typical aliphatic polyamine curing agents.

Polyamines including aliphatic, cycloaliphatic and aromatic types, are the most widely used curing agents for epoxy adhesives. Active hydrogen in aliphatic polyamines can rapidly react with epoxy resins at room temperature. Figure 1.7 shows chemical structures of typical aliphatic polyamines, i.e., DETA, TETA, MXDA and polyether amine. Cycloaliphatic amines can also react with epoxy resins at room temperature but need post thermal curing to achieve full cure with higher Tg suitable for high temperature resistant applications. Figure 1.8 shows chemical structures of typical cycloaliphatic polyamines, IPDA and PACM. Aromatic amines such as DDM (diamino diphenyl methane) normally require thermal curing with epoxy resin, offering high Tg and better mechanical properties.

Modified polyamines are prepared to improve handling properties and to lower toxicity as well as provide further improved performance of polyamines. Various grades of modified polyamines have been commercialized and supplied by major epoxy curing agent suppliers. Table 1.2 lists typical modified polyamines, their synthesis methods and key features. Polyamide type curing agent is a condensation reaction product of dimeric acid with excessive polyamine. Polyamide has high viscosity, low toxicity and long pot-life, offering good adhesion, good flexibility and toughness. Amidoamine type curing agent is the condensation reaction product of monobasic fatty carboxylic acid with excessive polyamine. Amidoamine has viscosity between polyamide and polyamine, offering good adhesion and good chemical resistance. Amine adduct is an adduct reaction product of epoxy resin with excessive amine. Amine adduct possesses good reactivity, low volatility and

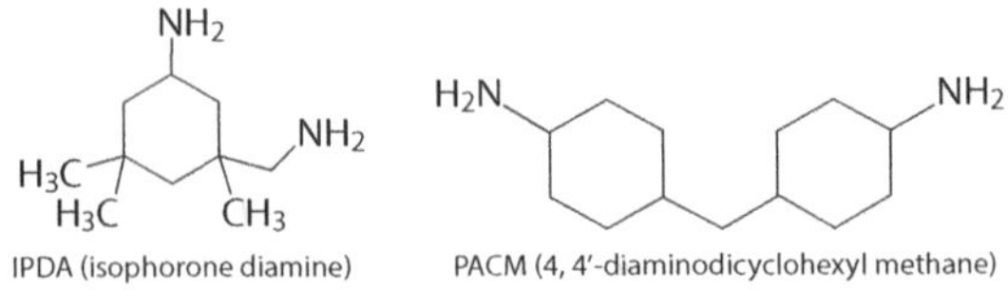

Figure 1.8 Chemical structures of typical cycloaliphatic amine curing agents.

Table 1.2 Typical modified polyamine type curing agents.

Modified polyamine	Synthesis mechanism	Key features
Polyamide	Condensation reaction of polyamine with dimer acid	Low toxicity Long pot-life Good adhesion and flexibility
Amidoamine	Condensation reaction of polyamine with monobasic fatty carboxylic acid	Good adhesion Good chemical resistance
Amine adduct	Adduct reaction of polyamine with epoxy resin	Good reactivity Low toxicity
Phenalkamine	Mannich reaction of polyamine with phenol and aldehyde	Fast curability Good chemical resistance

high mixing ratio for easy handling. Phenalkamine is a Mannich reaction product of polyamine with phenol and aldehyde. Phenalkamine possesses fast curability and good chemical resistance.

Mercaptans can cure epoxide very fast in the presence of basic accelerator even at room temperature and thus have been mainly used as curing agents for fast room temperature epoxy formulations. Phenols and anhydrides are mainly used for electrical castings, encapsulants or molding compounds, normally needing thermal cure.

Tertiary amines and imidazole compounds are a type of Lewis base that can be used as catalysts to cure epoxy resin via anionic polymerization. They can also be used as accelerators for polyamines, anhydrides and phenols to speed up curing reaction with epoxy resins. Typical chemical structures of tertiary amine and imidazole type curing agents are shown in Figure 1.9 and Figure 1.10, respectively.

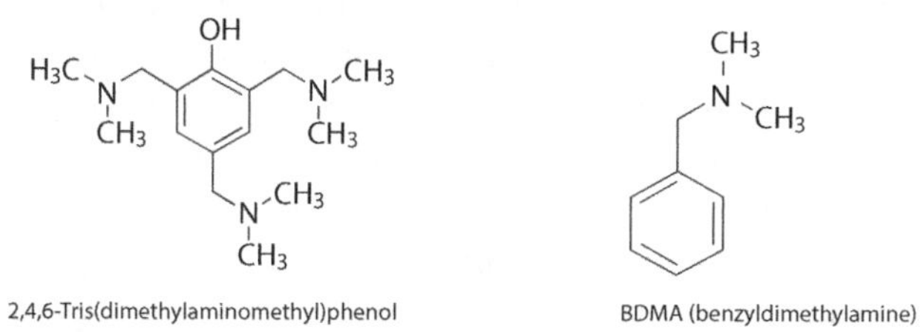

Figure 1.9 Chemical structures of typical tertiary amine curing agents.

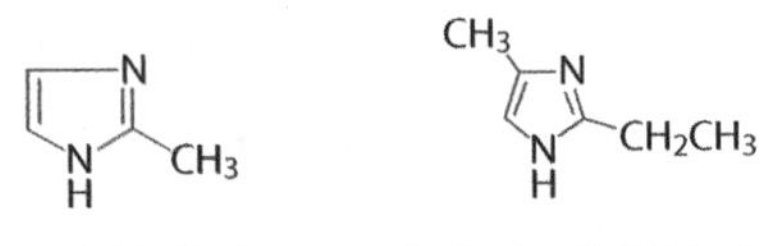

Figure 1.10 Chemical structures of typical imidazole curing agents.

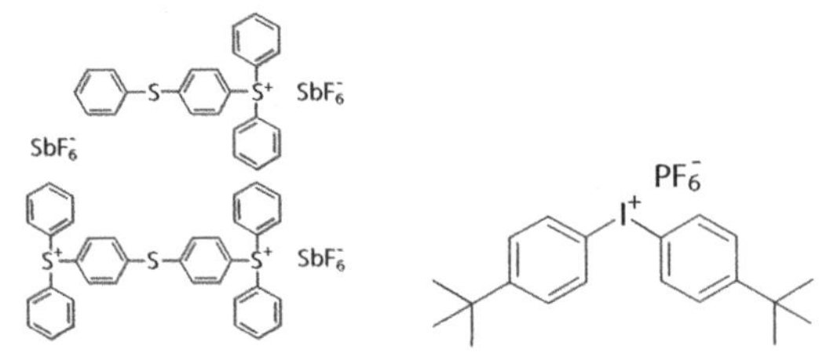

Figure 1.11 Chemical structures of onium salt and iodonium salt photoinitiators.

Certain onium salts and iodonium salts can generate strong acids via light radiation or heating that can initiate cationic polymerization of epoxy resin. Figure 1.11 illustrates typical structures of onium salts and iodonium salts suitable for use in UV cationic epoxy adhesives.

1.2.3 Formulating Epoxy Adhesives

Epoxy adhesives are typically formulated in the form of a compound containing epoxy resin and curing agent, catalyst with various modifiers and additives to achieve useful handling properties, good physical and mechanical properties as well as satisfactory adhesion performance required by the specific application purpose [23–26]. Key constituent, ingredients and their main functions are summarized in Table 1.3.

Epoxy adhesives can be formulated by selecting suitable curing agent/catalyst as either one-component or two-component packages. Two-component epoxy adhesive is prepared by packaging epoxy composition, commonly called resin part or part A, and curing agent composition, commonly called hardener part or part B, separately before use. The two-component adhesive materials will cure after mixing. Almost all room temperature cure epoxy adhesives are supplied in two-component packages. Epoxy adhesives can also be formulated in one-component package where all ingredients including epoxy resin and curing agent have been mixed in advance by use of latent curing agents. One-component epoxy

Table 1.3 Epoxy adhesive composition.

Function	Component	Main role
Primary	Epoxy resin Epoxy diluent Curing agent/catalyst Accelerator	Adhesive base Viscosity adjustment Curability, stability Cure speed enhancement
Modifier	Filler Toughener Plasticizer	Property enhancement, cost down Toughness enhancement Flexibility enhancement
Additive	Colorant Coupling agent Thixotropic agent Others	Coloring Adhesion promotion Rheology control

adhesives usually need elevated temperature cure and require chilled or even frozen storage condition at lower temperatures to ensure enough long shelf-life. UV cationic epoxy adhesives are normally formulated also as one-component type by selecting cationic photoinitiators that can cure quickly at UV radiation conditions.

1.3 Properties, Testing and Characterization

Typical properties, test methods and characterization techniques for epoxy adhesives in uncured state, curing stage, cured and adhered conditions are summarized in Table 1.4 [27–30]. Appearance, viscosity, and thixotropic value are the most important properties of uncured epoxy adhesives, determining primarily on how to apply them on the adherend substrates. Appearance is measured usually by visual inspection. Viscosity and thixotropic values can be easily measured by a viscometer.

Curing properties are important for handling and cure condition determination. Pot-life is defined as the length of time it takes for an initial mixed viscosity to double or quadruple for lower viscosity products at certain temperature, usually at room temperature around 25°C. Pot-life can act as a guide for epoxy adhesive use by providing a rough worktime range. Pot-life differs a lot for epoxy adhesives, ranging from a few minutes for fast room temperature curing two-component type to weeks for thermal cure one-component type. Gel time is determined by observing when it

Table 1.4 Properties, testing and characterization of epoxy adhesives.

Stage	Property	Test method	Test equipment
Uncured state	Appearance	ASTM E 284-17	Visual inspection
	Viscosity	ISO 3219, ASTM D2196	Viscometer
Curing stage	Pot-life	ISO 10364, ASTM 2471	Viscometer
	Gel time	ISO 2535, ASTM D2471	
	Cure behavior	ISO 11357, ASTM E 968	DSC
	Cure shrinkage	ISO 1675, ASTM D 792	Density meter
Cured adhesive	Hardness	ISO 868, ASTM 2240	Hardness gauge
	Modulus	ISO 527-3, ASTM D 882	Tension tester
	CTE	ISO 11359-2, ASTM E 1545	TMA
	Tg	ISO 11359-2, ASTM E 1545	DSC, TMA, DMA
Adhesion performance	Shear strength	ISO 4587, ASTM 1002-94	Tension tester
	Tensile strength	ISO 6922, ASTM D2095-92	Tension tester
	Peel strength	ISO 11339, ASTM D1876-93	Tension tester
	Impact strength	ISO 9653, ASTM 950-94	Impact tester

starts to become stringy, or gel-like, though not quite fully cured. Cure behavior of one-component epoxy adhesives is measured more precisely and quantitatively by DSC (differential scanning calorimetry). Cure behavior of UV cure epoxy adhesive can be measured using UV-DSC. Low cure shrinkage is one key feature of epoxy adhesives. It is normally calculated from density measurement of uncured and cured epoxy adhesives.

Physical, thermal and mechanical properties of cured epoxy adhesives can be measured by preparing suitable cured samples using the specific test and analytical methods. Hardness is measured by a hardness tester. CTE (coefficient of thermal expansion) is measured by TMA (thermal mechanical analysis). Tg (glass transition temperature) is the key factor, representing temperature resistant range of epoxy adhesives. Tg can be measured by various thermal analysis methods such as DSC, TMA and DMA (dynamic mechanical analysis). Mechanical properties such as tensile modulus, strength can be measured by a tension tester. The storage modulus is often measured by DMA. Other properties of cured epoxy adhesives such as electrical properties, optical properties can also be measured using specific test methods.

There are three main tests to determine the adhesion strength for epoxy adhesives – tensile, shear, and peel. Tensile strength is the resistance of a material to breaking under tension testing, representing adhesion behavior of an adhesive material while an axial stretching load is applied. Shear strength, commonly called as lap shear strength, is the load that a material can withstand in a direction parallel to the face of the material, representing the maximum shear stress in the adhesive prior to failure under torsional loading. Peel strength is the average force required to debond two adhered substrates. Peel strength is the average load per unit width of bondline while shear and tensile strengths are peak loads per adhesion area.

1.4 Typical Epoxy Adhesives

1.4.1 Room Temperature Cure Epoxy Adhesives

Room temperature cure epoxy adhesives are normally prepared and supplied in two-component packages by packaging epoxy resin component in the resin part while packaging curing agent component in the separate hardener part. Epoxy resin will react with the curing agent at room temperature conditions after mixing these resin part and hardener part together to become cross-linked strong thermoset structure that can bond adherend substrates tightly. By selection and combination of suitable curing agents, pot-life and cure time can be adjusted as required.

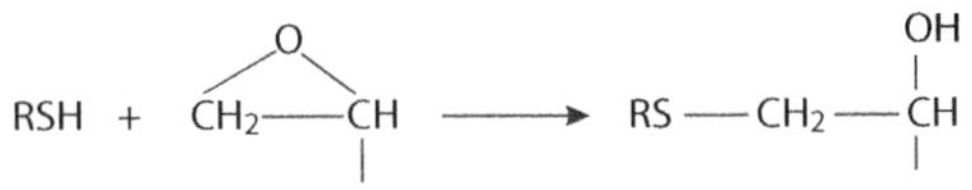

Scheme 1.1 Polyaddition reaction of epoxy resin and mercaptan.

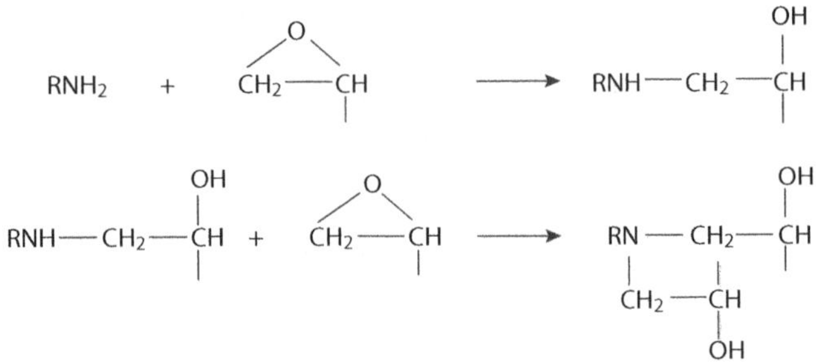

Scheme 1.2 Polyaddition reaction between epoxy resin and polyamine.

Mercaptan compounds are usually selected as curing agents for fast room temperature cure epoxy adhesives because their reaction with epoxy resin is very fast in the presence of small amount of basic chemicals such as tertiary amine or imidazole as accelerator. As illustrated in Scheme 1.1, epoxy resin reacts with mercaptan group via polyaddition reaction mechanism [31]. Gel time can be less than 15 minutes or even 10 minutes at room temperature. Full cure time will still need hours at room temperature conditions. Precautions need to be taken because of its very short worklife, often less than 10 minutes or even 5 minutes. Mercaptan based epoxy adhesives bond well to various substrates, and thus are suitable for use in general purpose structural bonding applications. Due to its aliphatic structure base, Tg of mercaptan based epoxy adhesives is usually less than 50°C so its thermal resistance is relatively limited.

Aliphatic polyamines are the most widely used curing agents in epoxy resin technology. Lots of modified polyamine type curing agents with adjustment for curability, handling or other properties for easy use have been commercialized in the market by various epoxy curing agent suppliers. As illustrated in Scheme 1.2, active hydrogen of primary and secondary amines reacts with epoxide via polyaddition mechanism [32]. Pot-life and gel time have been adjusted by selecting suitable curing agents and accelerators.

1.4.2 Thermal Cure Epoxy Adhesives

Thermal cure epoxy adhesives are prepared and supplied in both one-component and two-component packages depending mainly on curing agent type used. Compared to room temperature cure type, thermal cure two-component epoxy adhesives usually have higher glass transition temperature, and thus are suitable for applications requiring high temperature resistance. One-component epoxy adhesives do not need pre-mixing in use and thus can be handled much easily. Many new one-component epoxy adhesives have been commercialized and have become more and more important in recent years.

When cycloaliphatic amine or aromatic amine is used as curing agent, post thermal cure process is usually required to achieve full cure as their reactivity, especially of aromatic amine and secondary amine in cycloaliphatic amine, with epoxide is much lower as compared to aliphatic amines applicable for room temperature cure. Thermal cure epoxy adhesives have much stronger and more rigid structure and normally possess higher glass transition temperature as compared to room temperature cure epoxy adhesives which are mainly based on aliphatic amines or mercaptans. Two-component thermal cure epoxy adhesives are mainly used for applications requiring high temperature resistance, such as automobile production and aerospace assembly.

One-component epoxy adhesives do not require pre-mixing before use since all components have been mixed together and there is no concern for insufficient mixing problem as often is the case in two-component use. Potlife of one-component epoxy adhesives is usually long and one-component adhesives are thus suitable for automatic dispensing systems. Compared to two-component type, one-component epoxy adhesives can be handled much easily. On the other hand, one-component epoxy adhesives usually need to be cured at higher temperature because of long enough room temperature stability needed for adhesive preparation and storage. Most one-component epoxy adhesives require chilled storage condition at lower temperatures in a refrigerator or even freezer.

Recently one-component thermal cure epoxy adhesives have become more and more important especially in electronics assembly and automotive production where high production efficiency is required. With selection of suitable latent curing agents, lots of one-component epoxy adhesives have been developed and commercialized by epoxy adhesive suppliers for various applications. Typical commercial latent curing agents are summarized in Table 1.5.

DICY (dicyandiamide), chemical structure shown in Figure 1.12, is the oldest and widely used latent curing agent for epoxy resin technology. It is a solid chemical with a melting point of 208°C. DICY formulated epoxy composition is very stable, up to 6 months at room temperature. Latency mechanism is a combination of physical separation and chemical blocking with epoxide group. DICY cured epoxy resin shows very high adhesion and possesses high glass transition temperature and thus is especially suitable for applications requiring high performance such as bonding

Table 1.5 Typical commercial latent curing agents.

Latent curing agent	Latency mechanism	Curing agent state	Typical curing temperature
DICY	Chemical blocking/ physical separation	Solid	≧150°C
Dihydrazines			≧120°C
Modified imidazoles	Physical separation	Fine powder	≧80°C
Modified polyamine			≧80°C
Onium salts	Chemical blocking	Solid	≧80°C
Amine-BF_3 complex		Liquid	≧130°C

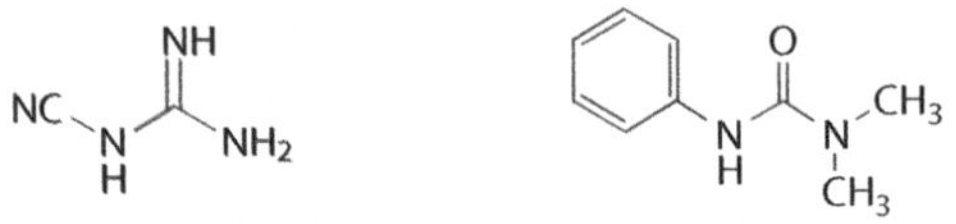

Figure 1.12 Chemical structures of DICY (left) and substituted urea (right) accelerators.

vehicle parts in automobile production. Cure temperature of DICY alone with epoxy resin normally needs to be at least 150°C to achieve full cure. By adding small amount of accelerator such as modified urea compounds and imidazole compound, cure temperature can be lowered to120°C [33].

1.4.3 UV Cure Epoxy Adhesives

Ultra-violet light (UV) curable epoxy adhesives can be cured quickly and have been very successfully used in several new electronics assembly and general bonding applications such as image sensor module assembly, display panels and modules assembly where fast production speed and high adhesion performance are required. Various UV cationic epoxy adhesive and UV acrylate hybrid thermal cure epoxy adhesives have been commercialized in recent years. As compared in Table 1.6, UV cure epoxy adhesives have no oxygen inhibition issue, have low curing shrinkage and show better adhesion as compared to common UV acrylate adhesives.

UV cationic epoxy adhesives are primarily composed of epoxy resin and cationic photoinitiator [34–39]. Cycloaliphatic type epoxy resins are usually used for UV cationic epoxy adhesives because of their faster cationic

Table 1.6 Comparison of UV acrylate, cationic epoxy and hybrid epoxy adhesives.

Adhesive type	**UV acrylate**	**UV cationic epoxy**	**Hybrid thermal cure epoxy**
Main components	Acrylate Photoinitiator	Epoxy resin Cationic photoinitiator	Acrylate Photoinitiator Epoxy resin Curing agent
<u>Polymerization</u>			
UV cure	Radical	Cationic	Radical
Thermal cure	N.A.	Cationic	Polyaddition, anionic
<u>Key features</u>			
Oxygen inhibition	Yes	No	Partial
Alkali inhibition	No	Yes	No
UV curability	High	Medium	High
Post thermal cure	No need	Partial	Need
Shadow cure	No	Low	Yes
Cure shrinkage	High	Low	Medium
Adhesion	Moderate	Good	Good

polymerization rate than that of normal bisphenol A diglycidyl ether type epoxy resin. As illustrated in Scheme 1.3 [40], cationic photoinitiator formulated in UV epoxy adhesives absorbs UV energy to generate strong acid that will react with epoxy to produce cationic species which can initiate homo-polymerization of epoxy resin. Compared to common acrylate based UV adhesives, UV cationic epoxy adhesives have lower cure shrinkage because of epoxy structure and have no surface cure issue resulting from oxygen inhibition from free radical polymerization since they cure via cationic polymerization. On the other hand, UV cationic epoxy adhesives are not suitable for bonding basic substartes which terminate cationic polymerization. UV cationic epoxy adhesives need longer cure time. In real use, a post thermal cure of UV cationic epoxy adhesives after the UV radiation is commonly combined for full cure to assure satisfactory adhesion performance. UV cationic epoxy adhesives have been commercialized and used in optical parts bonding, camera module sensor packaging and OLED panel aseembly applications [41–45]. The authors have found that adhesion perfromance of UV cationic epoxy adhesives can be much improved by using combination of cationic photoinitiator with thermal cationic initiator [46].

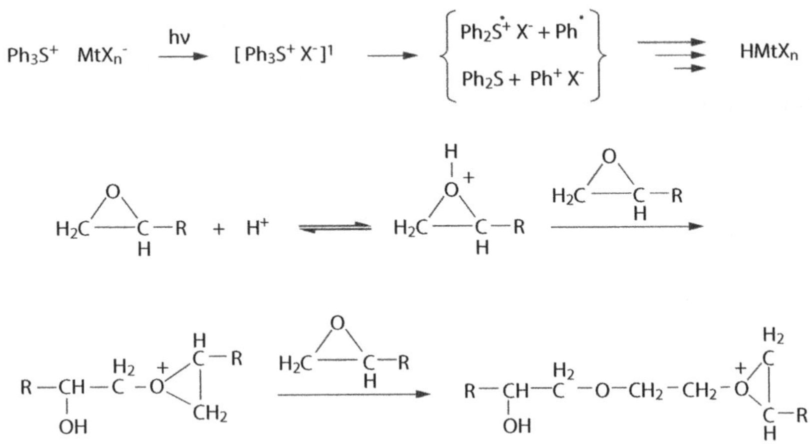

Scheme 1.3 UV cationic polymerization of epoxy adhesives.

1.5 Recent Developments and New Trends

1.5.1 High Performance Toughened Epoxy Adhesives

Cured epoxy resin polymer has highly cross-linked thermoset structure. It is rigid and strong but relatively brittle, not tough enough, tending to crack easily in actual uses. Flexible rubbers can be used to improve toughness performance of epoxy adhesives [47–50]. Several types of toughened epoxy resins such as isocyanate modified, NBR modified and CTBN modified epoxy resins have been commercialized and supplied in the market [51–53]. Figure 1.13 illustrates the chemical structure of isocyanate modified toughened epoxy resin that shows much improved toughness and flexibility as compared to standard bisphenol A epoxy resins. On the other hand, Tg becomes lower due to introduction of flexible chemical structure, resulting in decreased thermal resistance.

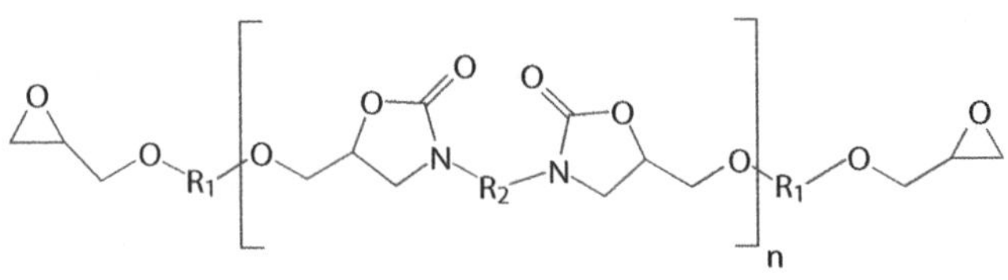

Figure 1.13 Chemical structure of isocyanate modified epoxy resin.

In recent years, various core shell rubber tougheners have been developed and commercialized. Core shell rubber bead is typically in the form of very fine powder with particle size in the range of sub-micrometer or nano size. Its core part is a synthetic rubber such as polybutadiene, polybutadiene-co-styrene or poly-siloxane rubber providing improved mechanical and toughness while the shell part out-layer is crosslinked polymer which can disperse easily in matrix epoxy resin. Core shell rubber toughener can improve not only toughness property significantly but also adhesion strength and thermal shock resistance. Unlike conventional flexible rubber toughener, there is almost no obvious Tg drop because of core shell rubber's crosslinked shell structure protection. Core shell rubber toughener formulated epoxy adhesives are especially suitable for use in high performance structural bonding applications [54–60].

1.5.2 Low Temperature Cure One-Component Epoxy Adhesives

In recent years, new types of latent curing agents have been developed and commercialized by several curing agent suppliers [61–65]. These latent curing agents are supplied as fine powder with average particle size well controlled within a few micrometers or as a premix of the fine powder latent curing agent and liquid epoxy resin. They are manufactured by grinding specially synthesized modified polyamine or imidazole solid with a softening point from 80 to 150°C. Latency mechanism is mainly physical separation between curing agent and epoxide. Curing temperature has been lowered significantly to as low as 80°C and its formulated epoxy composition can still be quite stable at room temperature. Many one-component epoxy adhesives commercialized recently are based on these new types of latent curing agents because of their lower temperature curability which is desirable for use in bonding heat-sensitive substrates such as plastics.

By combining small amount of liquid phenol compound, special stabilizer with low temperature cure latent curing agents, we found that cure time of one-component epoxy adhesives could be further shortened significantly [66, 67]. As shown in Table 1.7, cure time at 99% conversion rate of 5% liquid phenol added composition (sample No. 2) was measured as 6.75 minutes at 100°C as compared to 16.1 minutes for original latent curing agent-based composition (sample No. 4). Its pot-life remained very long, over 14 days while for non-stabilized composition (sample No. 5) it was shortened to 7 days.

Table 1.7 Stability, cure behavior and property comparison.

Sample No.	1	2	3	4	5
Component, %					
Bisphenol A epoxy resin[1]	48.5	48.5	48.5	49.0	49.0
Latent curing agent[2]	48.0	45.0	40.0	50.0	45.0
Liquid phenol resin[3]	2.0	5.0	10.0	-	5.0
γ-Glycidoxypropyltrimethoxy-silane	1.0	1.0	1.0	1.0	1.0
Stabilizer	0.5	0.5	0.5	-	-
Viscosity, mPa.s/25°C	24700	21900	25600	27200	25500
Pot-life @25°C[4], days	>14	>14	>14	>14	7
DSC cure time @100°C[5], min @ 99% conversion rate	7.38	6.75	11.7	16.1	10.9
Modulus[6], GPa@25°C	3.41	3.25	3.36	3.42	3.49
Tg, °C	154	151	146	159	152
Lap shear strength[7], MPa	14.0	13.9	13.6	13.2	13.8

*1. RE-310S, manufactured by Nippon Kayaku Co., Ltd.
2. Novacure HX-3722, manufactured by Asahi Kasei Corp.
3. MEH 8000H, manufactured by Meiwa Plastic Industries, Ltd.
4. Pot-life measured by monitoring viscosity change at 25°C storage.
5. Conversion rate calculated according to DSC isothermal method.
6. Measured by DMA method. Sample cure condition: 100°C for 60 min.
7. Adherend substrate: glass-epoxy substrate. Sample cure condition: 100°C for 60 min.

1.5.3 Instant Bonding Epoxy Adhesives

Epoxy adhesives usually need relatively long cure time, ranging from several days at room temperature to at least tens of minutes at elevated temperature, to achieve useful bonding performance. In recent decades, several new types of epoxy adhesives have been successfully developed and commercialized to meet high production efficiency required for semiconductor packaging and electronics assembly industry. These new epoxy adhesives can bond different substrates instantly within a few seconds at the specified curing conditions while still possessing satisfactory adhesion and reliability performance as structural epoxy adhesives.

Anisotropic conductive adhesives (ACAs) including anisotropic conductive film (ACF) and anisotropic conductive paste (ACP) are conductive epoxy adhesives used to bond and connect the driver electronics with the glass substrates in electronics device assembly [68–72]. ACF technology is used in chip-on-glass (COG), flex-on-glass (FOG), flex-on-board (FOB), flex-on-flex (FOF), chip-on-flex (COF) and chip-on-board (COB) assemblies. ACP is mainly used in chip-on-flex (COF) applications with low densities and cost requirements, such as for RFID antennas, or in FOF and FOB assemblies in handheld electronics. Anisotropic conductive adhesives are typically one-component epoxy compositions formulated from epoxy resin, latent curing agent and fine conductive particles coated with a polymer. There are mainly three steps in the ACAs bonding process: (1) apply anisotropic conductive adhesives, by using lamination for ACF or by dispensing or printing for ACP, on the base substrate; (2) mount the secondary substrate or device to the base substrate, and (3) cure the adhesives and complete the bonding process at certain high temperature with a press, usually at 170 to 200°C for 10 to 5 seconds.

By combination of UV acrylate with thermal cure epoxy composition, UV and thermal cure hybrid epoxy adhesives have been developed and commercialized for over two decades [73–78]. Acrylate monomer, epoxy resin, photoinitiator and epoxy curing agent are primarily formulated in the UV and thermal cure hybrid adhesives. UV hybrid epoxy adhesives combine advantages from both UV acrylate proportion and thermal cure epoxy part. Adhesion reliability performance could be much improved by introduction of epoxy composition as compared to normal UV acrylate adhesives. In the meantime, production efficiency could be much improved by shortening the bonding time to seconds via UV radiation. Successful development and industrialization of so-called ODF (One Drop Fill) process for large size LCD (liquid crystal display) panel production was an important technology revolution in the early 2000's that has made a big impact on our daily life. Development and commercialization of LCD ODF main sealant, an UV hybrid epoxy adhesive, played a key role in its mass production [79–80]. LCD ODF main sealant is an adhesive material that is used to bond two glass substrates and seal liquid crystal material between them. It is a UV hybrid epoxy adhesive, typically composed of acrylate monomer, photoinitiator, partially acrylated epoxy resin and latent curing agent. As illustrated in Figure 1.14 [81], the main steps for the adhesive use in this process are: 1) dispensing LCD main sealant on either color filter (CF) or thin film transistor (TFT) substrate; 2) dropping off liquid crystal material in each cell; 3) alignment and assembly; 4) UV cure of sealant; and 5) thermal post cure of the sealant. The author has invented initiator-free UV hybrid thermal cure epoxy adhesive by combining with bismaleimides that shows much better compatibility with liquid crystal material and high performance as new LCD ODF main sealant [82–85].

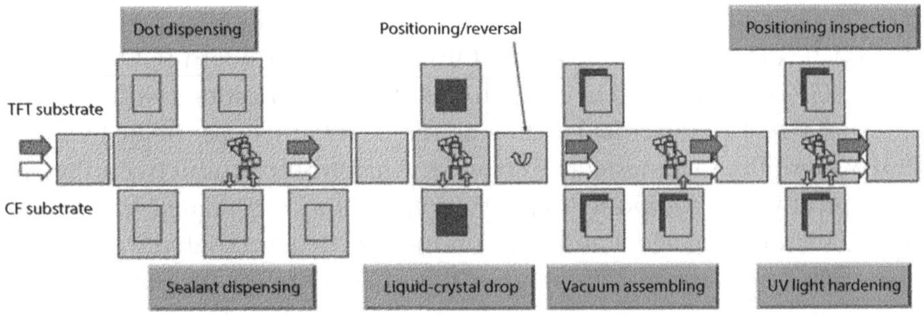

Figure 1.14 LCD ODF assembly process.

1.5.4 Sustainable Epoxy Adhesive Development

Sustainable epoxy adhesive development has become more and more important nowadays to meet the expectations from global environmental protection, government regulation restriction and customer requirements with mainly focusing on health and safety improvement, energy saving and circular economy contribution [86–90]. Some conventional raw materials, especially low viscosity epoxy resin diluent, aliphatic and aromatic amine curing agents and nonyl phenol additive, have certain health and safety concerns. Health and safety friendly sustainable epoxy adhesives need to be developed by carefully selecting green raw materials and optimizing the formulation to achieve satisfactory performance [91–93]. Sustainable biobased epoxy adhesives are prepared by using biobased epoxy resin and biobased curing agent. Biobased epoxies are usually produced from epoxidation of renewable precursors such as unsaturated vegetable oils, saccharides, tannins, cardanols, terpenes, rosins, and lignin. Biobased epoxies available commercially or under commercial investigation include epoxidized linseed oil, liquid epoxidized natural rubber, terpene-maleic ester type epoxy, diglycidyl ether of isosorbide and epoxidized cardanol. Typical biobased curing agents available commercially include vegetable oil derived polyamides, cardanol-derived phenalkamines and terpene-based anhydrides [94–105].

1.6 Summary

Epoxy adhesives show very good adhesion to various substrates and are the most widely used structural adhesives. Epoxy adhesives can be cured at room temperature conditions, at elevated temperature or via UV light radiation depending mainly on the curing agent used. Many epoxy adhesives, either supplied in one-component or two-component packages, have been commercialized and are widely used for bonding metals, concrete, glass, ceramics, plastics, wood, etc. in various industrial applications. Toughened high performance epoxy adhesives, low temperature cure one-component epoxy adhesives, instant bonding epoxy adhesives and sustainable epoxy adhesives are recent progressions and new trend in epoxy adhesive technology developments.

References

1. C.A. May, *Epoxy Resins Chemistry and Technology*, second edition, pp 1-8, Marcel Dekker, New York (1988).
2. H.Q. Pham and M.J. Marks, Epoxy resins, in: *Encyclopedia of Polymer Science and Technology Vol. 13*, pp 155-244, Wiley, New York (2004).
3. E.M. Petrie, *Handbook of Adhesives and Sealants*, pp 355-373, McGraw-Hill, New York (2006).
4. H. Panda, *Epoxy Resins Technology Handbook*, Asia Pacific Business Press, New Delhi (2016).
5. T.M. Goulding, Epoxy resin adhesives, in: *Handbook of Adhesive Technology*, second edition, A. Pizzi and K.L. Mittal (Eds.), pp 809-824, CRC Press, Boca Raton, FL (2003).
6. M. Bauer and J. Schneider, Adhesives in the electronics industry, in: *Handbook of Adhesive Technology*, second edition, A. Pizzi and K.L. Mittal (Eds.), pp 875-886, CRC Press, Boca Raton, FL (2003).
7. E.H. Cordes, Adhesives in the automotive industry, in: *Handbook of Adhesive Technology*, second edition, A. Pizzi and K.L. Mittal (Eds.), pp 983-999, CRC Press, Boca Raton, FL (2003).
8. A. Rudawska and Epoxy adhesives, in: *Handbook of Adhesive Technology*, third edition, A. Pizzi and K.L. Mittal (Eds.), pp 415-442, CRC Press, Boca Raton, FL (2018).
9. E. Sancaktar and L. Bai, Electrically conductive epoxy adhesives, *Polymers*, 3, 427-466 (2011).
10. C. Severijin, S. Teixeira de Freitas and J.A. Poulis, Susceptor-assisted induction curing behavior of a two-component epoxy paste adhesive for aerospace applications, *Int. J. Adhesion Adhesives*, 75, 155-164 (2017).
11. T. Vidil, F. Tournilhac, S. Musso, A. Robisson and Leibler, Control of reactions and network structures of epoxy thermosets, *Prog. Polym. Sci.*, 62, 126-179 (2016).
12. A. Zotti, S. Zuppolini, M. Zarrelli and A. Borriello, Fracture toughening mechanisms in epoxy adhesives, in: *Adhesives – Applications and Properties*, A. Rudawska (Ed.), pp 237-269, IntechOpen, London, UK (2016).
13. A. Kwakernaak, J. Hofstede, J. Poulis and R. Benedictus, Improvements in bonding metals for aerospace and other applications, in: *Welding and*

Joining of Aerospace Materials, M. Chaturvedi (Ed.), pp 235-319, Woodhead Publishing, Oxford, UK (2012).
14. S. Ebnesajjad and A.H. Landrock, *Adhesives Technology Handbook*, third edition, pp 110 – 113, Elsevier, London, UK (2015).
15. T.R. Kucklick, *The Medical Device R&D Handbook*, pp 34-35, CRC Press, Boca Raton, FL (2013).
16. A.F. Lewis, Epoxy resin adhesives, in: *Epoxy Resins – Chemistry and Technology*, second edition, C. A. May (Ed.), pp 653-718, Marcel Dekker, New York (1988).
17. F. Jin, L. Xiang and S. Park, Synthesis and applications of epoxy resin: a review, *J. Ind. Eng. Chem.*, 29, 1-11 (2015).
18. C.D. Wright and J.G. Muggee, Epoxy structural adhesives, in: *Structural Adhesives Chemistry and Technology*, S.R. Hartshorn (Ed.), pp 113-179, Springer, Boston (1988).
19. J.J. Licari and D.W. Swason, *Adhesive Technology for Electronics Applications: Materials, Processing, Reliability*, second edition, pp 77-87, Elsevier, London, UK (2011).
20. P. Jojibabu, Y. Zhang and B. Prusty, A review of research advances in epoxy-based nanocomposites as adhesive materials, *Int. J. Adhesion Adhesives*, 96, 10254 (2020).
21. J. Liu, C. Bai, D. Jia, W. Liu, F. He, Q. Liu, J. Yao, X. Wang and Y. Wu, Design and fabrication of a novel superhydrophobic surface based on a copolymer of styrene and bisphenol A diglycidyl ether monoacrylate, *RSC Advances*, 4, 18025-18032 (2014).
22. S. Muroi and H. Ishimura, *Epoxy Resin Introduction*, pp 2-3, Polymer Publishing Association, Tokyo (1988).
23. E.M. Petrie, *Epoxy Adhesive Formulations*, pp 85-110, McGraw-Hill, New York (2006).
24. B. Dewprashad and E.J. Eisebraun, Fundamentals of epoxy formulation, *J. Chem. Educ.*, 71, 290-294 (1994).
25. R.Junid, J. Siregar, N. Endot, J. Razak and A. Wilkinson, Optimization of glass transition temperature and pot life of epoxy blends using response surface methodology, *Polymers*, 13, 3304 (2021).
26. B. Muller and W. Rath, *Formulating Adhesives and Sealants*, pp 137-161, Vincentz Network, Hanover, Germany (2010).
27. Plastics-epoxy-test method, ISO 18280:2010 (2010).
28. Standard guide for testing epoxy resins, ASTM D 4142-89 (2009).
29. Standard specification for epoxy adhesive for bonding metallic and nonmetallic materials, ASTM D 6412/D6412M-99 (2012).
30. M. Biron, *Thermosets and Composites*, second edition, pp 211-212, Elsevier, London, UK (2014).
31. X. Fernandez-Frncos, A. Konuray, A. Belmonte, S. Flor, A. Serra and X. Ramis, Sequential curing of off-stoichiometric thiol-epoxy thermosets with a custom-tailored structure, *Polym. Chem.*, 7, 2280-2290 (2016).

32. R. Thomas, C. Sinturel, S. Thomas and E. Akiaby, Introduction, In: *Micro- and Nanostructured Epoxy/Rubber Blends*, S. Thomas, C. Sinturel and R. Thomas (Eds.), p 3, Wiley-VCH Verlag, Weinheim, Germany (2014).
33. T. Guthner and B. Hammer, Curing of epoxy resins with dicyadiamide and urones, *J. Appl. Polym. Sci.*, 50, 1453-1459 (1993).
34. V. Voytekunas, F.L. Ng and M. J.M. Abadie, Kinetics study of the UV-initiated cationic polymerization of cycloaliphatic diepoxide resins, *Eur. Polym. J.*, 44, 3640-3649 (2008).
35. B. Golaz, V. Michaud, Y. Leterrie and J.-A.E. Manson, UV intensity, temperature, and dark-curing effects in cationic photo-polymerization of a cycloaliphatic epoxy resin, *Polymer*, 53, 2038-2048 (2012).
36. M. Atif, R. Bongiovanni and J. Yang, Cationically UV-cured epoxy composites, *Polym. Rev.*, 55, 90-106 (2015).
37. C. Chen, B. Li, C. Wang, S. Iwasaki, M. Kanari and D. Lu, UV and thermal cure epoxy adhesive, in: *Painting and Coatings Industry*, F. Yalmaz (Ed.), pp 71-85, IntechOpen, London, UK (2019).
38. M. Fazal-ur-Rehman, Photo-curing schemes to cure the epoxy resins and their impacts on curing process, *J. Chem. Applications*, 4(1), 1-5 (2018).
39. C. Dekker, UV-radiation curing of adhesives, in: *Adhesives and Sealants*, P. Cognard (Ed.), pp 309-340, Elsevier, London (2006).
40. C. Corcione, G. Malucelli, M. Frigione and A. Maffezzoli, UV-curable epoxy systems containing hyperbranched polymers: kinetics investigation by photo-DSC and real-time FT-IR experiments, *Polym. Testing*, 28, 157-164 (2009).
41. T. Chang and T. Hsieh, A study of monomer's effect on adhesion strength of UV curable resins, *Int. J. Adhesion Adhesives*, 26, 520-531 (2006).
42. Y. Gan, C. Chen and K. Terada, Cationically photocurable epoxy resin composition, US Patent 7456230, assigned to Henkel (2002).
43. S. Velankar, J. Pazos and S.L. Cooper, High-performance UV-curable urethane acrylates via deblocking chemistry, *J. Appl. Polym. Sci.*, 62, 1361-1376 (1996).
44. J. Fourassier and J. Lalevee, *Photoinitiator for Polymer Synthesis*, pp 41-72, Wiley-VCH Verlag, Weinheim, Germany (2012).
45. T. Iida, Trouble-shooting of UV adhesives for opto-electronics, *J. Adhesion Society Japan*, 43, 72-78 (2007).
46. C. Chen and Y. Gan, Cationically curable epoxy resin composition, US Patent 7795744, assigned to Henkel (2004).
47. A.J. Kinloch, Toughening epoxy adhesives to meet today's challenges, *MRS Bulletin*, 28, 445-448 (2003).
48. D.J. Stamper, Toughened acrylic and epoxy adhesives, in: *Synthetic Adhesives and Sealants*, W.C. Wake (Ed.), pp 59-88, John Wiley & Sons, New York (1987).
49. K.P. Unnkrishnan and E.T. Thachil, Toughening of epoxy resins, *Designed Monomers Polymers*, 9, 129-152 (2006).

50. R. Bagheri, B.T. Marouf and R.A. Pearson, Rubber-toughened epoxies: a critical review, *Polym. Rev.*, 49, 201-225 (2009).
51. D. Ratna and A.K. Banthia, Robber toughened epoxy, *Macromol. Res.*, 12, 11-21 (2004).
52. Z. Ge, W. Zhang, C. Huang and Y. Luo, Study on epoxy resin toughened by epoxidized hydroxy-terminated polybutadiene, *Materials*, 11, 932 (2018).
53. H. Kishi, K. Uesawa, S. Matsuda and A. Murakami, Adhesive strength and mechanism of epoxy resins toughened with pre-formed thermoplastic prepolymer particles, *J. Adhesion Sci. Technol.*, 19, 1277-1290 (2005)
54. S.R. Mousavi, S. Estaji, N.R. Javidi, N. Paydayesh, H.A., Khonakdar, M. Almand, E. Rostami and S.H. Jafari, Toughening of epoxy resin systems using core-shell rubber particles: a literature review, *J. Mater. Sci.*, 56, 18345-18367 (2021).
55. D. Baek, K. Sim and H. Kim, Mechnical characterization of core-shell rubber/epoxy polymers for automotive structural adhesives as a function of operating temperature, *Polymers*, 13, 734 (2021).
56. J. Chen, A.J. Kinloch, S. Springer and A.C. Taylor, The mechanical properties and toughening mechanism of an epoxy polymer modified with polysiloxane-based core-shell particles, *Polymer*, 54, 4276-4289 (2013).
57. W. Thitsartarn, X. Fan, Y. Sun, J. Yeo, D. Yuan and C. He, Simultaneous enhancement of strength and toughness of epoxy using POSS-Rubber core-shell nanoparticles, *Composites Sci. Technol.*, 118, 63-71 (2015).
58. G. Chae, H. Park, J. Lee and S. Shin, Comparative study on the impact wedge-peel performance of epoxy-based structural adhesives modified with different toughening agents, *Polymers*, 12, 1549 (2020).
59. S. Kenig, H. Dodiuk, G. Otorgust and S. Gomid, Nanocomposite polymer adhesives: a critical review, *Rev. Adhesion Adhesives*, 7, 93-167 (2019).
60. W. Tsang and A.C. Taylor, Fracture and toughening mechanisms of silica and core–shell rubber-toughened epoxy at ambient and low temperature, *J. Mater. Sci.*, 54, 13938-13958 (2019).
61. Ajinomoto Fine-Techno Co., Ltd. Latent curing agent "AJICURE". available from: *https://www.aft-website.com/en/chemistry/ajicure*
62. ADEKA Corporation, ADEKA Hardener EH Series (Latent Hardener). available from: *https://www.adeka.co.jp/en/chemical/products/functional/pro143c.html#sp05*
63. Evonik Corporation, Epoxy Additives and Polyamides. available from: *https://crosslinkers.evonik.com/product/crosslinkers/downloads/evonik-poxyguide_asiapacific.pdf*
64. Asahi Kasei Corporation, Novacure. available from: *https://www.asahi-kasei.co.jp/advance/en/business/plastics/plastics04.html*
65. T & K TOKA Corporation, Latent Hardener "FUJICURE". available from: *https://www.tk-toka.co.jp/product/jushi/epoxy/detail/files/fujicure_solid.pdf*
66. C. Chen, Thermal curable liquid resin composition, Japanese Patent 4204814, assigned to Henkel (2002).

67. C. Chen, One-component epoxy resin composition, European Patent 2640765, assigned to Henkel (2011).
68. M. Shiiki, J. Imaizumi, T. Miyata and A. Chinda, Materials and components for flat panel display applications, *Hitachi Rev.*, 55, 32-39 (2006).
69. H. Toshioka, K. Nakatsugi, M. Yamamoto, K. Sato, N. Shimbara and Y. Okuda, Development of anisotropic conductive film for narrow pitch circuits, *SEI Technical Rev.*, 73, 41-44 (2011).
70. N. Takano, T. Fujinawa and T. Kato, Film technologies for semiconductors and electronic components, *Hitachi Chemical Technical Report*, 55, 20-23 (2013).
71. J.J. Licari and D.W. Swason, *Adhesives Technology for Electronic Applications: Material, Processing, Reliability*, second edition, pp 6-8, Elsevier, New York (2005).
72. J.W.C. De Vries and J.F.J.M. Caers, Anisotropic conductive adhesives in electronics, in: *Advanced Adhesives in Electronics*, M.O. Alam and C. Bailey (Eds.), pp 53-104, Woodhead Publishing, Cambridge, UK (2011).
73. A. Vitale, G. Trusiano and R. Bongiovanni, UV-curing of adhesives: a ctitical review, in: *Progress in Adhesion and Adhesives, Volume Three*, K.L Mittal (Ed.), pp 101-154, Wiley-Scrivener, Beverly, MA (2018).
74. C. Dekker, UV-radiation curing of adhesives, In: *Handbook of Adhesive and Surface Preparation*, S. Ebnesajjad (Ed.), pp 221-243, Elsevier, New York (2011).
75. B. Goss, Bonding glass and other substrates with UV curing adhesives, *Int. J. Adhesion Adhesives*, 22, 405–408 (2002).
76. C. Park, S. Lee, J. Park and H. Kim, Preparation and characterization of dual curable adhesives containing epoxy and acrylate functionalities, *React. Funct. Polym.*, 73, 641-646 (2013).
77. M. Xiao, Y. He and J. Nie, Novel bisphenol A epoxide-acrylate hybrid oligomer and its photopolymerization, *Designed Monomers Polymers*, 11, 383-394 (2008).
78. Y. Park, D. Lim, H. Kim, D. Park and I. Sung, UV- and thermal-curing behavior of dual-curable adhesives based on epoxy acrylate oligomers, *Int. J. Adhesion Adhesives*, 29, 710-717 (2009).
79. M. Matsuda, Sealants for one drop fill (ODF) process, in: *The Liquid Crystal Display Story*, N. Koide (Ed.), pp 199-, Springer, New York (2014).
80. C. Chen and K. Iida, Adhesive for flat-panel display manufacture, In: *Ullmann's Encyclopedia of Industrial Chemistry, Adhesives, 2. Applications*, Henkel Expert Team (Ed.), pp 519-520, Wiley, Weinheim, Germany (2012).
81. A. Hirai, I. Abe, M. Mitsumoto and S. Ishida, One drop filling for liquid crystal display panel produced from larger-sized mother glass, *Hitachi Rev.*, 57(3), 144-148 (2008).
82. C. Chen, Sealing agent for liquid crystal dropping technology and method of manufacturing liquid crystal display, Japanese Patent 5592081, assigned to Henkel (2009).

83. C. Chen, Sealant composition, US Patent 10108029, assigned to Henkel (2009).
84. C. Chen, Method of manufacturing liquid crystal display panel, Japanese Patent 5522938, assigned to Henkel (2007).
85. C. Chen, S. Iwasaki, M. Kanari, B. Li, C. Wang and D. Lu, High performance UV and thermal cure hybrid epoxy adhesive, *IOP Conf. Series: Mater. Sci. Eng.*, 213, 012032 (2017).
86. S. Kumar, S. Samal, S. Mohanty and S. Nayak, Recent development of biobased epoxy resin: a review, *Polym. Plast. Technol. Eng.*, 57, 133-155 (2018).
87. S. Cailoll, B. Boutevin and J. Pascault, Bio-sourced epoxy monomer and polymers, in: *Handbook of Adhesive Technology*, third edition, A. Pizzi and K.L. Mittal (Eds.), pp 443-470, CRC Press, Boca Raton, FL (2018).
88. L.A Heinrich, Future opportunities for bio-based adhesives – advantages beyond renewability, *Green Chem.*, 21, 1866-1888 (2019).
89. G.M. Roudsari, A.K. Mohanty and M. Misra, Green approaches to engineer tough biobased epoxies: a review, *ACS Sustainable Chem. Eng.*, 5, 9528-9541 (2017).
90. V. Hemmila, S. Adamopoulos, O. Karlsson and A. Kumar, Development of sustainable bio-adhesives for engineered wood panels – a review, *RSC Advances*, 7, 38604-38630 (2017).
91. E. Hase and M. Kitano, The role and place of risk assessment in chemical management especially in Japan, Europe and U.S.A., *Japanese J. Risk Analysis*, 22, 63-72 (2012).
92. D.E. Packham, The environmental impact of adhesives, in: *Eco-Efficient Construction and Building Materials Life Cycle Assessment (LCA), Eco-Labelling and Case Studies*, F. Pacheco-Torgal, L. F. Cabeza, J. Labrincha and A. de Magalhiies (Eds.), pp 338-367, Woodhead Publishing, Cambridge, UK (2014).
93. R. Rijk and R. Veraart, *Global Legislation for Food Packaging Materials*, pp 13-15, Wiley VCH, Weinheim, Germany (2010).
94. E. Ramon, C. Sguazzo and P. Moreira, A review of recent research on bio-based epoxy systems for engineering applications and potentialities in the aviation sector, *Aerospace*, 5, 110 (2018).
95. F. Ng, G. Couture, C. Philippe, B. Boutevin and S. Caillol, Bio-based aromatic epoxy monomers for thermoset materials, *Molecules*, 22, 149 (2017).
96. M. Benegaa, R. Rajab and J. Blake, A preliminary evaluation of bio-based epoxy resin hardeners for maritime application, *Procedia Eng.*, 200, 186-192 (2017).
97. K. Hambleton and J. Stanzione, Synthesis and characterization of a low-molecular-weight novolac epoxy derived from lignin-inspired phenolics, *ACS Omega*, 6, 23855-23861 (2021).
98. Z. Wang, P. Gnanasekar, S. Nair, S. Yi and N. Yan, Curing behavior and thermomechanical performance of bio-epoxy resin synthesized from vanillyl alcohol: effects of the curing agent, *Polymers*, 13, 2891 (2021).

99. S. Nikafshar, J. Wang, K. Dunne, P. Sangthonganotai and M. Nejad, Choosing the right lignin to fully replace bisphenol A in epoxy resin, *ChemSusChem*, 14, 1184-1195 (2021).
100. Z. Zhang, J. Li, Y. Zhang and Z. Jin, Kinetics of partially depolymerized lignin as co-curing agent for epoxy resin, *Int. J. Biol. Macromol.*, 150, 786-792 (2020).
101. T. Aziz, H. Fan, X. Zhang, F. Khan, S. Fahad and A. Ullah, Adhesives properties of bio-based epoxy resin reinforced by cellulose nanocrystal additives, *J. Polym. Eng.*, 40, 314-320 (2019).
102. Q. Ruiz, S. Pourchet, V. Placet, L. Plasseraud and G. Boni, New eco-friendly synthesized thermosets from isoeugenol-based epoxy resin, *Polymers,* 12, 229 (2019).
103. X. Yang, C. Wang, S. Li, K. Huang, M. Li, W. Mao, S. Cao and J. Xia, Study on the synthesis of bio-based epoxy curing agent derived from myrcene and castor oil and the properties of the cured products, *RSC Advances*, 7, 238-247 (2017).
104. S. Merighi, L. Mazzocchetti, T. Benelli and L. Giorgini, Evaluation of novel bio-based amino curing agent systems for epoxy resins: effect of tryptophan and guanine, *Processes,* 9, 42 (2021).
105. C. Ding and A. Matharu, Recent developments on biobased curing agents: a review of their preparation and use, *ACS Sustainable Chem. Eng.*, 2, 2217-2236 (2014).

2

Biological Reinforcement of Epoxies as Structural Adhesives

Anna Rudawska*, Jakub Szabelski, Izabela Miturska-Barańska and Elżbieta Doluk

Lublin University of Technology, Faculty of Mechanical Engineering, Nadbystrzycka, Lublin, Poland

Abstract

Epoxies as structural adhesives are used in a wide variety of structures, used in a wide variety of environments, for joining components made of different materials. However, there are limitations to the use of epoxies as structural adhesives in certain areas due to their delamination, low impact strength, inherent brittleness and low crack resistance. The mentioned limitations can be overcome by incorporating various modifying agents and their modification prior to their industrial application. Nowadays, modified epoxy resins are widely used for the manufacture of natural fiber-reinforced composites and various industrial products due to their excellent mechanical, thermal and electrical properties.

The chapter includes an introduction to the subject of modification of adhesives prepared on the basis of epoxy resins and curing agents characteristics and their properties, description of curing types, and properties and types of fillers. A separate section is dedicated to biofillers. It provides an overview of biofillers, their characteristics, and description of the properties of various biofilled epoxy adhesives.

***Keywords*:** Epoxy structural adhesive, epoxy resin, modification, biological reinforcement

2.1 Introduction

The modern concept of materials engineering places increasing demands on structures and materials. The increasing performance of the resulting

**Corresponding author*: a.rudawska@pollub.pl

K.L. Mittal and S.K. Panigrahi (eds.) Structural Adhesives: Properties, Characterization and Applications, (31–104) © 2023 Scrivener Publishing LLC

materials must meet the increasingly rigorous environmental aspects of modern society, including ecological and economic factors. Biological materials are a constantly growing area. In recent years, there has been an increasing interest in the use of natural fibers as reinforcements in polymer composites to produce low-cost materials such as construction materials. In addition, the use of biofillers to modify epoxy resins used in the automotive industry, among others, contributes to the creation of an interesting environmentally friendly material while minimizing production costs and ecological impact.

Natural fibers are potential reinforcing materials. They have long served many useful purposes, but the application of materials using natural fibers as reinforcement in a polymer matrix has been quite recent [1–7]. Many studies have been conducted on natural fibers such as kenaf [8–11], bamboo [12–15], jute [16, 17], hemp [18–20], coconut and coir [21, 22], pineapple [23], and oil palm [24–28].

The advantages of these natural resources are lightweight, low cost, low density, high strength, acceptable specific strength, recyclability and biodegradability [25, 28]. Natural fibers can be divided into four different types: leaves, phloems, fruits, and seeds [24, 29–31].

Epoxy resins have been known for a long time [32–36]. The oldest method of obtaining epoxy compounds of technical significance dates back to 1860. It involved the reaction of ammonia with epichlorohydrin. The introduction of epoxy resins as an industrial product in the second-half of the 20th century raised a great deal of interest due to a number of valuable properties that distinguish them from other polymers. These resins have found a place in various fields of technology, e.g. in electronics and electrical engineering, where they have become essential for further progress in this industry [37–45].

In the 1940s, solid epoxy resins were obtained by reacting low molecular weight epoxy resins with bisphenol A. This type of resin is used for coating purposes such as powder coatings and varnishes. In the early 1960s, cycloaliphatic epoxy resins were introduced as an industrial product. These resins are used where very high demands are made, for example in electrical engineering (as overhead insulation). However, due to complicated synthesis processes with many steps and high costs, they are produced only in small quantities. For several decades they have been competing with epoxy resins, which are obtained in a simpler way by reacting epichlorohydrin with cyclic or heterocyclic compounds that do not contain aromatic rings [44].

Diane epoxy resins are the basic component of numerous compounds cured with various curing agents or accelerators and with the addition of modifiers (such as fillers, flexibilisers, thinners, etc.). These resins have a number of advantages which make them increasingly used in industry. They are characterised by very attractive properties and relatively simple processing methods. Their advantage is their polar character, which provides good adhesion of these resins to various materials (such as glass, metals, ceramics or concrete). During the curing process of epoxy resins, only little shrinkage occurs, which makes the castings follow the shape of the mould very accurately. Internal stresses are therefore very low, and the addition of elasticising agents eliminates unwanted stresses [46].

Despite a number of advantages, epoxy resins also have some disadvantages. They are characterised by limited strength at elevated temperatures. Therefore, research is constantly being conducted to improve the heat resistance of cured resins. One of the directions of research is the introduction of proper fillers to the compounds, which will be able to provide better mechanical strength at elevated temperature [32–34].

Cured epoxy resins are characterised by high mechanical strength and are resistant to water, chemicals and weather conditions. Their advantage is also very good dielectric properties. They are characterised by high specific resistance and low dielectric loss factor. Fillers can be used to produce polymers with a wide range of properties (such as electrically conductive adhesives, moulding compounds or reinforced concrete). By adding glass fibers, engineering plastics can be produced that are as strong as metals but lighter and more corrosion resistant.A certain limitation in the use of epoxy resins may be their relatively high price. However, the introduction of fillers, e.g. mineral fillers, contributes to its significant cost reduction [32–34] One of the applications of epoxy resins is their use as structural adhesives [34–36, 41].

2.2 Epoxy Resins and Curing Agents

2.2.1 Epoxy Resins

Epoxy resins are low molecular weight polymeric and oligomeric compounds that contain more than one epoxy group in the molecule, also called an oxirane group [47]. The epoxy group is a three-membered ring consisting of two carbon atoms and one oxygen atom (Figure 2.1). Epoxy resins have the ability to react by curing to produce cross-linked, insoluble and non-fusible materials [32, 33, 48]. Depending on their molecular weight and structure, they are thermoplastic solids or viscous liquids with a density of 1.15-1.21 g/cm^3. They dissolve well in toluene, benzene, acetone and other organic solvents [32, 33].

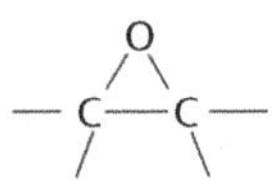

Figure 2.1 Epoxy group configuration [47].

Of the various methods of obtaining epoxy compounds, only two methods are used in industry that has practical applications. The first is the epoxidation with epichlorohydrin of chemical compounds that contain hydrogen atoms. The second method is the epoxidation of double bonds in unsaturated compounds by [32, 33]:

- direct attachment of oxygen,
- attachment of chlorine and a hydroxyl group to the double bond, followed by elimination of hydrogen chloride.

The number of studies on new methods of manufacturing epoxy resins is constantly increasing. However, an ingredient such as epichlorohydrin still remains the main component used in resin production. There is a classification of epoxy resins that divides them according to their chemical structure. An example of the classification is shown in Figure 2.2.

Epoxy resins find many applications due to their good adhesion to most materials. Depending on the purpose, resins can be distinguished in the form of solvent-free powder paints or varnishes, water dispersions for surface protection, as construction adhesives or as components of reactive resin concretes [34, 41, 47, 48].

2.2.2 Curing Agents

The reaction of curing epoxy resins into a non-melting and insoluble product is triggered by the addition of a crosslinking agent called curing agent [33, 49]. The choice of the appropriate curing agent depends mainly on the curing temperature as well as the curing conditions and the desired properties of the cured resin [32, 49–53]. Epoxy resins can be cured at room temperature (cold), at temperatures between 80-100°C (warm) as well as at elevated temperatures above 100°C (hot).

There are three basic types of curing agents: anhydride curing agents, amine curing agents and curing agents that cause ionic polymerisation of epoxy groups [32, 33, 48, 49, 54–57]. The first two classes mentioned above are most commonly used in industrial applications. Figure 2.3 shows examples of curing agents used to crosslink epoxy resins.

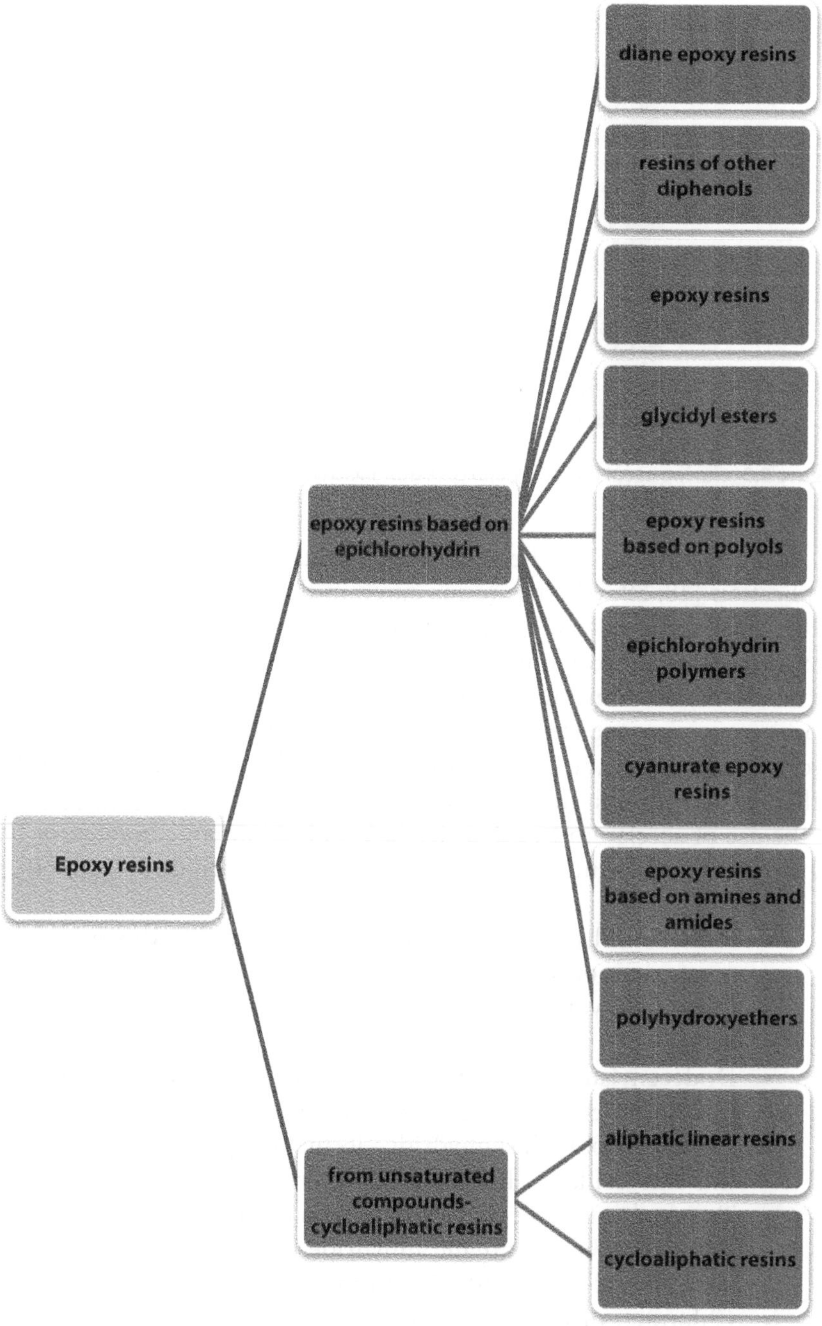

Figure 2.2 Scheme of classification of epoxy resins according to their chemical structure, based on [32, 33, 46, 49].

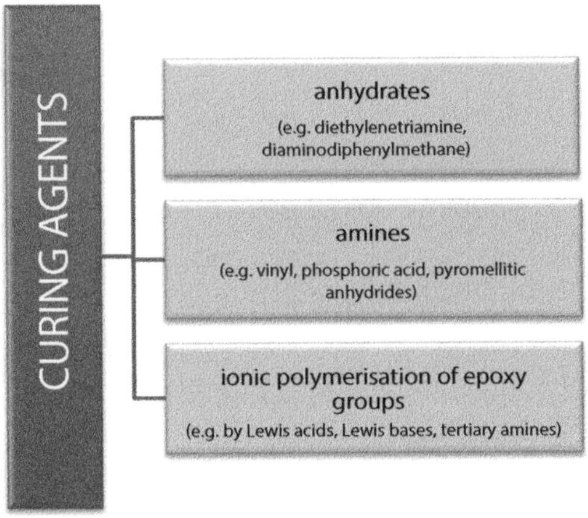

Figure 2.3 Curing agents used for epoxy resins, based on [32, 33, 48, 49, 58–60].

Acid anhydrides were the first curing agents which had found practical application [48, 58–60]. The resins cured in this way are most often used in the electrical industry. This group of curing agents include cyclic dicarboxylic or tetracarboxylic anhydrides of organic acids. They do not contain active hydrogen atoms that could react with epoxy groups However, these atoms are a consequence of the reaction of the anhydride groups with the hydroxyl groups of the epoxy resins. The crosslinking reaction with acid anhydrides requires a suitable temperature - from 100°C to 180°C, as well as the use of accelerators such as tertiary amines, organic borates, or lithium and tin compounds. Anhydride curing agents include [48, 58–60]:

- Phthalic anhydride - known as curing agent F; this is the most commonly used anhydride;
- Maleic anhydride - highly toxic and therefore very rarely used;
- Dodecenylsuccinic anhydride - this is a low viscosity liquid, mixes very easily with the resin and it easy to use;
- Hexahydrophthalic anhydride - its melting point is very low; this anhydride has low reactivity, therefore, it is often used;
- Tetrahydrophthalic anhydride - it is a crystalline substance;
- Methyltetrahydrophthalic anhydride - it occurs as a low viscosity liquid or in crystalline form;
- Chlorotetrahydrophthalic anhydride - it is a crystalline material, very rarely used at this time.

The following anhydrides are also known: endomethylenetetrahydrophthalic, methylendomethylenetetrahydrophthalic, trimellitic, pyromellitic and also adducts of terpenes and colophony with maleic anhydride. Selected examples of anhydride curing agents are shown in Figure 2.4.

The group of amine curing agents includes aliphatic polyamines and diamines, cycloaliphatic and aromatic diamines, adducts of amines with propylene or ethylene oxide, and other amine compounds. The most common curing agents are aliphatic polyamines, which were the first to be used to crosslink epoxy resins. Ethyleneamines such as ethylenediamine (EDA), diethylenetriamine (DETA), triethylenetetramine (TETA) are the most common resin curing agents [48, 61].

During the curing reaction with primary amines, the presence of compounds containing hydroxyl groups is essential. This is because a small amount of water, ethanol or methanol greatly accelerates the reaction. In addition, accelerators such as lactic or salicylic acids are often used.

Curing with tertiary amines, compared to the two groups discussed previously, is the least used. Lewis-base type compounds, belonging to tertiary amines, lead to the polymerisation of epoxy compounds. They are most often used as curing agents or accelerators for the resin crosslinking reaction [61].

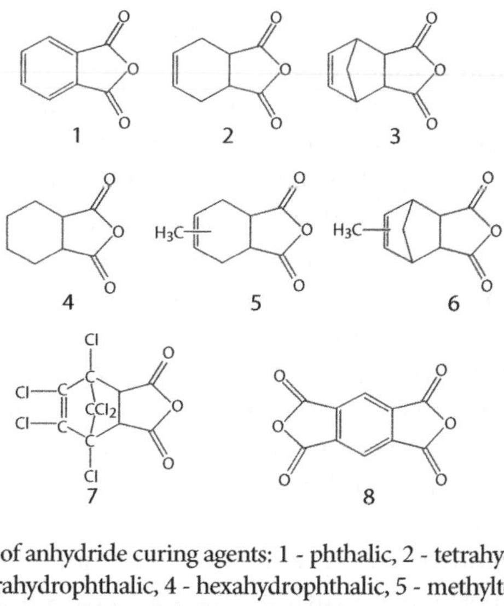

Figure 2.4 Examples of anhydride curing agents: 1 - phthalic, 2 - tetrahydrophthalic, 3 - endomethylenetetrahydrophthalic, 4 - hexahydrophthalic, 5 - methyltetrahydrophthalic, 6 - methylendomethylenetetrahydrophthalic, 7 - hexachloroendomethylenetetrahydrophthalic, 8 - pyromellitic [61].

2.2.3 Curing Methods

Uncured epoxy resins are viscous liquids that have no practical use. To cure them, a crosslinking agent, called a curing agent, is added to make an insoluble and non-melting product. This process gives the resins their proper working properties [33, 49].

The curing of resins is a chemical reaction of the active groups of the curing agents with the functional groups that are present in the resins. Their structure contains two types of functional groups that are involved in crosslinking. These are epoxy groups, located at the ends of the chain, and hydroxyl groups, which are attached along the chain of macromolecules [44, 58–60].

By adding a reactive chemical compound (curing agent), the epoxy resin takes the form of a hard solid. During this process, significant changes in physical properties occur. The reaction is exothermic, and thus a temperature increase can be observed. Examples of curing epoxy resins are shown in Table 2.1 [46, 49, 58–60, 62].

Another simple classification of curing methods for epoxy resins is the division into "cold", "warm" and "hot" curing. The most commonly used curing agents are amines, carboxylic acids, Lewis acids, acid anhydrides and diphenols [48, 58, 59, 64].

Cold curing of epoxy resins takes place at room temperature. Aliphatic polyamines are most commonly used to cure epoxy resins at room temperature. However, the adhesive compounds of such cured resins are characterised by low heat resistance [46].

Tertiary amines and primary aromatic amines are used for the hot cure. Aromatic amines provide long working times and good thermal and chemical resistance, but higher temperature curing is required to achieve increased heat resistance. For this process a suitable temperature between 80°C and 100°C is required. Such conditions allow to obtain increased thermal strength of the prepared adhesive compounds of such. The use of aromatic amines allows also to increase chemical and thermal resistance.

When using amine curing agents, their toxicity must also be considered. Aliphatic amines are characterised by strong caustic properties and may cause allergies. They are very irritating to eyes, skin and mucous membranes. Aromatic amines, on the other hand, can cause serious poisoning and are carcinogenic.

The third method of curing epoxy resins requires the use of curing agents in the form of acid anhydrides and novolac resins. The process temperature is above 100°C. The material obtained in this way is characterised by good dielectric properties and high heat resistance [58–60]. The ability

Table 2.1 Methods of curing epoxy resins, based on [33, 45, 48, 57–59, 61].

Method	Characteristics of the method
Curing with organic acid anhydrides	• Long high temperature heating required; • Low exothermic effect; • Low volume shrinkage; • Resins free of internal stresses and mechanical defects; • High thermal stability; • Application in electrical engineering as insulating materials;
Curing with primary and secondary amines	• Crosslinking at room temperature; • High strength of cured material; • Good dielectric properties; • Resistance to chemical agents; • Ability to obtain the desired flexibility; • Resistance to thermal ageing;
Curing with tertiary amine	• Polymerisation of epoxy compounds occurs; • Tertiary amines used as standalone curing agents or as cure accelerators; • Crosslinking at room temperature;
Curing with boron fluoride adducts	• Crosslinking at elevated temperature; • High shear strength;
Curing with phenolic compounds	• The reaction of phenols with resins takes place at high temperature; • Long curing time; • Need to use accelerators; • Very good chemical resistance; • High mechanical strength; • Application as anti-corrosive powder paints; • High viscosity of the compound;
UV curing	• Wavelength used for excitation; 200 - 300 nm; • Limited to curing of thin films; • Advantageous application for curing of commercial epoxy resin grades; • Fast crosslinking process; • Low water absorption; • High mechanical strength; • Low volume shrinkage;

(*Continued*)

Table 2.1 Methods of curing epoxy resins, based on [33, 45, 48, 57–59, 61]. (*Continued*)

Method	Characteristics of the method
Curing in microwave radiation field	• Possibility of accelerating the reaction by microwave heating; • High glass transition temperature.

of the acid anhydrides to react with both the hydroxyl groups of the resin and the epoxy groups allows obtaining higher crosslink density. As a result, hot cured epoxy resin products are characterised by high heat resistance and good dielectric properties [58–60]. Furthermore, they are also characterised by low volume shrinkage and low exothermic thermal effects. Lewis acid-type compounds are also used during hot curing. These adhesives can be cured within a few seconds, but they are characterised by low mechanical strength [63].

Most often curing agents are mixed with resin in stoichiometric ratio, i.e. in such a way that there is one hydrogen atom per one epoxy group [46]. Both the deficiency and the excess of the amine curing agent deteriorate the adhesive property. The curing reactions of epoxy resins are strongly exothermic; they are accompanied by heat release which increases the temperature in the resin-curing agent system and thus speeds up the reaction between the resin and curing agent.

The choice of curing agent depends, among other things, on the conditions under which the adhesive compound will be cured, i.e., the time and temperature of the process, and the amounts of curing agent and resin depend on the stoichiometric ratio of the specific components of the compound. The most commonly used curing agents are polyamides (dicyandiamide, polyamide), polyamines (diethylenetriamine, triethyleneamine, trimethyleneamine, triethylenetetramine, aliphatic amine), methyl-5-norbornene-2,3-dicarboxylic acid anhydrides, strong acids and bases (e.g. monomethylamine of boron trichloride, Mannich base), imidazoles and thiols [37, 42, 65].

2.2.4 Epoxy Structural Adhesives

For many years, epoxy adhesives have been the main family of adhesives used for structural bonding of metals and composite materials [41, 48, 66, 67]. The versatility of epoxy adhesives results from a large number of combinations of epoxy resins and curing agents with different chemical compositions and various curing methods, which results in a variety of molecular structures of the polymer obtained [35, 36, 40, 41, 48].

Many epoxy adhesives included in the group of structural adhesives can be defined as load-bearing materials with high modulus of elasticity and strength, which can transfer stresses without loss of structural integrity. The properties of adhesives prepared based on epoxy resins described above also depend on the curing agent used.

Epoxy adhesives have the following properties [38, 39, 54, 68, 69]:

- Adhesion - epoxy-based structural adhesives exhibit good adhesion to most materials,
- Mechanical strength - epoxy adhesives are characterised by high Young's modulus and strength, and their tensile strength may exceed 80 MPa,
- Chemical resistance - epoxy adhesive compounds are resistant to most chemicals, particularly alkalis,
- Diffusion - epoxy coatings generally have a relatively high resistance to vapour transmission, but can be 'opened' to diffusion by special techniques
- Watertightness - epoxy materials are considered watertight and are often used for waterproofing
- Electrical insulation property - epoxy adhesives are excellent electrical insulators,
- Shrinkage - adhesives based on epoxy resins are characterised by very low longitudinal shrinkage during curing,
- Modifiability - there are unlimited possibilities to fully modify the properties of epoxy adhesives to meet specific requirements.

Due to their properties, epoxy adhesives are used mainly in the aviation and automotive industries, as well as in civil engineering. They are also widely used in construction, marine industry, mechanical engineering and as protective coatings, among others [32, 39, 42, 62, 65, 66, 69–72].

2.3 Modification of Epoxies, and Modifying Agents

2.3.1 Epoxy Modification Methods

The polymer products, including adhesives, are modified to improve and/or impart new functional and performance properties [73, 74]. Modification of epoxy resins is aimed at improving certain properties of the cured polymer materials [58, 59]. Most commonly, it contributes to extending the life of the compound or reducing the exothermic effect of the crosslinking reaction. Additionally, it can lead to a reduction in manufacturing costs. An important step in modification is the selection of a suitable modifying substance and the determination of quantitative proportions in relation to the resin [32, 49, 54, 56, 75].

The main purposes of modification are to [33, 56–59, 64, 76–79]:

- reduce the brittleness of the cured resin,
- increase mechanical strength,
- improve chemical and dielectric resistance,
- minimise flammability,
- broaden the range of applications.

The use of modification can also lead to undesirable effects. The most important of them is the increase in viscosity of the compound, which complicates its processing.

There are three types of modifications: physical, chemical and physico-chemical. In physical modification the most commonly used agents are fillers, stabilizers, plasticizers or thinners. Chemical modification, on the other hand, consists in changing the composition of the hydroxyl groups and in changing the epoxy equivalent. Physico-chemical modification combines these two modification methods.

Physical modification of polymers is carried out by physical phenomena, for example by mixing. Modified polymers (including adhesives) differ from those prior to modification in structure, physical properties, usability, visual appearance, etc. The most common methods of physical modification are the addition of the various types of fillers [59, 74, 80–83]. The performance properties of modified materials significantly depend on the type of filler used (particle shape and size, specific surface area, dispersed phase concentration) [59, 74, 84, 85]. The best properties are obtained when the smallest possible fillers are incorporated, preferably with particle sizes in the nanometer scale [74, 84, 86–88].

Chemical modification usually involves the attachment of other chemical compounds to all or part of the mers or the elimination of specific functional groups from the mers. Epoxy resins can also be modified using other polymers. The chemical compounds introduced into the resin-curing agent system are intended, among other things, to make the resin more flexible. Compounds that are often subject to chemical modification are polymers of natural origin, e.g. cellulose or natural rubber. Chemical modification, in addition to changing the chemical nature of the modified compound, it also causes a change in primary characteristics (resulting from the chemical nature of the molecule, its structure and size) and secondary characteristics (resulting from the method of polymer synthesis, processing, or improvement of specific properties) [58–60].

An epoxy adhesive compound consists of an epoxy resin and a curing agent, while a modified epoxy compound may also contain various additives (Figure 2.5) in the form of fillers, stabilisers, plasticisers or diluents (Figure 2.6) [49, 54, 59]. The general composition of an epoxy resin-based compound is shown in Figure 2.6.

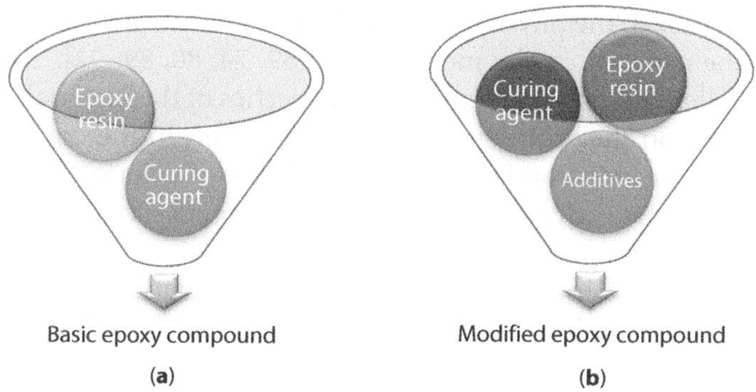

Figure 2.5 General compositions of adhesive compounds.

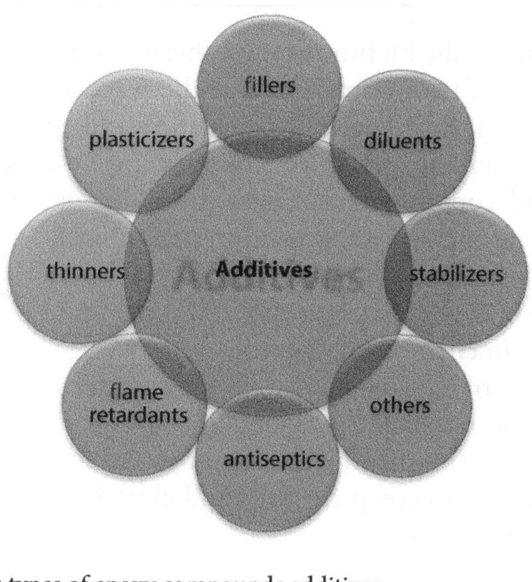

Figure 2.6 Various types of epoxy compounds additives.

2.3.2 Fillers Properties

The main purpose of using fillers is to improve certain functional properties of polymeric materials. Furthermore, fillers also simplify processing and reduce the exothermic effect. The desired properties of polymeric materials, which can be obtained by using fillers, depend mainly on two factors: the selection of an appropriate filler for the polymer and the selection of an appropriate amount of filler. When choosing a filler, its properties, form and price should be taken into account. In addition, the environmental aspect is also an important consideration [49, 59].

Fillers are solids which when added to plastics give them desirable properties but also reduce the price of the finished material. Depending on the type of filler used, the physical, mechanical, electrical and chemical properties of the material can be modified [49, 59, 74, 80, 89–92]. There is no universal filler that will ensure optimal properties of the material. Its selection depends on the requirements set for the resin in the cured state, as well as on processing conditions. The possibility of deterioration of some resin properties at the cost of improvement of others should be taken into account.

When fillers are introduced, the following benefits are obtained as a result of physical modification [45, 56–59, 73, 78, 79, 83, 87, 89–91, 93–95]:

- increase in heat conductivity of the cured resin,
- decrease in thermal expansion of the cured resin,
- increase in heat resistance, fire resistance and thermal stability of the material,
- extension of the lifetime of the adhesive compound,
- reduction of the thermal effect of the reaction,
- reduction of shrinkage during gelation and curing,
- improvement of some mechanical properties, such as hardness and compressive strength, modulus of elasticity, and significant increase in flexural strength, tensile strength and impact strength,
- reduction of water absorption,
- increase in chemical resistance,
- reduction of internal stresses in the cured resin,
- increase in resistance to ageing,
- possibility to modify dielectric properties, especially increase in resistance to creep currents and electric arc and increase in dielectric strength.

The introduction of fillers into adhesive compounds can also cause the following undesirable effects [47, 59, 73]:

- an increase in viscosity of the compound, which makes processing more difficult (e.g. it requires the additional operation – removal of air bubbles from the polymer castings, frequently required for an elevated temperature process or the addition of a solvent or diluent),
- tendency of fillers to sediment, as a result of which the compound of polymer is not homogeneous,
- deterioration of certain mechanical properties due to using coarse-grained fillers,
- unfavourable influence of moisture adsorbed on the surface of filler grains on dielectric properties,
- possibility of agglomeration of fillers, resulting in deterioration of mechanical properties of the compound,
- increased weight which is not desirable, especially in space and aerospace structures.

Some authors [96] revealed that the addition of certain type of filler, e.g. calcium carbonate, can have a negative effect on the mechanical properties of the modified polymers. Sai Sravani *et al.* [97] similarly have shown that there is a decrease in tensile and flexural strengths with the addition of this filler material.

Fillers should not react with the resin and curing agent and should be neutral or weakly alkaline. Some filler should not be combined with certain curing agents. An example of this is mica, which if added in large quantities can react with some anhydrides or white zinc which speeds up the reaction of resins with amine curing agents. The moisture content of the fillers should not exceed 0.5% [44]. Fillers can be added to resins in varying quantities, e.g. from 25 to 50 wt %. In the case of liquid resins, mainly those with added thinners to reduce viscosity, fillers can be as much as 95% of the total weight of the material. In this case, the resin acts as a binder.

2.3.3 Fillers Types

A general classification of fillers is given in Figure 2.7.

Depending on the type of filler, the physical, mechanical, electrical and chemical properties of the polymer can be adjusted to a large extent [40, 44, 56, 57, 73].

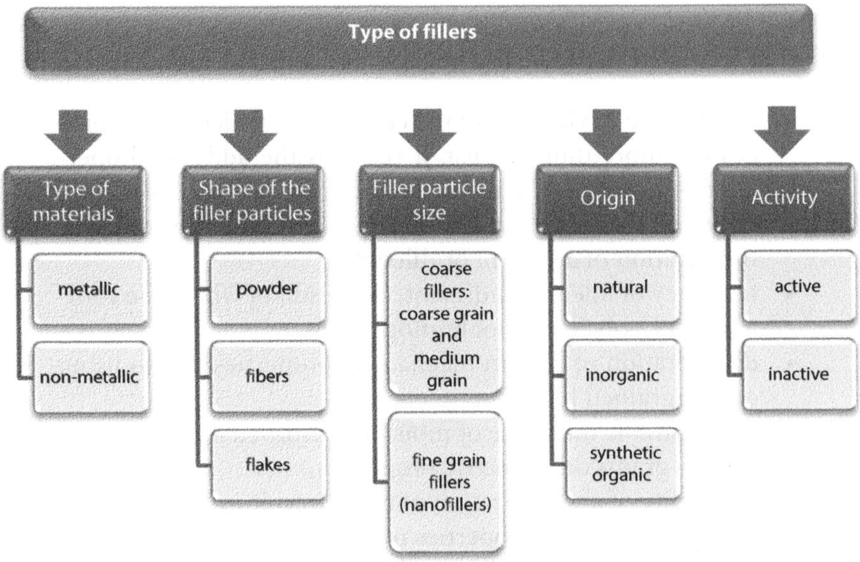

Figure 2.7 Categorization of fillers, based on [40, 44, 56, 59, 73].

2.3.3.1 *Filler Classification Criterion: Type of Material*

Fillers can be divided according to the type of material. One can distinguish between metallic and non-metallic fillers (Figure 2.7).

Metallic fillers include tin, copper, lead, aluminium, bronze, brass, magnesium, zinc or copper oxides [59, 91, 94]. Zinc oxide powder, used for filling polypropylene, elastomers and unsaturated polymer resins, improves their resistance to weathering. Aluminium oxide is also used for filling unsaturated polyester resins, which when added to polymers leads to an increase in their resistance to flame [59, 98, 99]. Other examples of metallic fillers are bronze and aluminium. These powders when introduced into polyamides or polyacetals form compounds that easily conduct electricity. The use of magnesium oxide filler in polymer compounds increases their hardness, stiffness and resistance to flow under load.

In the group of non-metallic fillers, other polymers (e.g. polytetrafluoroethylene (PTFE) or polyethylene (PE)) and mineral fillers such as mica, chalk, silica, and other auxiliary substances such as graphite, carbon black, carbon fiber or glass fiber are used [87, 91, 92, 96, 97]. By using a silica filler, the chemical resistance as well as the electric arc resistance can be improved. Most non-conductive fillers increase the dielectric constant, while they do not significantly change the dielectric strength [40, 91]. Carbon black is a reinforcing filler. The enhancement effect is greater the smaller the particle size of this filler is. It is mainly used as a filler for rubber used in tyre production. Carbon black also acts as a pigment and light stabiliser.

2.3.3.2 *Filler Classification Criterion: Shape of the Filler Particles*

Another classification of fillers is based on the shape of the filler particles. According to this criterion, fillers are distinguished as powder, fiber and flake (Figure 2.7).

Powder fillers are used as additives in polymers. They improve mechanical, thermal, dielectric, chemical and processing properties, mainly of duroplastics (e.g. they facilitate flow, reduce shrinkage and the exothermic effect of the crosslinking process). Such fillers are solid chemical compounds in the form of particles, usually spherical, with grain sizes in the range of 0.5-10 μm. Powder fillers are mainly inorganic fillers (chalk-calcium, silica, graphite, talc, glass, etc. [84, 87–89, 92], but also organic fillers (wood flour) [100, 101].

Powder fillers are mainly used in the preparation of sealings and expansion joints, as well as for lacquered coatings and polymeric plastic linings. Talc is the basic powder filler. It is magnesium silicate with a density of 2.4 g/cm^3 and a hardness of 1 on Mohs scale. When introduced into thermoplastic polymers, it reduces their susceptibility to creep under load. When applied to polypropylene, it significantly increases its stiffness and heat resistance, while its introduction to poly(vinyl chloride) (PVC) carpets increases their shear strength [40].

Fibrous fillers are another type of fillers. They may take the form of continuous or chopped fibers, as well as ropes. The common feature of such fillers is that the length of the fibers is much greater than their thickness. The fibers are used either with or without coating, and sometimes are impregnated, which modifies their surface so as to aid polymer adsorption and chemical reactions between the fiber filler and the polymer matrix. The occurrence of this adsorption or the formation of chemical bonds is an inherent condition for high adhesion [103]. Fibrous fillers are based on natural inorganic fibers (e.g. asbestos fibers) and organic fibers (e.g. cellulose fibers) as well as synthetic fibers [59, 78, 80, 95, 97, 102].

A classification of synthetic fibers is as follows [59, 78, 91, 97, 103–106]:

- fibers based on boron nitride, boron carbide, aluminium oxide,
- glass fibers, quartz fibers, basalt fibers, boron fibers, silicate fibers,

- metallic fibers - aluminium, steel, silver,
- graphite fibers, carbon fibers,
- monocrystalline fibers - fibers based on curable and thermoplastic polymers.

Among the mentioned fibers, glass fibers are the most commonly used.

Another type of fillers is flake fillers. They are available in the form of fabrics, sheets, mats and films. Flake fillers are produced from similar materials as the fibrous fillers. One of these types is polymer-coated flake fillers.

2.3.3.3 Filler Classification Criterion: Filler Particle Size

There is a variety of requirements that relate to the particle size of the fillers [84, 104, 107–109]. The following types of fillers are specified:

a) coarse fillers (particle size in micrometers):
   - coarse fillers (average grain size above 250 μm),
   - medium grain fillers (average grain size: 10-50 μm),
b) fine grain fillers - nanofillers (average grain size below 10 μm, characterised by at least one dimension on the nanometric scale not exceeding 100 nm).

The reason for using particles with small grain sizes is to increase the specific surface area and decrease the tendency to sedimentation. A certain critical size of the particles used for modification has been identified, below which a significant change in the performance of the adhesive is observed. This size ranges from 5 nm for catalytic properties to 100 nm for mechanical properties. The introduced filler even in a small amount (less than 10 wt. %) allows to improve the structural and functional properties. Sosiati *et al.* [107] found that the $CaCO_3$ particle size and the ratio of kenaf to $CaCO_3$ content influenced the impact strength of the hybrid composite based on an epoxy resin.

The nanofillers used in adhesive compounds are usually in the form of platelets, fibers or tubes. Tile-like nanofillers have a larger specific surface area than spherical fillers and so increase the viscosity of modified adhesive compounds.

Fillers and nanofillers, which in addition to the classification above, can also be divided into the following three groups [87, 94, 104]:

- 3D – "powdery" and spherical having all 3 dimensions in the nanoscale. These include materials in the form of: carbon

black, silica, silicates, chalk, aluminium hydroxides, carbides, oxides, borides, metal nitrides, various salts, as well as metals themselves (e.g. Au, Ag, Pt, Cu) and their oxides (e.g. SiO_2, Al_2O_3, ZnO, TiO_2). Powder nanofillers added to the polymer matrix modify many properties including electrical and magnetic; and increase in hardness, abrasion and flame resistance, and can even activate antibacterial features,
- 2D – lamellar, characterized by two dimensions in the nanoscale. 2D fillers include layered silicates and alumina-silicates. These are naturally occurring clay materials which are the main component of bentonite. The most common is montmorillonite, as well as saponite, hectorite, beidelite and mica,
- 1D – linear, characterised by one dimension on the nanoscale. 1D nanofillers include structures formed, among others, by carbon in the form of nanofibers and nanotubes. They are characterised by a high ratio of length to diameter.

In addition to particle shape and dimension, fillers and nanofillers can differ in physical structure (crystalline, amorphous, gaseous inclusions - nanofoams) and chemically and can be divided into [58, 103, 104]:

- inorganic fillers, which include:
  - silicates, talc, chalk, kaolin, mica,
  - oxides: aluminium, titanium, zinc, magnesium,
  - powders of metals or alloys: copper, aluminium, steel, tin, silver, bronze,
  - glass: microballoons, beads, fibers, etc,
- organic fillers, which include:
  - proteins: keratin,
  - wood flour, sisal, cellulose fibers,
  - synthetic fibers: polyamide, polyacrylonitrile,
  - hydrocarbon fillers: carbon black, graphite, carbon fiber, etc.

2.3.3.4 *Filler Classification Criterion: Origin*

Fillers can also be divided according to their origin. A distinction is made between natural, inorganic, and synthetic organic fillers [2, 3, 7, 29, 59, 95] (Figure 2.7).

Natural organic fillers include e.g. wood flour [100, 101, 110–113], cellulose fibers [22, 114–116], sisal fibers [1, 117–119] or flax fibers [120, 121]. Wood flour is one of the cheapest fillers used for polymers. It is obtained by grinding wood sawdust. It is mainly used to fill phenol-formaldehyde resins or polypropylene and polyethylene composites [101, 110, 112, 122, 123].

However, inorganic fillers are most commonly used, which include quartz flour, graphite, talc, silica, porcelain flour, chalk, kaolin, alumina, quartz, mica, powdered metals or their oxides, etc. [40].

These fillers are mainly of mineral origin. They are obtained usually by grinding natural materials, which are typically cheaper than synthetic substances. The grinding can be carried out either wet (with the addition of water) or dry. The method used to produce the filler determines the fineness, purity, shape and particle-size distribution (PSD), which have a considerable influence on the properties of the cured resin. Quartz flour, due to its good abrasion resistance, is used to obtain poly(vinyl chloride) (PVC) floor coverings. It has good dielectric properties and its thermal conductivity and dimensional stability can be modified.

The criterion of origin also divides the fillers into synthetic fillers, which include the already mentioned glass fibers, carbon fibers, graphite fibers, glass beads, etc. [59, 103].

2.3.3.5 *Filler Classification Criterion: Activity*

An additional division of fillers is the classification into active and inactive fillers. By adding a filler to a polymer, a number of properties can be influenced, such as [59]:

- increase in mechanical strength,
- increase or decrease of the friction coefficient,
- increase in chemical and thermal resistance,
- improvement of thermal conductivity,
- change of electric properties.

Such a group of fillers is called active fillers. Fillers which do not change the properties of the material are called inactive fillers. They can only be used to reduce the cost of the material.

Chalk is an example of an inactive filler. This is mainly used to reduce the price of the material but it generally reduces the mechanical properties of the material.

2.4 Biological Reinforcement of Epoxy Adhesives

2.4.1 Introduction

More and more research, as well as work on the implementation into mass production, is devoted to composites of synthetic polymers with natural materials [123–131]. The production of new polymer composites filled with organic materials is gaining more and more interest due to their numerous advantages. These composites are gradually replacing the glass fiber-reinforced laminates. The natural fillers are replacing the mineral fillers in many new materials [2, 5, 95, 117, 132–134].

One class of the additives to modify epoxy adhesive compounds are modifiers of natural origin. Biodegradable polymers are also often used as matrices in composites [135, 136]. Moreover, biodegradable polymers are reinforced with fibers of natural origin in order to maintain their characteristics of an environmentally friendly material [7, 29–31, 115, 137–140].

Composites containing natural filler materials have very good functional properties; moreover, they can be recycled [112, 123, 141–144]. Currently, the introduction of organic fillers into a polymer matrix such as starch [145, 146], wood flour or wood fibers [147, 148], coconut fibers [134, 149] or straw [126, 150, 151], is now a widely used industrial practice. It should be emphasized that composites with organic fillers, even in the case of using a very high proportion of the filler (up to 80 wt %), often do not lose the functional properties of the finished product. Frequently used natural reinforcing and filling materials used to obtain composites with polymers include fibers and wood. Currently, composites containing flax, hemp, sisal, kenaf, ramie, jute and wood with various degrees of fragmentation are commonly used.

2.4.2 Types of Biological Reinforcements

Biological reinforcement materials for epoxy adhesives and polymer composites, including those based on epoxy resin, include [2, 3, 132, 142, 152–170] (Figure 2.8):

- natural fibers,
- wood, which although is classified as natural fibers, is often presented as a separate group due to its characteristics,
- plant oils,
- fungi,

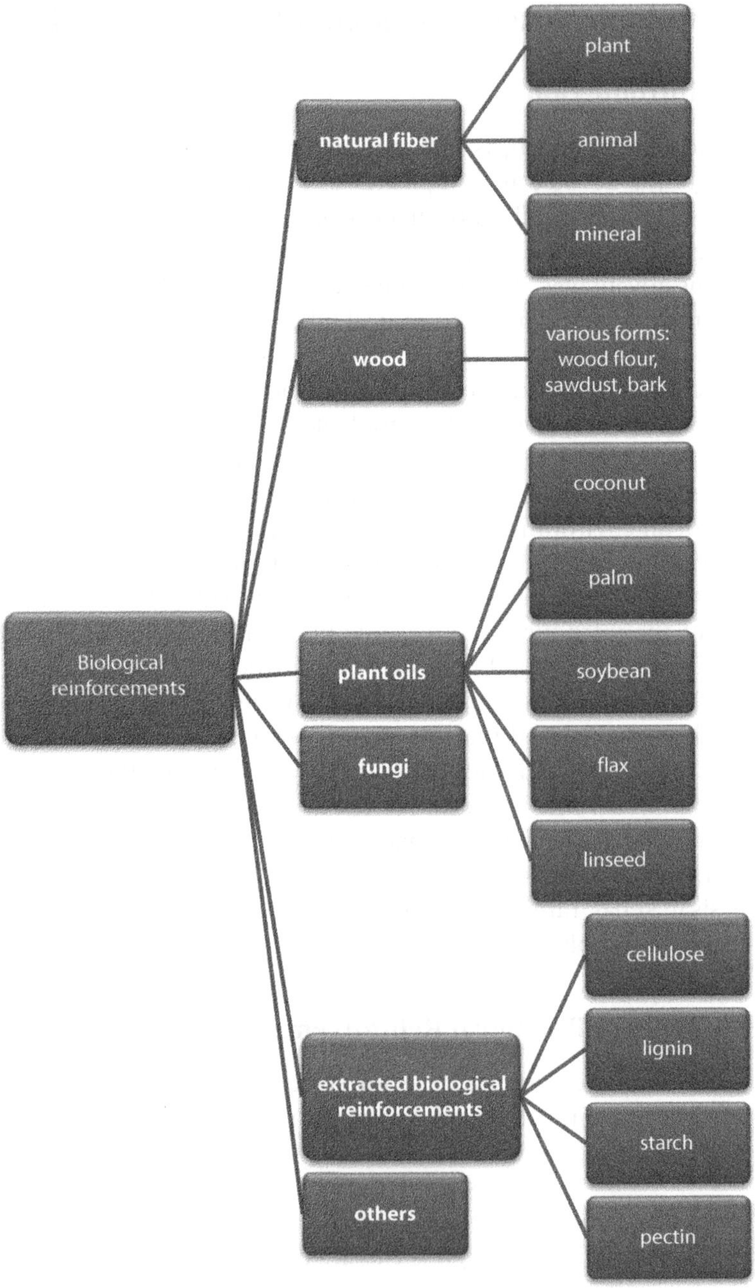

Figure 2.8 Biological reinforcements for polymers, based on [2, 3, 132, 141, 152–170].

- extracted biological reinforcements, which are components of plant structure, such as:
  - cellulose,
  - lignin,
  - starch,
  - pectins,
- others (e.g. pine cones).

2.4.2.1 Natural Fibers

Natural fibers occur in nature in a ready-to-process state. They can be provided by plants, animals and in the form of minerals [30, 133, 140–143, 170]. The characteristics of natural fibers are given in section 2.4.3, and an overview and characterization of fibers of plant origin is given in section 2.4.4.

2.4.2.2 Wood

Wood is a type of natural composite made of numerous fibers. It contains cellulose (the organic compound that binds the cell walls of wood) and lignin (the substance that binds cellulose in plants) [101, 109, 112]. The structure of the wood is anisotropic - the fiber is relatively difficult to cut, but the fibers can be separated easily along the cross section.

There are many types of wood and each type has its own properties (they differ not only in hardness but also in other parameters) and a characteristic structure. Different types of wood are selected for different uses. The properties of wood depend on its species (and, in the case of mechanical properties, on the plane at which these are determined). Most often, the density of wood at a humidity of 15% is lower than the maximum water density (1 g/cm^3) and in the case of wet wood it does not exceed 1.5 g/cm^3. However, in the case of balsa, very low densities of 40-180 kg/m^3 (0.04-0.18 g/cm^3) are found [171].

The strength properties of wood depend primarily on the decomposition of cellulose polymers in the cell wall, which forms the supporting skeleton of the wood, and on the amorphous lignin filling the free spaces of the skeleton. Cellulose influences elastic properties, and lignin - plastic properties [158, 172–180].

Wood under the influence of long-term load shows rheological features. Creeping and relaxation phenomena occur. When the wood grain is stretched, stress corrosion cracking also develops over time. That is why wood is distinguished by temporary and long-term durability.

Wood (especially waste wood from the furniture industry) in various forms (fractions) is used as a filler in polymer composites, thus obtaining wood-polymer composites (WPCs). For the production of wood-polymer composites, thermoplastic polymers based on polyolefins (polypropylene, polyethylene) and poly(vinyl chloride) are used [101, 109, 110, 112, 180, 181]. Wood as a waste material is relatively cheap and may constitute even 70% of the composite filling [182].

The following forms (fractions) of wood are used as reinforcement for polymer materials [100, 101, 110, 111, 123, 182]:

- pieces of wood,
- bark (cortex),
- shavings,
- wood flour,
- wood dust.

The particle sizes of the mentioned fractions depend on the technology used and the type of machines. For the production of elements made for WPCs with thin walls, it is recommended to use fine fractions of sawdust or wood dust.

Selected properties and applications of frequently used reinforcements of polymer composites in the form of wood flour and cork are presented below.

2.4.2.2.1 Wood Flour

Wood flour plays the most important role as the filler and polymer reinforcement [177–179]. Wood flour is a material composed of particles of wood crushed and segregated using special sieves. The basic elements of wood flour are carbon (49.5%), oxygen (43.8%), hydrogen (6.0%), nitrogen (0.2%) and others. Chemical composition of wood flour includes about 50% cellulose, 20-25% lignin, 20-25% hemicellulose. Wood flour also contains simple carbohydrates, protein, starch, tannins, essential oils, natural rubber and mineral salts. The chemical composition depends on the type of tree, climate, soil, etc.

It is one of the cheapest fillers that are used for polymers. It is obtained by grinding wood sawdust. It is mainly used for filling phenol-formaldehyde resins (phenolic molding compounds).

Wood flour is obtained from raw wood and is characterized by a loose and very fine consistency. The material is characterized by very good mechanical properties (including water insolubility, durability, resistance to external factors, low thermal conductivity, the ability to easily bind with other substances, plasticity) which makes it a more and more commonly used prefabricated product in various industries, mainly in furniture and construction. The low price and the ease of obtaining this material are also important.

Wood flour can be used as a filler for various polymer materials [177–179]. In addition to epoxy additives, commonly used wood-polymer composites (WPCs) also appeared on the plastics market already in the nineties of the twentieth century. Wood composites are widely used in industry as a substitute for wood. They are resistant to weather conditions, especially moisture. Composites based on polypropylene (PP), high-density polyethylene (HDPE) and poly(vinyl chloride) (PVC) are of major importance [122, 123, 182]. In the case of biodegradable polymers, poly(lactic acid) (PLA) is the most promising polymer for this type of application.

2.4.2.2.2 Cork

Cork is an impermeable, water-resistant, flexible and fire-resistant material obtained from the bark of cork oak. It consists predominantly of suberin-saturated cork tissue. The cork can be used for reinforcement as a powder or as a granule [183–187]. Cork powder and granules are the main by-products of the cork industry [187].

Cork granules are widely used in many industries. These can be used for the production of [130, 149, 166, 183, 187]:

- composite cork products,
- surfaces of sports fields,
- soil improvers,
- agglomerated cork stoppers,
- cork flooring,
- shoe soles,
- insulating cork blocks,
- insulating mass.

Pintor *et al.* [187] presented the properties of cork and cork powder and their application in adsorption technologies. Many applications for this product are envisaged, from cork stoppers to incorporation into agglomerates and briquettes for use as an adsorbent in the treatment of gaseous emissions, water and wastewater.

Cork biomass was used in its original form as a biosorbent for heavy metals and oils, and is also a precursor of activated carbon for the removal of organic pollutants in water and volatile organic compounds in the gas phase.

The use of cork to solve environmental problems, namely oil spills, is the first commercial application of the cork as a potential sorbent. The variety of raw materials that can be produced or extracted from cork has attracted the attention of many researchers.

The chemical composition of cork has been studied since the 18th century [188], but a full knowledge of the chemical properties of cork, and even less of all its potential uses and transformations, is not yet complete.

According to the Portuguese standards NP-114 and NP-273, cork powder is a material with dimensions smaller than 0.25 mm [189]. There are different types of cork powder depending on the origin: grinding, granulating or pre-grinding powder; cleaning powder, no pollution; finishing powder from cutting and grinding operations; finishing powder from agglomerated cork slabs; finishing powder from agglomerated cork and discs; insulating powder from cork slabs [190].

A mixture of these powders is considered "flaming powder" as it is used to feed boilers due to its high calorific value [189]. Other applications include use as a filler, mixed with adhesives, to improve cork quality: in linoleum production, in agglomerates, and briquettes [191].

Being an environmentally friendly and cheap material, cork powder can also be used to produce activated carbon with high specific surface, comparable to commercially available activated carbon, or it can be used directly for the adsorption of pollutants, as a biosorbent. It is also used commercially as an absorbent in oil spills.

Barbosa *et al.* [166] emphasized in their work that there is a growing interest in developing methods to improve the strength of the epoxy adhesives. These authors presented an overview of current advances in the use of reinforcement materials and findings on the use of microparticles to increase the strength of adhesives. They also discussed the use of materials of natural origin as reinforcing materials, with particular emphasis on the use of cork particles as a hardening material for epoxy adhesives. The cured epoxy adhesive generally contains elastic or thermoplastic domains dispersed discontinuously in the resin matrix to increase resistance to crack initiation.

Barbosa *et al.* [130] in another work also indicated that the modification of epoxy resins, which are the basic component of epoxy adhesives with particles (nano or micro), was an effective method for improving the strength of the structural adhesives. In the presented research, natural cork microparticles were used to increase the strength of brittle epoxy adhesives. The concept is for the cork particles to act as a plug for the cracks, which leads to greater energy absorption. The influences of cork particle size, quantity and surface treatment were investigated. Cork particles ranging from 38 to 53 and 125 to 250 μm were admixed with Araldite 2020. The amount of cork in the adhesive varied from 0.25 to 1% by volume. It was possible to conclude that cork can improve strength, and the quantity and size of cork and the use of plasma surface treatment affect the mechanical properties.

Fernandes *et al.* [191] presented research on biocomposites made of a combination of various biodegradable aliphatic polyesters with cork (30% by weight). Physico-mechanical and thermal properties of matrices and biological cork composites were investigated. Observations of the fracture morphology exhibited good physical adhesion in the cork–matrix system. Cork increases the degree of crystallinity of biocomposites. Cork biomass contributes to produce lightweight and sustainable biocomposites. The results obtained indicated that cork-polymer biocomposites are a viable alternative to developing more sustainable composite materials such as car elemenents and wine bottle caps.

2.4.2.2.3 Bark (Cortex) Fibers

Bark tissues amount to 10–20 wt % of woody vascular plants and are composed of various biopolymers, tannins, lignin, suberin and polysaccharides [192]. Up to 40% of the bark tissue is made of lignin, which forms an important part of a plant, providing structural support by crosslinking between different polysaccharides, such as cellulose [192]. Industrial bark was found to contain a high amount of wood (up to 21%), a sufficient amount of tannin for industrial extraction (10.7% of wood-free bark), and a high amount of non cellulosic glucose, varying according to the felling season (7.7–11.5% of wood-free bark) [193]. The bark (cortex fibers) is used, *inter alia*, as a biological reinforcement of polymer composites [194].

Stepczyńska *et al.* [194] conducted research to determine the possibility of using cortical fibers as reinforcement in the poly(lactic acid) (PLA) matrix and the production of biocomposites that may cause bacterial lethality and increase the rate of biodegradation. The evaluation of the mechanical properties of the composites showed that the tensile modulus improved with increasing cortical fiber content, but the tensile strength and elongation at break decreased. The addition of natural fibers reduced the impact strength of PLA composites. The addition of cortical fibers caused two competing effects: the plasticizing effect associated with the extraction of oils and other plant substances present in the *Lapacho* cortex during the processing of the composite, and the mechanical strengthening effect typical of cellulose fillers. However, the enhancement effect was only observed when the filler content was significant. Thermal stability increased due to the addition of cortical reinforcement. The oils and other substances found in *Lapacho's* bark presumably increased the durability of the composites, leading to an increase in the temperature at which their thermal decomposition began. The sample biodegradation studies showed that the biocomposites had a greater weight loss compared to pure PLA. In addition, the bark fibers support bacterial/fungal colonization by stimulating extensive invasion and microbial colonization. Moreover, cellulose, which is the main structural compound of the cerebral cortex, increased the efficiency of the enzyme activity. The results presented indicate that *Lapacho* bark is not an active biocide against *Escherichia coli* and *Staphylococcus aureus* strains. The biocomposites produced caused bacterial mortality, but still do not meet the requirements of ISO 22196: 2011 and the biocidal activity of oils and other substances present in *Lapacho* bark was insufficient.

2.4.2.3 Vegetable Oils

Vegetable oils are liquid fats of vegetable origin which retain a liquid consistency at room temperature. In terms of chemistry, they are a combination of triglycerides of higher saturated and unsaturated fatty acids. Vegetable oils, depending on the percentage of individual acids in the fat molecule, exhibit different properties. Most often, oils are obtained from seeds, sprouts and fruits, by extraction or pressing.

Vegetable oils are used, among others, in [134, 139, 160, 195]:

- food industry,
- cosmetology,
- construction,
- furniture industry,
- automotive as a biofuel,
- for the production of epoxidized vegetable oils used in the manufacturing of biomaterials, including epoxy biobased,
- adhesives.

Vegetable oils can be an essential component of bioresins [139, 159, 160, 195, 196], also used in composite materials, or serve to modify the properties of the polymer, e.g. its plasticization. The use of such bio-resin and natural reinforcement, among other fibers, allows the carbon footprint to be reduced.

Epoxides based on soybean oil, which is one of the most easily available vegetable oils in the world, have a high potential for use in the synthesis of polymeric materials [195, 196]. In addition, linseed and castor oils are also of great importance in this area [197].

Epoxidized linseed oil (ELO) is one of the more intensively researched compounds in terms of obtaining new epoxy resins. A method of obtaining thermosetting polymer networks based on ELO and dicarboxylic acid anhydrides as cross-linking compounds was developed. The influence of anhydride and various types of catalysts, such as tertiary amines and imidazoles, on the cross-linking mechanism, structure and thermomechanical properties of the obtained material were investigated [197]. In addition, mixtures of ELO with epoxidized soybean oil (ESO) were used with phthalic anhydride and maleic anhydride as the crosslinking agent and with benzyldimethylamine as the catalyst. The properties of the products obtained were compared with materials based solely on ELO or ESO. The most favorable physicochemical and mechanical properties were demonstrated by the materials obtained from ELO and the ELO: ESO mixture (80:20% by weight) [160].

Epoxidized linseed oil and cyclo-carbonated linseed oil (CLO) have proven to be effective reactive diluents for use in epoxy resin based compounds, coatings and adhesives [198]. Contrary to epoxidized soybean oil, the most frequently tested reactive bio-diluents ELO and CLO did not have a negative effect on the mechanical properties of the obtained thermosetting materials. Their addition resulted in an improvement in tensile strength, fracture toughness and adhesion to aluminum.

2.4.2.4 Fungi

Fungi are found all over the globe. They live mainly on land, although they are also found in inland and marine waters. They are also commonly present in forests, soil, dead tissues, living plants and animal organisms. There are many types of fungi in the world. So far, over 100,000 species have been identified. Most of them are small and even microscopic organisms [199–201].

Fungi play a very large and varied role in nature. One example of fungi is saprophytes that live on dead organic plant or animal remains. They break down organic matter and prevent the build-up of dead debris. They also participate in soil-forming processes - they enrich and fertilize the soil, creating humus. Among the common macrofungi that decompose forest litter, the fungiflies, the ceps, andthe fungusflies can be mentioned. A large group of saprophytic fungi are selective with respect to trees. The sick and the weak trees are removed and the healthy ones are left alive. They also cause the wood to rot. Microscopic fungi also have this ability and cause grey rot. This type of fungus can cause discolouration of the wood, which is particularly unfavourable from a utilitarian point of view. They are food for many forest animals and provide shelter for many insects. Some fungi live in symbiosis with tree roots. In this way, they supply them with valuable nutrients (mycorrhiza).

Also, the importance of fungi in the human economy and in the industry is very diverse both positive and negative [199–202]. Among the positive features, fungi are used for the production of antibiotics and vitamins A, B2, B12, and vaccines, in the baking industry (yeast, in the dairy industry, can be used in the fight against pests and are food for humans and animals. such as: causing human, animal and plant diseases, destruction of food and industrial materials, the formation of indoor molds.

But they can also be used as an additive to polymer compounds, including epoxy [157, 158]. An example of the use of fungi as an additive to epoxy adhesive compounds can be found in section 2.5.

2.4.2.5 Extracted Plant Ingredients

All plant fibers are made of cellulose [172]. In addition, plant fibers contain small amounts of lignin (a polymer whose monomers are organic compounds that are derivatives of phenolic alcohols) and plant adhesives (pectins) [173–176, 202]. Plant components often used to reinforce adhesives, resins and epoxy materials are characterized below.

2.4.2.5.1 Cellulose

Cellulose $(C_6H_{10}O_5)$ n (n is the number of glucose residues) is one of the most abundant organic polymers found in the world. It is a polysaccharide made of β-glucose linked by β-1,4-glycosidic bonds [203].

A cellulose molecule is an unbranched polysaccharide chain consisting of 2,500-10,000 β-glucose residues interconnected by glycosidic bonds. The basic structural unit of cellulose is cellobiose. Long chains form bunches called micelles. In the cell wall, micelles combine into fibrils by connecting by hydrogen bonds. Long fibrils made of cellulose form many layers. Between the bundles of cellulose are the remaining polysaccharides that connect them [204].

Cellulose as a carbohydrate is an organic compound consisting of carbon, hydrogen and oxygen that functions as energy source for living organisms. Plants are able to make their own carbohydrates, which they use to produce energy and build cell walls. There are several different types of carbohydrates depending on how many atoms they have, but glucose is the simplest and most common in the plants. Plants produce glucose, use it as an energy source, or store it as starch for later use. The plant uses glucose to produce cellulose when it combines many simple glucose units to form long chains. These long chains are called polysaccharides, which are long molecules that plants use to build their structures.

Some of the foods that the plant produce when converting light energy into chemical energy (photosynthesis) are used as fuel, and some are stored. The rest are turned into cellulose to serve as the main building material for the plant. Cellulose is ideal as a construction material because its fibers give strength to the leaves, roots and stems of the plant.

Cellulose is a significant structural component of the primary cell wall of green plants, various forms of algae and oomycetes, as well as fungi and bacteria. For example, wood consists of about 50% cellulose and cotton contains about 90% cellulose. The structure of the cell wall depends on the function of a given cell and its location. The wall of the crumb cells will be different from that of the scleroderma cells. Cellulose can make up to 40 to 60% of the dry weight of the cell walls. Along with cellulose, there are other supporting substances, such as lignin, pectin or hemicellulose, which build the transverse connections between cellulose fibrils.

Due to long chains, cellulose does not dissolve in cold and hot water and in organic solvents. It is also resistant to dilute acids. Hydrolysis only takes place in the presence of concentrated acids. Cellulose is dissolved in Schweitzer and Cross-Bewen reagents. Its bonds are broken by enzymes called cellulases. In the course of the reaction, cellobiose (disaccharide) is formed and the end product is glucose.

Cellulose can be obtained by various extraction procedures, using various processes such as oxidation, etherification, and esterification, which convert the prepared celluloses into cellulose derivatives. Due to the fact that it is a non-toxic, biodegradable polymer with high tensile and compressive strengths, it is widely used in various fields such as nanotechnology, pharmaceutical, food, cosmetic, textile and paper industries, drug delivery systems for cancer treatment and other diseases.

Microcrystalline cellulose in particular is one of the most commonly used cellulose derivatives in the food, cosmetic, pharmaceutical industries and is an important excipient due to its binding and tablet properties, characterized by plasticity and wet cohesiveness. Due to its high availability, tasteless and odorless nature, bacterial cellulose has many industrial applications.

In the field of pharmaceuticals and food technology, modification of the cellulose structure with other chemical groups leads to the production of structures offering better biocompatibility, flexibility, stability, and emulsifying action. Moreover, cellulose, being indigestible for humans, usually has a zero calorific value and can therefore be added to foods for many purposes. Compounds such as hydroxypropyl methyl cellulose (HPMC), sodium carboxymethyl cellulose, hydroxyethyl cellulose and others are widely used in the pharmaceutical industry and food technology [205–207].

2.4.2.5.2 Lignin

Lignin is a naturally derived polymer that is part of the wood cell together with cellulose and hemicellulose. The lignin content in the lignocellulosic raw materials remains at an average level of 15-30%, while the share of cellulose is about 50% [174, 176].

About 20% of aromatic compounds and CH_3-O- methoxy groups are found in lignin, from which methyl alcohol is formed by dry distillation of wood. The quantitative share of lignin in wood ranges from 26% to 30%, depending on the tree species and the degree of lignification of the cell membrane [208].

As a binder, lignin makes the wood cell structure compact, increases the compressive strength of the wood and maintains its rigidity. Lignin can be found in the walls of water-conducting elements: coils and vessels, and saturates the walls of sclerenchyma cells. It provides these tissues with mechanical resistance and durability. Due to its chemical composition, it also has a protective function, i.e., it is insensitive to microorganisms. The elimination of lignin (by delignification), by adding sodium hydroxide (a cooking liquor with a strongly alkaline reaction), from the wood leads to the softening of the wood substance, which is a necessary process in the production of paper.

Lignin is a polymer whose monomers are organic compounds derived from phenolic alcohols (including coniferyl alcohol, sinapine alcohol, and coumaryl alcohol). The chemical structure of lignin is cross-linked with ether and carbon-carbon (C-C) covalent bonds. Lignin is decomposed by fungi, causing the wood to rot white. The most common monomeric decomposition products of lignin are vanillin and vanillic acid [176, 202].

Lignin is a natural polymer that is also available in modified forms as industrial by-product. Industrial lignins cannot be directly used for the production of biomaterials with acceptable product parameters. Hence, there is a need for pre-treatment to reduce the sulfur content and odor and to improve the properties of the lignin so that it can be used as a filler to reinforce composites and as plasticizers [69, 209–211].

Wood *et al.* [211] noticed an increase in impact properties of the fabricated composites with the energy absorbed by the composite containing 5 wt % lignin being 145% higher than the composite with no lignin added. Both flexural and tensile moduli showed increases when lignin was added up to 2.5 wt %, although there was a drop in both when the lignin was increased to 5 wt %, attributed to poor mixing and infusion due to the increased viscosity of the resin. In all cases, the addition of lignin increased the structural properties of the composites to some degree when compared with composites with no additional lignin.

Yamini *et al.* [212] characterized epoxy matrix composites modified with fillers based on lignin. From lignin-based industrial wastes, a reactive filler in the form of cyclocarbonate lignosulfonate (CLS) was prepared. Then CLS (10, 20 and 30 wt %) was mixed with diglycidyl ether of bisphenol A (DGEBA) epoxy resin followed by amine curing to produce biocomposites. The spectral, thermal, thermomechanical, mechanical and morphological properties of the CLS-epoxy composites were investigated by FTIR, DSC/TGA, DMTA, tensile/bend tests and SEM, respectively. Compared to the pure DGEBA network, by increasing the CLS% in biocomposites, the tensile strength was reduced, the flexural modulus was well preserved, and the Young's modulus and the storage modulus were significantly improved. This has been attributed to the stiffening effect of the modified lignin particles and the formation of urethane bonds and intermolecular interactions such as hydrogen bonds. Incorporating carbonated lignin into epoxy can provide an eco-friendly approach to low-cost, high-performance biocomposites.

2.4.2.5.3 Starch

Starch is a white, granular and organic chemical product. Starch is produced in the green leaves of plants from excess glucose produced during photosynthesis [213].

The basic chemical formula of the starch molecule is $(C_6H_{10}O_5)_n$. Starch is a polysaccharide comprising glucose monomers joined in α-1,4 linkages. The simplest form of starch is the linear polymer amylose; amylopectin is the branched form. Starch is stored in chloroplasts in the form of granules and in such storage organs as the root of the cassava plant; the tuber of potato; the stem pith of sago; and the seeds of corn, wheat, and rice.

Most commercial starch is made from corn, although wheat, tapioca, and potato starches are also used [16, 145, 146]. Commercial starch is obtained by crushing or grinding starch-containing tubers or seeds and then mixing the pulp with water; the resulting paste is free of its remaining impurities and then it is dried. Aside from their basic nutritional uses, starches are used in brewing and as thickening agents in baked goods and confections. Starch is used in paper manufacturing to increase the strength of paper and is also used in the surface sizing of paper. Starch is used in the manufacture of corrugated paperboard, paper bags and boxes, and gummed paper and tape. Large quantities of starch are also used in the textile industry for warp sizing, which imparts strength to the thread during weaving.

One of the uses of starch is as a biological reinforcement of polymer composites [146].

2.4.2.6 Nut Shells

Nut (walnut, hazelnut and almond) shells are hard, light and natural abrasives. Abrasive grains are produced from crushed, cleaned and sieved walnut shells. Nut shells are classified as reusable abrasives. Nut shells are 100% natural product and contain no free, crystalline, chemically unbound SiO_2 silica. Nut shells can be used as biological reinforcement in polymer composites [22, 128, 165, 214–216].
Characteristics of the walnut shells are as follows [217]:

- shape: sharp-edged,
- hardness: Mohs scale 2.5 - 3.5,
- specific gravity: 1.0 - 1.2 g/cm^3,
- bulk density: approx. 0.7 g/cm^3,

Prabhakar *et al.* [165] presented the results of their research on composites madeusing the waste peanut shell powder (PSP) as natural filler in an epoxy resin matrix. The natural filler extracted by the manual process was treated with an alkaline solution at concentrations of 2,5 and 7 wt %. Composites were made by varying the weight fraction of filler in the range from 5 to 15% by weight. The influence of bio-filler content in composites on tensile and thermal properties was assessed using Fourier transform infrared spectroscopy (FTIR), universal testing machine, scanning electron microscopy (SEM) and thermogravimetric analysis (TGA). The research showed that the tensile strength and the tensile modulus increased with the increase of the biofiller content. The highest mechanical properties of the epoxy composite loaded with 7% alkali treated PSP were achieved with the biofiller mass content of 15% by weight. The morphology of the composites shows a better binding of the filler to the resin, which leads to an improvement in the mechanical properties. TGA results showed that the polymer composites had increasing thermal resistance with an increase in NaOH concentration and filler content.

Singh [214] presented the problems of epoxy resin modification using walnut particles. 10, 15, 20 and 25 wt % of walnut particles were mixed with epoxy resin. Scanning electron microscopy (SEM) showed that the walnut particles were well dispersed in the epoxy resin matrix. The addition of walnut particles increased the elasticity modulus of the biocomposite. The addition of walnut particles in biocomposites lowered the compressive and tensile strengths. However, the addition of walnut particles in biocomposites increased the hardness. Flexural modulus also increased with an increase in the weight percentage of walnut particles, while the flexural strength and deformation decreased with an increase in the weight percentage of walnut particles.

2.4.2.7 Straw

Straws (of different forms) can be used as biological reinforcements in various types of polymer composites [121, 126, 150, 151, 218]. The following types of straws are used as reinforcement in polymer composites:

- rape (rapeseed) straw [219–222],
- rice straw [151, 221, 223],
- flax straw [121, 224],
- wheat straw [126],
- triticale straw [150],
- legume straw [221].

Rapeseed straw has a high carbon to nitrogen ratio (C:N) of 84:1. By comparison, the C:N ratio is 15-20:1 [219]. Therefore, there is a widespread belief that organic compounds readily available in straw stimulate the growth of soil microorganisms that use nitrogen in the soil and are introduced in mineral fertilizers for their development.

The high content of hemicellulose and cellulose in straw makes it a suitable raw material for production of oligo- and monosaccharides and other useful products from their degradation like aldehydes, ketones carboxylic acids during their hydrothermal depolymerization [22]. The cellulose in the woody part of the rapeseed straw stem provides the cell walls with high stiffness and tensile strength.

About 40% of cellulose and 20% of lignin are found in the lignified fragments of the stalks of rapeseed cultivars. The cellulose content is comparable to the cellulose content in selected types of wood, both coniferous and deciduous. Based on the results of the presented research and analogy with lignocellulosic material from various tree species, it can be concluded that woody rapeseed stalks can be a valuable source of lignocellulosic material intended for the production of thermoplastic polymer composites [222].

2.4.3 Natural Fibers

Fibers are strongly elongated cells with cell walls containing high amount of cellulose. They are found in all organs of all plants [225]. Cellulose is present in fibers with a share of over 50%, reaching the highest shares (over 90%) in cotton and white brush fibers (ramie), over 70% of cellulose in fibers of flax, hemp, and kapok [16, 18, 142]. In addition to cellulose, the fibers contain the so-called encrusting substances: hemicelluloses, pectins, lignins, waxes, and to a lesser extent tannins, resins, proteins and fats. The different proportions and compositions of these substances have an impact on the technological processes and properties of the fibers. For example, pectins are responsible for bonding of fiber bundles with adjacent tissues. The extraction of the fibers requires prior decomposition of the pectins [142, 225]. Fibers of various plants differ in their chemical, physical, morphological and anatomical properties, which translates into their different length, color, shape, composition, strength, durability and water resistance [142, 170, 225, 226].

Among the natural fibers, the following fibers are distinguished (Figure 2.9) [142, 170, 225–227]:

- plant,
- animal,
- mineral.

Plant fibers are obtained from different parts of plants (Figure 2.9) [6, 23, 142, 225, 226]:

- seeds (e.g. cotton),
- stalks (e.g. flax, hemp, jute, ramie),
- leaves (e.g. sisal),
- fruit (e.g. coconut fibers).

Fibers of animal origin are obtained, for example, from the hair of mammals (wool, bristles), and also from animal secretions (silk) [170, 224].

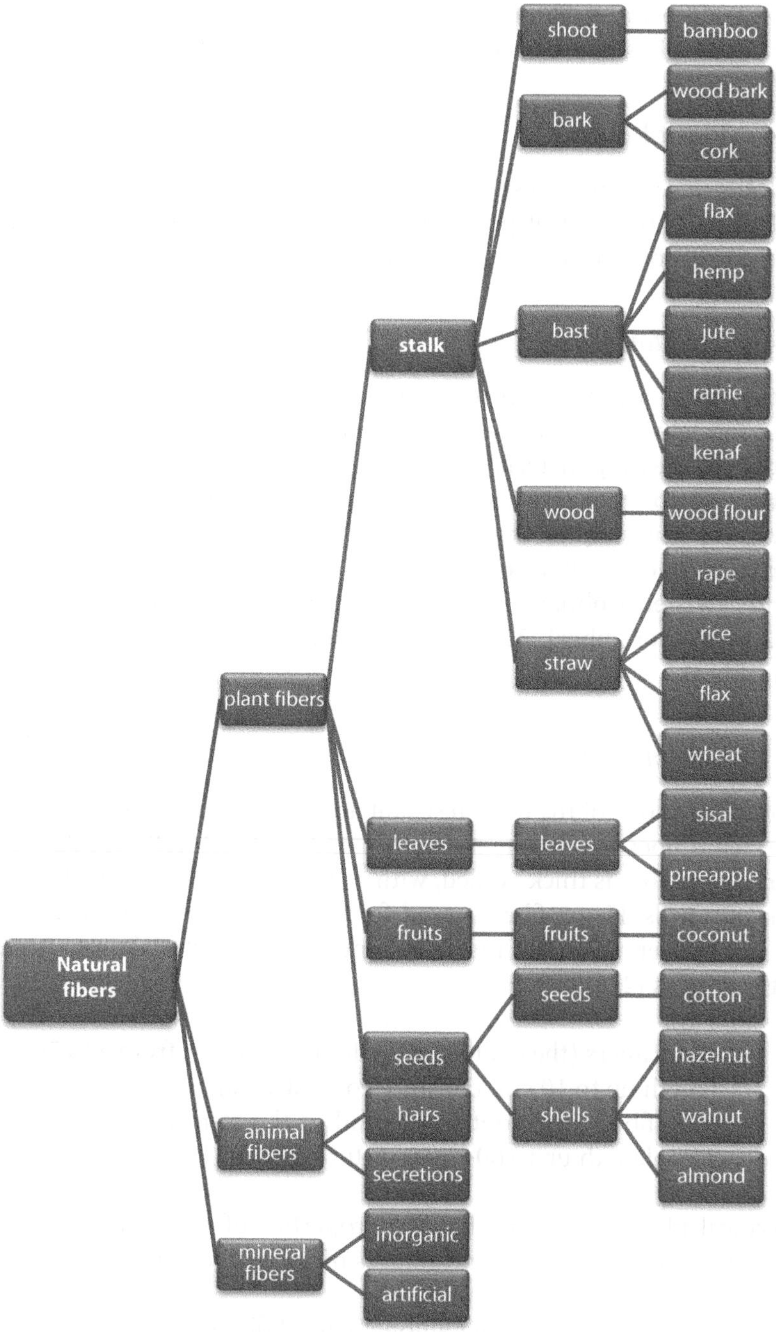

Figure 2.9 Classification of natural fibers, based on [142, 225, 226].

Linen fiber is used to produce a fabric that is characterized by high resistance to friction and stretching. It is characterized by high hygroscopicity, strong wet swelling, high crumbling and stiffness, but also high durability. Flax fiber is used in its original form to make paper and even banknotes.

Linen fiber as used in composites is utilized in various industries: ecological construction, recreational equipment, car equipment, furniture and also as biological reinforcement of polymer composites [239–243].

Linen fibers are used for the production of yarns, ropes as well as tows, non-woven materials, etc. In the form of flax yarn, it is used as a component of a composite that can be used in furniture as a replacement for glass fibers.

Short flax fiber is used for the production of low-count yarns by the dry-spinning method, as well as for the production of tarpaulins. Low-grade short fiber is used to make nonwovens, which are widely used in the production of paper. Cotton fiber is used in the manufacture of twisted products: twine, cotton wood. In addition, it is used in the production of fiber-reinforced polymer composites and fiberboards. The latest area of application of short flax fiber in combination with composite materials is the production of a large number of parts for car bodies, airplanes, ships, cars. Tayfun *et al.* [243] underlined that flax fiber-reinforced polymer composites have become popular in textile, transport, and construction markets and also these composites have many modern applications, and are used effectively in the manufacture of footwear, packaging, protective coatings, cables, wires, and tubes (mainly for the automotive and construction industries).

Long linen fibers are used to make high-quality yarn, which is used to make awnings, bedding and fabrics. The length of the fibers ranges from 45 to 80 cm. In the production process, the straw is dried and passed through water using toothed rollers, and then the chaff is separated from the fibers using special devices.

The advantages of flax fiber include [239–243]:

- environmental friendliness,
- no static electricity,
- high resistance to friction and stretching,
- high moisture absorption,
- very low shrinkage,
- bacteriostatic and bactericidal properties,
- flax fiber contains lignins, which are an excellent natural absorber of UV radiation,
- resistance to sea water,

- linen fabrics have the ability to thermoregulate,
- not attacked by moths, fungi or bacteria.

Müller *et al.* [81] discussed the problem of the use of biological reinforcement in the adhesive layer. The advantage of using biological reinforcement is the simplification of the subsequent regeneration of adhesive bonds compared to reinforcements based on carbon and glass fibers. Biocomposites combining biological reinforcement with inorganic adhesive can be used in the area of connecting materials by means of bonding technology. Research has focused on analyzing the strength of the adhesive bond when the bonds have been reinforced with biological fabric such as jute, linen and cotton. The aim of this experiment was to describe the effect of biological reinforcement in the form of a fabric on the adhesive bond strength at various loading speeds from 0.5 to 300 mm/min. The experimental part is devoted to research on the effect of the added natural fabrics on the shear strength of the adhesive joint produced with the use of a polymer binder. The results of the experiment showed the benefit of including the biological fabric in the adhesive layer. Strengthening the adhesive layer with linen or cotton fibers showed a positive result with the increase in the adhesive strength by approx. 50%. The biological reinforcement was not completely wetted by the resin, which did not significantly affect the adhesive bond strength.

Liang *et al.* [239] and Coroller *et al.* [240] emphasized the need to analyze the mechanism of reinforcing polymer composites with plant fibers. The paper [240] presents a study of the influence of mechanical properties, dispersion and surface properties of flax fibers on the stretching of unidirectional composites. Based on the research results, it was indicated that the mechanical properties of unidirectional composites could be related to the properties of the used flax fibers, as well as the microstructure of the composites. The orientation of flax fibers, related to the homogeneity of the microstructure, is strongly dependent on the fiber extraction process.

2.4.4.4 *Jute Fibers*

Jute is a stalk fiber produced from a plant of the genus jute. Jute fiber, after cotton, is the most widespread because of its low cost and high efficiency [16, 17, 244]. It is used in the production of bags, ropes and fabrics, in gardening, interior decoration, and jute waste is used in the paper industry and also as reinforcement for polymer composites [17, 245–247] used in e.g. automotive application [248]. The jute fibers have been used as reinforcement in epoxy resin [245–249] and also polyester resin [248].

Gopinath *et al.* [248] based on the test results underlined that the jute reinforced epoxy composite exhibited better mechanical properties than jute-polyester composite.

Abdellaoui *et al.* [17] determined the mechanical properties of a laminated composite based on natural jute fibers and epoxy resin. The laminate was prepared by using a compression system according to two jute fiber directions (0° and 45°), and two sample cutting directions (0° and 45°). This allowed to reduce the anisotropy of the produced epoxy composites. The experimental results show that the mechanical properties increase as the number of layers increases. The maximum Young's modulus of the laminate with 1, 3, 5 and 7 layers was found to be 5264, 5902, 6400 and 5562 MPa for 0° fiber direction and 0° cut direction, respectively. Moreover, the predicted Young's modulus, shear modulus and Poisson's ratio for each fiber direction turned out to be close to the experimental ones. The difference probably lies in the errors in the considered assumptions about the perfect bonding of the fibers with the matrix. It was also observed that the use of different stacking orientations reduces the anisotropy of the obtained composites.

The issue discussed in many studies is to increase the adhesion of the fibers used as reinforcement of composites to the polymer matrix in order to obtain a composite with better strength properties. This applies to virtually any type of reinforcement. This trend in relation to jute fibbers was also presented by some researchers [245, 248]. Doan *et al.* [245] studied the alkaline treatment of jute fibers in order to enhance the interfacial interaction between jute natural fibers and an epoxy matrix.

2.4.4.5 Sisal Fibers

Sisal is a hard and very durable fiber obtained from a special species of agave - sisal agave. This perennial plant requires a tropical climate. Ripe agave leaves are 2 meters long, 4-6 centimeters thick and 15 centimeters wide in the middle. Sisal is considered to be the third most important textile plant after cotton and jute [1, 117].

The fiber is obtained from the leaves using the manual or manual-semi-mechanical method. This process must take place as soon as the leaves are cut, as extraction is then easiest and gives the best results. Sisal is used for making ropes, braids, mats, bags, etc. [250].

Sisal fibers are characterized by high strength and tear resistance. Sisal is the hardest plant fiber, it successfully replaces coconut, e.g. in the production of mattress inserts, and is definitely cheaper than coconut. The sisal itself does not absorb moisture and does not retain it, so the mattress is "breathable".

Sisal is used in the production of polymer composites [117, 118, 251–253], coatings [119], also floor coverings. Due to its hardness and strength, sisal flooring will not bend or deform. Sisal is also characterized by a very low thermal resistance, which allows it to be used for underfloor heating; it will be great for transmitting heat [1, 250].

Due to its high strength [1, 117], sisal is perfect for the production of strings and ropes. Scratchers for animals are made of sisal cords. Sisal is one of the few materials for which the environmental conditions are not a threat; therefore sisal cords are perfect for agriculture and gardening. Due to their properties, i.e. resistance to stretching and resistance to loads, sisal cords are used to maintain the stability of various objects, where other natural fibers cannot cope.

Different amounts of sisal fiber additives are found in polymer composites. For instance, in [251] amounts of woven sisal fiber in the composite were 3, 5 and 7 wt %. Srisuwana *et al.* [251] also noted that the flexural modulus of all composites was higher than that of the modified composite and increased with increasing the amount of fiber.

In the case of sisal fibers, also an important aspect is the good adhesion of the fibers to the polymer matrix in order to obtain a composite with better strength properties. This issue in relation to sisal fibers was investigated in [252]. Rong *et al.* [252] investigated the effect of fiber processing on the mechanical properties of unidirectional epoxy composites reinforced with sisal. Various treatments of the fibers have been used to modify the surface of the fiber and its internal structure. It has been shown that treating the fibers can significantly improve adhesion and also lead to the penetration of the matrix resin into the fibers, making it difficult to pull out the cells. However, it should be noted that as a result of the increase in adhesion, the dependence of the mechanical properties of the composite on the processing methods becomes more complicated.

2.4.4.6 Coconut (Coir) Fibers

Coconut fiber is obtained from the fibrous mass surrounding the coconut by soaking it in sea water for several months, mechanically separating the fibers, combing and drying. The length of the fibers ranges from 10 to 33 cm, the thickness is 0.05-0.3 mm. Due to the high content of lignin, the fiber is very flexible, durable and does not rot. It is also resistant to sea water. The fibers do not sink, even when they soak up water.

Coconut fibers are used in the production of mats, carpets, ropes, etc. and composites that are used in various industrial and utility areas [21, 22, 134, 149, 254], e.g. as mortar [255].

Romli *et al.* [254] studied the composite made from the combination of coir fibers and epoxy resin. The analyzed parameters were fiber volume fraction in the material, curing time and amount of compression load applied during the fabrication process.

It was found that the volume fraction of coir appears to be the most dominant factor influencing the tensile strength of the resulting composite. Moreover, the influences of the curing time and the volume fraction of coir are also important and should be taken into account in the composite production process.

2.4.4.7 Cellulose Fibers

The characteristics of cellulose are described in section 2.4.2.5. Selected properties, applications and tests on cellulose fibers are presented here.

Natural cellulose fibers derived from wood are a white, tasteless and odorless, water-insoluble material from which plant tissue is made. Cellulose is one of the most abundant biopolymers in nature. Cotton fibers are 98% pure cellulose, while the average wood contains 40-53% cellulose. The wood pulp polysaccharide has 300 to 1700 glucose units $(C_6H_{10}O_5)_{x.}$ In woody parts of plants, cellulose occurs with brittle substances that are useless in the textile industry (e.g. lignin, hemicellulose, pectin). Chemical separation of undesirable materials is done by boiling in sodium hydroxide solution (NaOH) and carbon disulfide (CS_2). In the final stage, the material is forced through small openings of spinning nozzles and subjected to an acidic chemical bath containing H_2SO_4, Na_2SO_4 and $ZnSO_4$. The result is a popular viscose fiber with properties similar to cotton and a silk-like feel. Pure wood cellulose is the basic material for making paper, artificial silk, celluloid and explosive nitrocellulose [256, 257].

Due to the boron content, cellulose fibers inhibit the growth of mold and fungi. They are also suitable for making acoustic insulation. Cellulose is a very good thermal and acoustic insulation due to its spongy structure and a large amount of "closed" air (70-80% by volume). As a result, its heat transfer coefficient λ is equal to mineral wool ($\lambda = 0.039$ W/mK according to one of the producers). Cellulose is, at least in one aspect, better than other thermal insulation materials - it is vapor-permeable, and at the same time does not require the use of a vapor barrier. This means that one can design and implement the so-called "breathing" partitions and create an appropriate microclimate in the interior. This is possible due to the fact that cellulose fibers have the ability to absorb and release water from the environment (up to 23%, with natural humidity 11-17%). Water does not displace air trapped in the pores and even a damp material does not lose its thermal insulation properties. Of course, this only works so well if it is ventilated and has the opportunity to dry (very quickly, due to the huge surface area of the cellulose fibers) [258].

The hygroscopicity of this material also means the possibility of capillary rise of water and therefore this insulation cannot be used below ground level or in floors on the ground. There are always some concerns about the flammability of cellulose (i.e. paper). However, it is wrong, because due to the impregnation and crystallization of the fibers, it is a material that is hard to ignite and does not spread fire. In this respect, it is classified like polyurethane foam, but it is much safer during a fire, because it does not emit toxic fumes.

Aziz *et al.* [158] analyzed the adhesive property of bio-based epoxy resin reinforced with cellulose nanocrystals (CNCs) additive. With the addition of CNCs, the composites significantly increased the tensile modulus at lower wt % (in the tested range) and maximum CNC crystallinity was obtained. The high porosity of CNCs was also obtained. Scanning electron microscopy (SEM) showed an even distribution of CNCs. The composite matrix with high CNCs produced material with good expansion, low crystallinity, and increased elongation. The original CNCs were more evenly distributed in the prepared biological epoxy resin, which facilitated the transformation, supported by the improved dispersion of the native CNCs. The presence of native CNCs greatly improved the bonding efficiency of the bioepoxy resin in the contact area. Improving the mechanical properties of native CNCs has broad application prospects. This suggests that native CNCs can be widely used in environmental engineering applications, especially in terms of adhesive property.

Saba *et al.* [256] presented the results of research on the properties of epoxy nanocomposites reinforced with cellulose nanofibers (CNFs). Various amounts of filler were used for the production of composites, i.e.: 0.5, 0.75 and 1% by weight) CNF additives significantly improve the mechanical properties of epoxy composites for 0.75% CNF compared to other epoxy nanocomposites. In addition, the electron micrographs revealed an excellent distribution and dispersion of CNF in the epoxy matrix for 0.75% CNF/epoxy nanocomposites, while the existence of voids and agglomerates was observed for above 0.75% of the CNF filler content. Overall analysis of the results clearly showed that the filler charge of 0.75% CNF was the best and effective in improving the mechanical and structural properties of epoxy composites.

Barari *et al.* [114] studied epoxy composites of plant-derived cellulose nanofibers (CNF)/bio-composites, which were produced by the process of liquid composite molding (LCM). The composites used CNFs with and without chemical modification. The microstructures, mechanical and tribological properties of the cured composites were studied in order to understand the structure-property relation of composites and the curing kinetics of the prepared composites. The results showed that the produced composites showed better mechanical and tribological properties compared to the pure epoxy resin samples. Moreover, the composites reinforced with chemically modified CNF outperformed the untreated composites. The modification of the fiber surface improved the curing of the resin by reducing the activation energy and led to an improvement in the mechanical properties. CNF/bio-based epoxy composites form a uniform tribo-layer when sliding, which minimizes direct contact between surfaces, thus reducing both friction and wear of the composites.

2.4.4.8 Bamboo Fibers

Bamboo fibers are a raw material of natural origin [12]. Bamboo is grass, more specifically a botanical species belonging to the family of grasses, just like wheat and rye; however, it has woody stems that look more like tree trunks than blades.

Bamboo stalks consist of more than 40% - 50% cellulose, approx. 30% lignin, 20% pentose, and the rest of the structure consists of ashes and silica. Their structure is often comparable to jute or linen, although the crystalline form of cellulose itself resembles cotton or ramie. One year after the appearance of new shoots, with the help of which bamboo multiplies intensively occupying more and more land, the stalks attain their final structure, but they still grow constantly in width and height [14].

Bamboo fiber, due to its specific cross section, has many micropores. As a result, it is flexible, delicate and has exceptional moisture absorption. It is also worth emphasizing that bamboo has antistatic property, due to which it does not accumulate dust and mites. Also due to its natural origin, it is hypoallergenic. In addition, bamboo fiber has natural protective barrier against microorganisms [14, 15].

Bamboo fibers are used for construction purposes, and production of interior furnishings, weapons and paper. In addition, clothing textiles as well as medical and hygienic textiles are produced. Bamboo fibers are also used in the production of composites which are used in many applications, including industrial [13, 15, 259–261].

The work [261] investigated composites composed of various types of fibers such as sisal, bamboo, banana and vakka. The composites were produced up to a maximum fiber volume fraction of 0.37 for the tensile test and 0.39 for bending and dielectric tests. It has been observed that the tensile strength increases in relation to the volume fraction of the fiber for the vakka fiber composite and is also greater than that of sisal and banana composites and is comparable with bamboo composites. This shows that the type of fiber used in the polymer composite is of significant importance.

2.4.4.9 *Kenaf Fibers*

One of the types of plant fibers used to make polymer composites are kenaf fibers [9–11, 262, 263]. Kenaf fiber is extracted from bast fiber of kenaf plant. Kenaf fiber composites have a bright future due to its renewability and being eco-friendly [262]. The work in [262] presents an overview on kenaf fiber reinforced composites in terms of their market, manufacturing methods and overall properties. Mahjoub *et al.* [264] and Manral and Bajpai [265] presented the characteristics and properties of kenaf fibers.

Kenaf fibers have great potential to replace synthetic fibers such as glass fiber. The use of kenaf fiber can provide mechanical properties, i.e. tensile strength, comparable to synthetic fibers with a lower density than traditional materials, and thus lightweight and environmentally friendly polymer composites can be made [266].

Fairuz *et al.* [267] underlined that the tensile strength of kenaf pultruded composites increased as the fiber percentage increased to 50%. Above 50% of the fiber load, the tensile strength decreased slightly. Apart from the fiber content, the strength properties of kenaf-reinforced polymer composites are greatly influenced by the orientation of the fiber.

Yong *et al.* [268] investigated the influence of fiber orientation on the mechanical properties of kenaf-polyester sandwich composites. From the tensile strength results, they found that the strength of the kenaf-polyester sandwich composite increased as the fiber orientation changed. A polyester/kenaf sandwich composite with kenaf fiber in anisotropie orientation achieved the highest mechanical properties. The kenaf fiber in anisotropic orientation can absorb the impact energy and allow the sandwich composite to withstand greater impact forces compared to composite with fiber in perpendicular or isotropic orientation. The polyester/kenaf sandwich composite also showed higher thermal stability compared to a conventional plywood sheet.

El-Shekeil *et al.* [269] determined the effects of fiber content on the mechanical (i.e. tensile, flexural and impact strengths, hardness and abrasion resistance) and thermal (i.e. TGA) properties of kenaf fiber reinforced thermoplastic polyurethane (TPU) composites. Different fiber contents in the composite were used: 20, 30, 40 and 50% by weight. The composite with 30% kenaf fibers had the highest tensile strength, with the modulus and deformation increasing with increasing fiber content. The bending strength and modulus increased with increasing fiber loading. Increasing fiber loading resulted in a decrease in the impact toughness. Hardness increased by the addition of 30% fiber content. The abrasion resistance decreases as the fiber loading increases. It has been found that increasing the fiber content reduces the thermal stability of the composite.

2.4.4.10 Other Fibers

In addition to the above-mentioned types of fibers, other types of plant fibers are also used to strengthen composites, such as [261, 270–276]:

- banana fibers [270, 271],
- ramie fibers [144, 272–274],
- vakka fibers, [261],
- jowar fibers [260],
- henequen fibers [275–277],
- piassava fibers [278],
- and others.

Banana fibers, which are concentrated near the outer surface, are extracted by hand scraping, chemically, by retting, or by means of scraper. They can also be extracted by boiling the leaves in a sodium hydroxide solution [271].

Ramie fiber is one of the oldest textile fibers of plant origin. Environmental awareness has increased interest in plant-derived ramie that is safe, biodegradable and recyclable. The ramie has excellent microbial resistance and hygienic properties. The physical properties of the ramie fibers exhibit high tensile strength, high gloss and brightness. This fiber has resistance to heat, light, acids and bases, etc. The primary use for the ramie fibers is fabric production and, for example, blending ramie with silk with different blend proportions provides excellent possibilities to produce various materials for different applications [272].

Another application of fibers is their use as a reinforcement for polymer composites and green composites used in various applications [144, 273, 274].

The work in [276] presents the literature on henequen plants and various properties of the henequen fibers. Moreover, the influence of selected types of fiber processing on the mechanical properties and the henequen/epoxy composite was determined. In the work [277] authors present the problem of obtaining henequen fiber with a surface layer with a magnetic property. Scanning electron microscopy showed that henequen fiber has a regular porous structure, making it compatible with ferrofluids and allowing a layer of magnetic nanoparticles and stearic acid to be deposited on the surface of the fiber. It is emphasized that the main advantage of these magnetic fibers is that they can be easily manipulated with an applied magnetic field.

Nascimento *et al.* [278] determined the morphology, physical and strength properties of the piassava fiber, which is very stiff and can be used as a composite reinforcement. Epoxy matrix composites were reinforced with continuous and stacked piassava fibers treated with alkali. Composites with fibers loading greater than 20% demonstrated effective reinforcement behavior in both bending and tensile tests, while the impact energy increased linearly with the amount of piassava fibers used. Fractographic examination showed relatively less fiber adhesion to the matrix, which was the point of crack initiation. There is also evidence of fiber crack stopping above 20% fiber. Fiber loading, along with the spiny protrusions of the surface in the piassava fibers, was found to be responsible for the reinforcement of epoxy composites.

Various types of fibers mentioned in this and in other subsections are used in hybrid composites. Hybrid composites i.e. composites containing several types of reinforcements are often used, and various research issues concerning such composites are included in numerous works [238, 247, 279–284].

Various properties of the natural fibers are utilized, which contribute to obtaining composites with more favorable or different properties than composites containing only one type of reinforcement.

Mohanavel *et al.* [284] tested hybrid composites that were prepared with ramie and natural jute as reinforcement, epoxy polymer as matrix, and SiC and cellulose as filler. Mechanical and thermogravimetric analyses were carried out to quantify the impact of reinforcement on hybrid composites and to evaluate their mechanical properties and thermal stability. The mechanical tests conducted showed an improvement in the tensile and bending strengths of hybrid composites with an increase in the content of bi-directional woven ramie fiber, and a decrease with an increase in the content of chopped jute fiber. The Izod impact test results showed that the chopped jute fibers can withstand greater impact loads compared to woven ramie yarns. Therefore, the impact energy increased with the increase in the content of chopped jute fiber in hybrid composites.

2.5 Fungi-Modified Adhesives

The epoxy resins can be modified using various modifying agents, other polymers or resins, and curing agents. They are used as agents to modify various properties, cross-linking and making it more flexible, which allows changing the existing properties or to obtain other new properties, including achieving a biodegradable material.

The works [156, 157] include studies and research results on the modification of epoxy adhesives with fillers in the form of fungal metabolites obtained from white wood rot fungi (WRF).

Wood rot fungi (WRF), especially white rot fungi, aroused the interest of several researchers because of their extremely effective biodegradation [201, 285, 286]. Given their effective production of a variety of secondary metabolites, including enzymes such as laccase, these organisms have long been used in various areas of biotechnology [210, 287]. The laccase is also used in research on the modification of adhesive property of plant biopolymers, such as lignin [209, 287], which can be found in resins or wood adhesives [69, 285, 286, 288]. Lignin has been used in epoxy resin, and many different formulation approaches have been investigated [298, 290]. Considering the metabolic biodiversity of wood-destroying fungi, their use has great promise for the modification of adhesive compounds [291, 292]. The composition and properties of *Pycnoporus sanquineus*, i.e., a low molecular weight fraction used as a modifying agent are precisely described in [293].

The main objective of the studies presented in [156, 157] was to determine the effect of biochemical modification of epoxy adhesive on some mechanical properties of cured epoxy adhesiveexposed to various climatic factors. Preliminary studies have been carried out on both the properties of the modified adhesive compounds and the adhesive joints made with these modified adhesives. Based on these studies, it can be assumed that the use of metabolites derived from fungal cultures as modifiers of epoxy compounds may have a positive effect not only on the strength of adhesive joints exposed to various climatic conditions, but also on aging and degradation of modified epoxy adhesives.

The work [156] describes the technology of preparing modified epoxy adhesive compounds with a lyophilized fungal formulation, i.e. technological parameters of the mixing and hardening process and the mechanical properties of modified epoxy adhesive compounds were determined. The resultant properties were then used as the basis for the development of the technology of making adhesive joints, which is included in the second paper [157].

The paper [157] presents the influence of biochemical modification of epoxy adhesive compounds on selected mechanical properties of adhesive joints of hot-dip galvanized steel sheet made with modified epoxy adhesive. In order to determine the influence of modifications on the mechanical properties of the produced adhesive joints, the study also conducted control experiments using the unmodified adhesive as a reference. In addition, the study also investigated the effect of aging on the mechanical properties of adhesive joints made using the epoxy modified adhesive.

The main conclusions from the research presented in [156] are:

- The method of preparing the adhesive mass and the modifier content used affect the mechanical properties of the cured adhesive compound. Depending on the method of preparation, either an increase or a decrease in the elasticity of the adhesive is observed, from which it can be concluded that the development of proper technology for the preparation of modified adhesive compounds is very important.
- Aging of the modified adhesives did not adversely affect their mechanical properties. On the other hand, the tested properties of unmodified adhesive decreased after aging. Although aging had a negative effect on the value of the breaking force during strength tests, it did lead to a significant increase in the flexibility of the modified adhesive. In most cases, the elongation at break of the modified adhesives subjected to aging increased by over 50% (even by over 70%) compared to the modified adhesives that were not aged. In modified epoxy adhesives, elongation at break increased after aging compared to unmodified adhesive cured for 7 days at ambient temperature.
- Summarizing the obtained results, it can be concluded that the high antioxidant potential of fungal modifiers, as well as their composition (the content of phenolic compounds and low molecular weight of active proteins or carbohydrates) can significantly change the properties of the tested adhesive mixtures. In addition to the typical chemical reactions, it is likely that the typical mechanical changes in adhesives directly affect their adhesive property. However, the

mechanisms of the described modifications require further research.

The primary conclusions from the research presented in [156] are:

- The strength of adhesive joints decreases with an increase in the amount of lyophilized fungi used as a filler compared to the results for adhesive joints made with unmodified epoxy adhesive.
- The results of the aging effect on the strength of the adhesive joint show that the aging has a positive effect on the mechanical properties of adhesive joints made of hot-dip galvanized steel sheets with modified epoxy adhesives. This was observed for all concentrations of the modifying agent used, i.e. lyophilized fungi. In contrast, aging has a negative effect on the strength of adhesive joints made with the unmodified epoxy adhesive.
- Moreover, the strength of the adhesive joints made with the modified adhesive is not affected by increased temperature or humidity. The modified samples placed in the climatic chamber show greater strength and elongation with the increase of the modifier concentration compared to the results of adhesive joints made using unmodified adhesives.

It can be assumed that the use of metabolites derived from fungal cultures as modifiers of epoxy compounds may delay degradation processes for the modified adhesive compounds. It can be concluded that the high antioxidant potential of fungal modifiers and their composition (the content of phenolic compounds, low molecular weight of active proteins or carbohydrates) can significantly change the properties of modified epoxy adhesives used for adhesive bonding.

In conclusion, it can be stated that the properties of low molecular weight fungal metabolites offer great promise in research on the modification of adhesive compounds. Apart from the unique composition of fungi preparations used, they seem to be noteworthy especially in the context of the modification of epoxy adhesives, and also due to their very high antioxidant and antibacterial potential.

The results obtained also have a practical aspect. The modification of the epoxy adhesive allows achieving industrial, economic and ecological benefits. Extending the service life of the adhesive joints allows them to be suitably designed for a given working environment. The economic aspect will be expressed in a lower consumption of the adhesives (extended lifetime - lower costs of adhesive materials) and the ecological aspect will be the introduction of natural, non-toxic and environmentally friendly bio-products, and the production of biodegradable materials will contribute to greater safety of the natural environment.

2.6 Prospects

The demand for polymer materials with new properties increases in line with economic development and constant technological progress. A constantly growing amount of plastic waste is polluting the natural environment. The legal regulations and the decreasing oil resources have contributed to the increased interest in biodegradable materials, as well as to the need for modification of existing polymer materials with natural (biological) fillers [2, 29, 56, 226, 270, 271, 294]. Yashas Gowda *et al.* [294] emphasized that the current emphasis is placed on materials that are environmentally friendly and biodegradable, and one of the main areas of research in this direction are composites of polymers with natural fibers. These composites with appropriately selected composition ensure composite materials with improved properties.

Biodegradable materials are commonly considered the polymers of the future, mainly because they are rapidly biodegradable. Due to their biocompatibility and biodegradability, they are used in various industries: biotechnology, building construction, automotive, packaging, as well as in medicine, pharmacy and horticulture and other engineering and biotechnological applications [7, 9, 69, 104, 295–298]. For example Veerasimman *et al.* [298] underlined that sisal fiber composites have been used in vast engineering applications like automobile, railways, building materials, electrical industry, geotextiles, defense and in the packaging industry and also that the increased strength of sisal fiber has attracted the design engineers in developing fiber composites. Yang *et al.* [297] presented research on the use of jute-epoxy composites with carbon nanotubes as electrically conductive materials. Moreover, by-products from the agri-food industry may be an interesting, competitive alternative to standard polymeric materials reinforced with wood fibers [180, 296].

The rapid development of specialized applications of polymer materials in various fields of technology and other aspects of life create more and more requirements for the quality of products made of such materials and requirements for their properties or use in specific conditions. The widespread use of polymeric materials is not without significance, and therefore the introduction of mass applications of biodegradable polymers seems justified in practically all areas of their applications.

Due to the growing global concern for the environment, the possibility of partial or complete replacement of fossil raw materials is intensively sought, and thus the introduction of materials based on renewable raw materials on an industrial scale. The chemical composition and oligomeric nature of vegetable oils enable their epoxidation and facilitate the development of new polymeric materials with different chemical characteristics and properties [24–27, 83, 247, 281]. Epoxidized oils are the chemical basis for polyethers, polyesters, polyurethanes and polyhydroxyurethanes, but also as reactive modifiers for natural and synthetic polymers [138, 139, 160]. Polymers based on epoxidized vegetable oils in combination with fillers, which are more and more often natural fibers, make it possible to obtain biocomposites and create both cross-linked and linear thermoplastics [82]. The ultimate competitiveness of such materials is due not only to the use of renewable raw materials, but also due to properties that are often comparable to or better than petrochemical polymer products that meet the requirements of various industries. Some have functional properties, including shape memory and self-healing properties, others, such as adhesives, have higher adhesion to the substrate.

Another aspect in the production of epoxy structured adhesives is the production of bio-adhesives and bio-epoxy resins based entirely on products of natural origin [114, 137, 158, 244, 286, 299–303] and also biocomposites or epoxy green composites [140, 144, 169, 175, 273, 295].

Based on the available literature and research results, it is expected that the interest in the use of products based on natural materials or synthetic products, but modified by natural reinforcements, of which natural fibers of plant origin are of great importance in various application areas.

2.7 Summary

In the present world, it is becoming necessary to use greener materials, which is why researchers around the world are focusing on developing new materials that would improve the quality of products in terms of ecology and economy.

The need for new ecological materials has led to the use of composites made of raw natural fibers and polymer matrices, which has become one of the most studied research topics in recent times. Natural fiber composites are an alternative to replacing environmentally harmful synthetic materials and help control pollution problems. Moreover, they are cheap, have better mechanical properties and require low energy consumption in the production technique.

Biobased raw materials contained in biomaterials including, for example, bio-adhesives, are obtained from renewable sources. The advantage is that biobased, renewable raw materials can contribute to reducing CO_2 emissions compared to fossil feedstock used in the supply chain. In this way, the end products make a positive contribution to climate change.

Biomaterials are used in many areas, such as:

- packaging industry,
- printing industry,
- paper industry,
- construction industry,
- automotive industry,
- building industry,
- medicine,
- agriculture and horticulture,
- others.

By using such materials, it is also possible to eliminate waste from various industries, because often reinforcing materials or raw materials for biomaterials are by-products of different production processes, e.g. wood flour, sawdust, or various types of plant and animal fibers.

References

1. K. Joseph, R.D.T. Filho, J. Beena, T. Sabu and L. Carvalho, A review on sisal fiber reinforced polymer composites. *Revista Brasileira de Engenharia Agr'icola e Ambiental* 3, 367-379 (1999).
2. A. Karimah, M.R. Ridho, S.S. Munawar, D.S. Adi, Ismadi, R. Damayanti, B. Subiyanto, W. Fatriasari and A. Fudholi, A review on natural fibers for development of eco-friendly bio-composite: characteristics, and utilizations. *J Mater Res Technol* 13, 2442-2458 (2021).
3. M.L. Ankit, P. Rinawa, P. Chauhan, D. Suresh, S. Kumar and R.S. Kumar, A review on mechanical properties of natural fiber reinforced polymer (NFRP) composites. *Mater Today Proc* (2021), in press, https://doi.org/10.1016/j.matpr.2021.07.275
4. C. Rajaravi, S. Kishore, M. Easwar and B. Vasanthkanna, Analysis of epoxy resin with natural materials. *J Adv Mater Eng* 9, 33-39 (2019).
5. R. Kumar and A. Anand, Tribological behavior of natural fiber reinforced epoxy based composites: A review. *Mater Today Proc* 18, 3247-3251 (2019).

6. S. Kumar, L. Prasad, V.K. Patel, V. Kumar, A. Kumar, A. Yadav and J. Winczek, Physical and mechanical properties of natural leaf fiber-reinforced epoxy polyester composites. *Polymers* 13, 1369 (2021).
7. S. Alsubari, M.Y.M. Zuhri, S.M. Sapuan, M.R. Ishak, R.A. Ilyas and M.R.M. Asyraf, Potential of natural fiber reinforced polymer composites in sandwich structures: A review on its mechanical properties. *Polymers* 13, 423 (2021).
8. C. Tezara, J. Siregar, H.Y. Lim, F.A. Fauzi, M.H. Yazdi, L.K. Moey and J.W. Lim, Factors that affect the mechanical properties of kenaf fiber reinforced polymer: a review. *J Mech Eng Sci* 10, 2159-2175 (2016).
9. I. Yusuff, N. Sarifuddin and A.M. Ali, A review on kenaf fiber hybrid composites: Mechanical properties, potentials, and challenges in engineering applications. *Prog Rubber Plast Recycl Technol* 37, 66-83 (2021).
10. N. Saba, M.T. Paridah and M. Jawaid, Mechanical properties of kenaf fiber reinforced polymer composite: A review, *Constr Build Mater* 76, 87-96 (2015).
11. F. Bajuri, N. Mazlan, M.R. Ishak, and J. Imatomi, Flexural and compressive properties of hybrid kenaf/silica nanoparticles in epoxy composite. *Procedia Chem* 19, 955-960 (2016).
12. S. Ch. Chin, K. F. Tee, F. S. Tong, H.R. Ong and J. Gimbun, Thermal and mechanical properties of bamboo fiber reinforced composites, *Mater Today Commun* 23, 100876 (2020).
13. A. Muhammad, M. Rahman, S. Hamdan and K. Sanaullah, Recent developments in bamboo fiber-based composites: a review. *Polym Bull* 76, 2655-2682 (2019).
14. M. Ramesh, L. RajeshKumar and V. Bhuvaneshwari, Bamboo fiber reinforced composites, in: *Bamboo Fiber Composites*, M. Jawaid, S. Mavinkere Rangappa and S. Siengchin (Eds), pp.1-13, Springer, Singapore (2021).
15. D. Liu, J. Song, D.P. Anderson, P.R. Chang, and Y. Hua, Bamboo fiber and its reinforced composites: structure and properties. *Cellulose* 19, 1449-1480 (2012).
16. J. Prachayawarakorn, S. Chaiwatyothin, S. Mueangta and A. Hanchana, Effect of jute and kapok fibers on properties of thermoplastic cassava starch composites. *Mater Design* 47, 309-315 (2013).
17. H. Abdellaoui, H. Bensalah, J. Echaabi, R. Bouhfid and A. Qaiss, Fabrication, characterization and modelling of laminated composites based on woven jute fibers reinforced epoxy resin. *Mater Design* 68, 104-113 (2015).
18. A. Shahzad, Hemp fiber and its composites - A review. *J Composite Mater* 46, 973-986 (2012).
19. D.M. Panaitescu, R. C. Fierascu, A. R. Gabor and C.A. Nicolae, Effect of hemp fiber length on the mechanical and thermal properties of polypropylene/SEBS/hemp fiber composites. *J Mater Res Technol* 9, 10768-10781 (2020).
20. R. Sepe, F. Bollino, L. Boccarusso and F. Caputo, Influence of chemical treatments on mechanical properties of hemp fiber reinforced composites. *Composites B* 133, 201-217 (2018).

21. S. Rajamuneeswaran, J. Vairamuthu, S. Nagarajan, B. Stalin and S. Jayabal, A comparative study on mechanical properties of coir fiber reinforced polymer composites filled with calcium carbonate particles. *Mater Today Proc* 33, 4790-4792 (2020).
22. A.K. Chaudhary, P.C. Gope and V.K. Singh, Water absorption and thickness swelling behavior of almond (Prunus amygdalus L.) shell particles and coconut (Cocos nucifera) fiber hybrid epoxy-based biocomposite. *Sci Eng Composite Mater* 22, 375-382 (2015).
23. R.M.N. Arib, S. Sapuan, S. Megat Ahmad, M. Mohamad Hamdan, P.M. Tahir, and K. Zaman, Mechanical properties of pineapple leaf fiber reinforced polypropylene composites. *Mater Design* 27, 391-396 (2006).
24. K.B. Khairiah and A.M.A. Khairul, Biocomposites from oil palm resources. *J Oil Palm Res*, Special Issue, 103-113 (April 2006).
25. K. Myrtha, O. Holia, A.A.H. Dawam and S. Anung, Effect of oil palm empty fruit bunch fiber on the physical and mechanical properties of fiber glass reinforced polyester resin. *J Biol Sci* 8, 101-106 (2008).
26. H. Ismail, N. Rosnah and H.D. Rozman, Effects of various bonding systems on mechanical properties of oil palm fiber reinforced rubber composites. *Eur Polym J* 33, 1231-1238 (1997).
27. N. Saba, M.T. Paridah, K. Abdan and N.A. Ibrahim, Physical, structural and thermomechanical properties of oil palm nano filler/kenaf/epoxy hybrid nanocomposites. *Mater Chem Phys* 184, 64-71 (2016).
28. H.Y. Sastra, J. Siregar, S. Sapuan, Z. Leman and M.M.H.M. Hamdan, Flexural properties of *Arenga pinnata* fiber reinforced epoxy composites. *Amer J Appl Sci* 21, 21-24 (2005).
29. M. Wollerdorfer and H. Bader, Influence of natural fibers on the mechanical properties of biodegradable polymers. *Ind Crops Prod* 8, 105-112 (1998).
30. A. Bismarck, S. Mishra and T. Lampke, Plant fibers as reinforcement for green composites, in: *Natural Fibers, Biopolymers and Biocomposites*, A.K. Mohanty, M. Misra and L.T. Drzal (Eds.), pp. 36-108, CRC Press, Boca Raton, FL (2005).
31. A.K. Błędzki, S. Reihmane and J. Gassan, Properties and modification methods for vegetable fibers for natural fiber composites. *J Appl Polym Sci* 59, 1329-1336 (1996).
32. C.A. May, *Epoxy Resins: Chemistry and Technology*. Marcel Dekker, New York (1988).
33. H.L. Lee and H. Neville, *Handbook of Epoxy Resins*. McGraw-Hill, New York (1988).
34. S. Ebnesajjad (Ed.), *Adhesives Technology Handbook*, second edition. William Andrew, Norwich, NY (2008).
35. A.J. Kinloch, *Adhesion and Adhesives*. Chapman and Hall, New York (1987).
36. E.M. Petrie, *Handbook of Adhesives and Sealants*. McGraw Hill, New York (1999).

37. L.F.M. da Silva, A. Öchsner and R.D. Adams (Eds.), *Handbook of Adhesion Technology*. Springer, Heidelberg (2011).
38. D.A. Dillard (Ed.), *Advances in Structural Adhesive Bonding*. Woodhead Publishing Limited, New York (2010).
39. D.J. Dunn, *Engineering and Structural Adhesives*. Rapra Technology, Shrewsbury, UK (2004).
40. C.V. Cagle (Ed.), *Handbook of Adhesive Bonding*. McGraw-Hill, New York (1973).
41. I. Skeist (Ed.), *Handbook of Adhesives*, third edition. Chapman and Hall, New York (1990).
42. A. Pizzi and K.L. Mittal (Eds.), *Handbook of Adhesive Technology*, third edition. CRC Press, Boca Raton, FL (2018).
43. A.V. Pocius, *Adhesion and Adhesives Technology: An Introduction*. Carl Hanser, Munich (2002).
44. Ch. Czaderski, E. Martinelli, J. Michels and M. Motavallli, Effect of curing conditions on strength development in an epoxy resin for structural strengthening. *Composites B* 43, 398-410 (2012).
45. R. Gomatam and K.L. Mittal (Eds.), *Electrically Conductive Adhesives*. CRC Press, Boca Raton, FL (2008).
46. A. Rudawska, Epoxy adhesives, in: *Handbook of Adhesive Technology*, third edition, A. Pizzi and K.L. Mittal (Eds.), pp. 415-441, CRC Press, Boca Raton, FL (2018).
47. W. Brockmann, P.L. Geiß, J. Klingen and B. Schröder, *Adhesive Bonding. Materials, Applications and Technology*. Wiley-VCH, Weinheim (2009).
48. R.D. Adams (Ed.), *Adhesive Bonding. Science, Technology and Applications*. Woodhead Publishing Limited, Cambridge, UK (2010).
49. L.S. Penn and H. Wang, Epoxy resins, *in*: *Handbook of Composites*, Chapter 3, S.T. Peters (Ed.), pp. 48-74. Chapman and Hall, London (1998).
50. H. Okuhira, T. Kii, M. Ochi and H. Takeyama, Characterization of epoxy resin hardening with ketimine latent hardeners. *J Adhesion Sci Technol* 18, 205-211 (2004).
51. O. Moussa, A.P. Vassilopulos and T. Keller, Effects of low-temperature curing on physical behaviour of cold-curing epoxy adhesives in bridge construction. *Int J Adhesion Adhesives* 32, 15-22 (2012).
52. M. Frigione, C. Naddeo and D. Acierno, Cold-curing epoxy resins: aging and environmental effects. I – Thermal properties. *J Polym Eng* 21, 23-51 (2001).
53. M. Frigione, C. Naddeo and D. Acierno, Cold-curing epoxy resins: aging and environmental effects. II – Mechanical properties. *J Polym Eng* 21, 349-367 (2001).
54. S.J. Shaw, Epoxy resin adhesives, in: *Chemistry and Technology of Epoxy Resins*, E. Blackie (Ed.), pp. 117-143, Chapman and Hall, London, UK (1994).
55. M. Frigione and A. Maffezzoli, Curing agents, in: *Encyclopedia of Composites*, second edition, Vol. 1, L. Nicolais, A. Borzacchiello and S.M. Lee (Eds.), pp. 643-648. John Wiley & Sons, Hoboken, New Jersey (2012).

56. J.-P. Pascault and R.J.J. Williams (Eds.), *Epoxy Polymers. New Materials and Innovations*, pp. 1-11. Wiley-VCH, Weinheim (2010).
57. K. Mimura and H. Ito, Characteristics of epoxy resin cured with *in situ* polymerized curing agent. *Polymer* 43, 7556-7559 (2002).
58. B Ellis (Ed.), *Chemistry and Technology of Epoxy Resins*, first edition, pp. 1-35, 37-68, 72-99, 121-126, 206-239. Springer Science+Business Media, Dordrecht (1993).
59. E.M. Pertie, *Epoxy Adhesive Formulation*, pp. 3-19, 26-41, 43-53, 71-82, 85-98, 207-225, 229-236, McGraw-Hill, New York (2006).
60. I. Hamerton, *Recent Developments in Epoxy Resins*, Rapra Review Reports, second edition, pp.7-8, 21-34, 49-55. Rapra Technology Limited, Shawbury, UK (1997).
61. B. Bereska, J. Iłowska, K. Czaja and A. Bereska, Hardeners for epoxy resins. *Chemical Industry* 4, 443-448 (2014).
62. G. Habenicht, *Applied Adhesive Bonding. A Practical Guide for Flawless Results*, pp. 23-26. Wiley-VCH, Weinheim (2009).
63. M.J. Gerace, Surface activated rubber particles improve structural adhesive. *Adhesives Age* 38, 26-31 (Dec. 1995).
64. K.P. Unnikrishnan and E.T. Thachil, Toughening of epoxy resins. *Designed Monomers Polymers* 9, 129-152 (2006).
65. J. Shields (Ed.), *Adhesives Handbook*. Butterworths, London (1984).
66. A. Rudawska (Ed.), *Adhesives - Applications and Properties*. InTech, Rijeka, Croatia (2016).
67. A. Rudawska, I. Miturska, J. Szabelski, A. Skoczylas, P. Droździel, E. Bociąga, R. Madleňák and D. Kasperek, Experimental research and statistic analysis of polymer composite adhesive joints strength. *J Phys Conf Ser* 842, 1-9 (2017).
68. G. Gierenz and W. Karmann, *Adhesives and Adhesive Tapes*. Wiley-VCH, Weinheim (2001).
69. A. Pizzi and K.L. Mittal (Eds.), *Wood Adhesives*. CRC Press, Boca Raton, FL (2011).
70. R.D. Adams, J. Comyn and W.C. Wake, *Structural Adhesive Joints in Engineering*. Chapman and Hall, London (1997).
71. M. Miravalles, The creep behaviour of adhesives, Ph.D. thesis, Chalmers University of Technology, Göteborg, Sweden (2007).
72. K.L. Mittal, Interfacial chemistry and adhesion: recent developments and prospects. *Pure Appl Chem* 52, 1295-1305 (1980).
73. J. Yang, Modification of epoxy resins by functional hyperbranched poly(arlene esters). Ph.D thesis, Virginia Polytechnic Institute and State University, Blacksburg, VA (1998).
74. I.-N. Yoon, Y. Lee, D. Kang, J. Min, J. Won, M. Kim, Y. Soo Kang, S. Kim and J.-J. Kim, Modification of hydrogenated Bisphenol A epoxy adhesives using nanomaterials. *Int J Adhesion Adhesives* 31, 119-125 (2011).
75. J. Szabelski, R. Karpiński and A. Machrowska, Application of an artificial neural network in the modelling of heat curing effects on the strength of

adhesive joints at elevated temperature with imprecise adhesive mix ratios. *Materials* 15, 721 (2022).
76. S.G. Prolongo, G. Del Rosario and A. Urena, Comparative study on the adhesive properties of different epoxy resins. *Int J Adhesion Adhesives* 26, 125-132 (2006).
77. A. Rudawska, The impact of seasoning conditions on mechanical properties of modified and unmodified epoxy adhesive compounds. *Polymers* 11, 804 (2019).
78. A. Rudawska, Experimental study of mechanical properties of epoxy compounds modified with calcium carbonate and carbon after hygrothermal exposure. *Materials* 13, 5439 (2020).
79. A. Rudawska, Mechanical properties of epoxy compounds based on Bisphenol A aged in aqueous environments. *Polymers* 13, 952 (2021).
80. I. Miturska I., A. Rudawska, M. Müller and P. Valášek, The influence of modification with natural fillers on the mechanical properties of epoxy adhesive compositions after storage time. *Materials* 13, 291 (2020).
81. M. Müller, P. Valášek and A. Rudawska, Mechanical properties of adhesive bonds reinforced with biological fabric. *J Adhesion Sci Technol* 31, 1859-1871 (2017).
82. A. Rudawska and M. Frigione, Aging effects of aqueous environment on mechanical properties of calcium carbonate-modified epoxy resin. *Polymers* 12, 2541 (2020).
83. P. Czub and I. Franek, Epoxy resin modified with palm oil derivatives – preparation and properties. *Polimery* 58, 135-139 (2013).
84. J. Suwanprateeb, S. Tiemprateeb, S. Kangwantrakool and K. Hemachandra, The role of filler volume fraction in the strain-rate dependence of calcium carbonate-reinforced polyethylene. *J Appl Polym Sci* 70, 1717-1724 (1998).
85. M. Müller, P. Valášek, A. Rudawska and R. Chotěborský, Effect of active rubber powder on structural two-component epoxy resin and its mechanical properties. *J Adhesion Sci Technol* 32, 1531-1547 (2018).
86. S.M. Zebarjad and S.A. Sajjadi, On the strain rate sensitivity of HDPE/$CaCO_3$ nanocomposites. *Mater Sci Eng A* 475, 365-367 (2008).
87. Y. Zhou, F. Pervin V.K. Rangari and S. Jeelani, Influence of montmorillonite clay on the thermal and mechanical properties of conventional carbon fiber reinforced composites. *J Mater Process Technol* 191, 347-351 (2007).
88. R. Jain, A.K. Narul and V. Choudhary, Studies on epoxy/calcium carbonate nanocomposites. *J Appl Polym Sci* 26, 2162-2168 (2009).
89. A.B. Al-Zubaidi, A.A. Al-Tabbakh, H.A. Al-Qaessy and R.N. Al-Kaseey, Mechanical, thermal and wear characteristics of polymer composite material reinforced with calcium carbonate powder. *Eng Tech J* 32, 519-532 (2014).
90. H. He, K. Li, J. Wang, G. Sun, Y. Li and J. Wang, Study on thermal and mechanical properties of nano-calcium carbonate/epoxy composites. *Mater Design* 32, 4521-4527 (2011).

91. C.L. Poh, M. Mariatti, M.N. Ahmad Fauzi, C.H. Ng, C.K. Chee and T.P. Chuah, Tensile, dielectric, and thermal properties of epoxy composites filled with silica, mica, and calcium carbonate. *J Mater Sci Mater Electronics* 25, 2111-2119 (2014).
92. A. Rudawska and M. Frigione, Cold-cured bisphenolic epoxy adhesive filled with low amounts of $CaCO_3$: Effect of the filler on the durability to aqueous environments. *Materials* 14, 1324 (2021).
93. A. Rudawska, P. Jakubowska and A. Kloziński, Surface free energy of composite materials with high calcium carbonate filler content. *Polimery* 62, 434-440 (2017).
94. J.F. Feller, I. Linossier and Y. Grohens, Conductive polymer composites: comparative study of poly(ester) - short carbon fibers and poly(epoxy) - short carbon fibers mechanical and electrical properties. *Material Letters* 57, 64-71 (2002).
95. N. Saba, M. Jawaid, O.Y. Alothman, M. Paridah and A. Hassan, Recent advances in epoxy resin, natural fiber-reinforced epoxy composites and their applications. *J Reinf Plast Composites* 35, 447-470 (2016).
96. L. Li, H. Zuo, L. Shao, G. Wang, and J. Chen, Study in mechanical property of epoxy composite fillers with nano-sized calcium carbonate particles. *J Mater Sci* 40, 1297-1299 (2005).
97. K. Sai Sravani, B. Ram Gopal Reddy and R. Mohammed, Effect of $CaCO_3$ and Al_2O_3 fillers on mechanical properties of glass/epoxy composites. *Int J Modern Trends Sci Technol* 3, 207-213 (2017).
98. M. Rakotomalala, S. Wagner and M. Döring, Recent developments in halogen free flame retardants for epoxy resins for electrical and electronic applications. *Materials* 3, 4300-4327 (2010).
99. U. Braun, B. Schartel, M.A. Fichera and C. Jäger, Flame retardancy mechanisms of aluminium phosphinate in combination with melamine polyphosphate and zinc borate in glass-fiber reinforced polyamide 6,6. *Polym Degrad Stab* 92, 1528-1545 (2007).
100. C. Prapruddivongs and N. Sombatsompop, Roles and evidence of wood flour as an antibacterial promoter for triclosan-filled poly(lactic acid). *Composites B* 43, 2730-2737 (2012).
101. E. Pérez, L. Famá, S.G. Pardo, M.J. Abad and C. Bernal, Tensile and fracture behaviour of PP/wood flour composites. *Composites B* 43, 2795-2800 (2012).
102. A. Zotti, A. Borriello, A. Martone, V. Antonucci, M. Giordano and M. Zarrelli, Effect of sepiolite filler on mechanical behaviour of a bisphenol A-based epoxy system. *Composites B* 67, 400-409 (2014).
103. S.J. Shaw, Additives and modifiers of epoxy resin, in: *Chemistry and Technology of Epoxy Resins*, B. Ellis (Ed.), pp. 117-143, Springer, Netherlands (1993).
104. H.S. Katz and J.V. Milewski (Eds.), *Handbook of Fillers for Plastics*, van Nostrand Reinhold, New York (1987).

105. A. Ghanbari, N. S. Jalili, S. A. Haddadi, M. Arjmand and M. Nofar, Mechanical properties of extruded glass fiber reinforced thermoplastic polyolefin composites. *Polym Composites* 41, 3748-3757 (2020).
106. S.P. Gumfekar, B. M. Amoli, A. Chen and B. Zhao, Polyaniline-tailored electromechanical responses of the silver/epoxy conductive adhesive composites. *J Polym Sci B Polym Phys* 51, 1448-1455 (2013).
107. H. Sosiati, C.T. Utomo, I. Setiono and C. Budiyantoro, Effect of $CaCO_3$ particles size and content on impact strength of kenaf/$CaCO_3$/epoxy resin hybrid composites. *Indonesian J Appl Phys* 10, 24-31 (2020).
108. F.-L Jin and S.-J. Park, Thermal stability of trifunctional epoxy resins modified with nanosized calcium carbonate. *Bull Korean Chem Soc* 30, 334-338 (2009).
109. W.-H. Wang, H.-B. Huang, H.-H. Du and H. Wang, Effects of fiber size on short-term creep behavior of wood fiber/HDPE composites. *Polym Eng Sci* 55, 693-700 (2015).
110. A.J. Nunez, P.C. Sturm, J.M. Kenny, M.I. Aranguren, N.E. Marcovich and M.M. Reboredo, Mechanical characterization of polypropylene–wood flour composites. *J Appl Polym Sci* 88, 1420-1428 (2003).
111. N.E. Marcovich, M.I. Aranguren and M.M. Reboredo, Modified wood flour as thermoset filler. Part I. Effect of the chemical modification and percentage of filler on the mechanical properties. *Polymer* 42, 815-825 (2001).
112. N.E. Marcovich and M.A. Villar, Thermal and mechanical characterization of linear low-density polyethylene/wood flour composites. *J Appl Polym Sci* 90, 2775-2784 (2003).
113. R. Liu, J. Cao, S. Luo and X. Wang, Effects of two types of clay on physical and mechanical properties of poly(lactic acid)/wood flour composites at various wood flour contents. *J Appl Polym Sci* 127, 2566-2573 (2012).
114. B. Barari, E. Omrani, A.D. Moghadam, P. L. Menezes, K. M. Pillai and P. K. Rohatgi, Mechanical, physical and tribological characterization of nano-cellulose fibers reinforced bio-epoxy composites: An attempt to fabricate and scale the 'Green' composite. *Carbohydrate Polym* 147, 282-293 (2016).
115. A. Alhuthali, I.M. Low and C. Dong, Characterisation of the water absorption, mechanical and thermal properties of recycled cellulose fiber reinforced vinyl-ester eco-nanocomposites. *Composites B* 43, 2772-2781 (2012).
116. H. Alamri, I.M. Low and Z. Alothman, Mechanical, thermal and microstructural characteristics of cellulose fiber reinforced epoxy/organoclay nanocomposites. *Composites B* 43, 2762-2771 (2012).
117. S. Rao, K. Jayaraman and D. Bhattacharyya, Micro and macro analysis of sisal fiber composites hollow core sandwich panels. *Composites B* 43, 2738-2745 (2012).
118. E.E.M. Ahmad and A.S. Luyt, Effects of organic peroxide and polymer chain structure on morphology and thermal properties of sisal fiber reinforced polyethylene composites. *Composites A* 43, 703-710 (2012).

119. N. Chand and R. Joshi, Analysis of mechanical, thermal, and dynamic mechanical behaviors of different polymer-coated sisal fibers. *J Natural Fibers* 7, 100-110 (2010).
120. V. Gopalan, V. Suthenthiraveerappa, A.R. Annamalai, S. Manivannan, V. Pragasam, P. Chinnaiyan, G. Mannayee and C.-P. Jen, Dynamic characteristics of woven flax/epoxy laminated composite plate. *Polymers* 13, 209 (2021).
121. A.K. Zykova, P.V. Pantyukhov, E.E. Mastalygina, Ch. Chaverri-Ramos, S.G. Nikolaeva, J.J. Saavedra-Arias, A.A. Popov, S.E. Wortman and M. Poletto, Biocomposites of low-density polyethylene plus wood flour or flax straw: biodegradation kinetics across three environments. *Polymers* 13, 2138 (2021).
122. D. Ndiaye, L.M. Matuana, S. Morlat-Therias, L. Vidal, A. Tidjani and J.-L. Gardette, Thermal and mechanical properties of polypropylene/wood-flour composites. *J Appl Polym Sci* 119, 3321-3328 (2010).
123. A.K. Błędzki, M.M. Letman, A. Vīksne and L. Rence, A comparison of compounding processes and wood type for wood fiber-PP composites. *Composites A* 36, 789-797 (2005).
124. N. Ayrilmis, U. Buyuksari and T. Dundar, Waste pine cones as a source of reinforcing fillers for thermoplastic composites. *J Appl Polym Sci* 117, 2324-2330 (2010).
125. M. Rivai, A. Gupta, M.R. Islam and M.D.H. Beg, Characterization of oil palm empty fruit bunch and glass fiber reinforced recycled polypropylene hybrid composites. *Fibers Polym* 15, 1523-1530 (2014).
126. Ch. Dong and I.J. Davies, Flexural properties of wheat straw reinforced polyester composites. *Amer J Mater Sci* 1, 71-75 (2011).
127. D. Paukszta, Investigations of lignocellulosic materials from rape for the purpose of producing composites with thermoplastic polymers. *Fibers Textiles in Eastern Europe* 13, 90-92 (2005).
128. Ch. Dong and I.J. Davies, Flexural properties of macadamia nutshell particle reinforced polyester composites. *Composites B* 43, 2751-2756 (2012).
129. C.X. Viet, H. Ismail, A. Masa and N. Hayeemasae, Silane-coupled kenaf fiber filled thermoplastic elastomer based on recycled high density polyethylene/natural rubber blends. *Polymers* 31, 2 (2021).
130. A.Q. Barbosa, L.F.M. da Silva, J. Abenojar, M. de Figueiredo and A. Öchsner, Toughness of a brittle epoxy resin reinforced with micro cork particles: Effect of size, amount and surface treatment. *Composites B* 114, 299-310 (2017).
131. B.F. Yousif, A. Shalwan, C.W. Chin and K.C. Ming, Flexural properties of treated and untreated kenaf/epoxy composites. *Mater Design* 40, 378-385 (2012).
132. T. Singh, Optimum design based on fabricated natural fiber reinforced automotive brake friction composites using hybrid CRITIC-MEW approach, *J Mater Res Technol* 14, 81-92 (2021).
133. P. Manimaran, P. Senthamaraikannan, M.R. Sanjay, M.K. Marichelvam and M. Jawaid, Study on characterization of *Furcraea foetida* new natural fiber as

composite reinforcement for lightweight applications. *Carbohydrate Polym* 81, 650-658 (2018).
134. S.B. Viswanath, S. Sathees Kumar, D. Sudarsan and R. Muthalagu, Mechanical attributes of coir fiber, rice husk and egg shell reinforced hybrid polyester composites. *Mater Today Proc* 46, 874-877 (2021).
135. P. Georgiopoulos, E. Kontou and M. Niaounakis, Thermomechanical properties and rheological behavior of biodegradable composites. *Polym Composites* 35, 1140-1149 (2013).
136. V. Fombuena, L. Sánchez-Nácher, M.D. Samper, D. Juárez and R. Balart, Study of the properties of thermoset materials derived from epoxidized soybean oil and protein fillers. *J Am Oil Chem Soc* 90, 449-457 (2012).
137. D. Lascano, D. Garcia-Garcia, S. Rojas-Lema, L. Quiles-Carrillo, R. Balart and T. Boronat, Manufacturing and characterization of green composites with partially biobased epoxy resin and flaxseed flour wastes. *Appl Sci* 10, 3688 (2020).
138. S. Kumar, S.K. Samal, S. Mohanty and S.K. Nayak, Epoxidized soybean oil-based epoxy blend cured with anhydride-based cross-linker: thermal and mechanical characterization. *Indust Eng Chem Res* 56, 687-698 (2017).
139. J.M. España, L. Sánchez-Nacher, T. Boronat, V. Fombuena and R. Balart, Properties of biobased epoxy resins from epoxidized soybean oil (ESBO) cured with maleic anhydride (MA). *J Am Oil Chem Soc* 89 2067-2075 (2012).
140. Y.-F. Shih, J.-X. Cai, C.-S. Kuan and C.-F. Hsieh, Plant fibers and wasted fiber/epoxy green composites. *Composites B* 43, 2817-2821 (2012).
141. M.R. Sanjay, P. Madhu, M. Jawaid, P. Senthamaraikannan, S. Senthil and S. Pradeep, Characterization and properties of natural fiber polymer composites: A comprehensive review. *J Cleaner Prod* 172, 566-581 (2018).
142. M. Milosevic, P. Valášek and A. Ruggiero, Tribology of natural fibers composite materials: an overview. Lubricants 8, 42 (2020).
143. P. Chandra and S.R. Benin, Study on the influence of barium sulphate on flexural and impact strength of simple natural fiber composite. *Mater Today Proc.*, in press.
144. N. Suizu, U. Takashi, K. Goda and J. Ohgi, Tensile and impact properties of fully green composites reinforced with mercerized ramie fibers. *J Mater Sci* 44, 2477-2482 (2009).
145. P. Marie Arockianathan, S. Sekar, S. Sankar, B. Kumaran and T.P. Sastry, Evaluation of biocomposite films containing alginate and sago starch impregnated with silver nano particles. *Carbohydrate Polym* 90, 717-724 (2012).
146. M.H. Izmar, M.M. Afiq and A.R. Azura, Effects of different additions of sago starch filler on physical and biodegradation properties of pre-vulcanized NR latex composites. *Composites B* 43, 2746-2750 (2012).
147. Y. Altun, M. Doğan and E. Bayramlı, Effect of alkaline treatment and pre-impregnation on mechanical and water absorption properties of pine wood flour containing poly(lactic acid) based green-composites. *J Polym Environ* 21, 850-856 (2012).

148. K. Kamau-Devers and S.A. Miller, The environmental attributes of wood fiber composites with bio-based or petroleum-based plastics. *Int J Life Cycle Assessment* 25, 1145-1159 (2020).
149. E.M. Fernandes, V.M. Correlo, J.F. Mano and R.L. Reis, Novel cork-polymer composites reinforced with short natural coconut fibers: effect of fiber loading and coupling agent addition. *Composites Sci Technol* 78, 56-62 (2013).
150. T. Ngo, M. Ton-That and W. Hu, Triticale straw and its thermoplastic biocomposites. *Sci Eng Composite Mater* 22, 343-351 (2015).
151. U. Tayfun, M. Dogan and E. Bayramli, Effect of surface modification of rice straw on mechanical and flow properties of TPU-based green composites. *Polym Composites* 37, 1596-1602 (2014).
152. N.-M. Barkoula and T. Peijs, Interface engineering through matrix modification in natural fiber composites, in: *Interface Engineering of Natural Fiber Composites for Maximum Performance*, N. Zafeiropoulos (Ed.), pp. 43-81, Woodhead Publishing (2011).
153. S. Indran and R.E. Raj, Characterization of new natural cellulosic fiber from *Cissus quadrangularis* stem. *Carbohydrate Polym* 117, 392-399 (2015).
154. Z. Belouadah, A. Ati and M. Rokbi, Characterization of new natural cellulosic fiber from *Lygeum spartum L. Carbohydrate Polym* 134, 429-437 (2015).
155. M. Jacob, S. Thomas and K.T. Varughese, Mechanical properties of sisal/oil palm hybrid fiber reinforced natural rubber composites. *Composites Sci Technol* 64, 955-965 (2004).
156. A. Rudawska, I. Haniecka, M. Jaszek and M. Osińska-Jaroszuk, The influence of biochemical modification on the properties of adhesive compounds. *Polymers* 9, 442 (2017).
157. A. Rudawska, I. Haniecka, M. Jaszek and D. Stefaniuk, The influence of adhesive compounds biochemical modification on the mechanical properties of adhesive joints. *Polymers* 10, 344 (2018).
158. T. Aziz, H. Fan, X. Zhang, F.U. Khan, S. Fahad and A. Ullah, Adhesive properties of bio-based epoxy resin reinforced by cellulose nanocrystal additives. *J Polym Eng* 40, 314-320 (2020).
159. T. Takahashi, K.-i. Hirayama, N. Teramoto and M. Shibata, Biocomposites composed of epoxidized soybean oil cured with terpene-based acid anhydride and cellulose fibers. *J Appl Polym Sci* 108, 1596-1602 (2008).
160. M.D. Samper, V. Fombuena, T. Boronat, D. García-Sanoguera and R. Balart, Thermal and mechanical characterization of epoxy resins (ELO and ESO) cured with anhydrides. *J Am Oil Chem Soc* 89, 1521-1528 (2012).
161. D.L. Motoc, J.M. Ferri, S. Ferrandiz-Bou, D. Garcia-Garcia and R. Balart, Dynamic–mechanical and decomposition properties of flax/basalt hybrid laminates based on an epoxidized linseed oil polymer. *Polymers* 13, 479 (2021).
162. E. Lizundia, M.H. Sipponen, L.G. Greca, M. Balakshin, B.L. Tardy, O.J. Rojas and D. Puglia, Multifunctional lignin-based nanocomposites and nanohybrids. *Green Chem* 23, 6698-6760 (2021).

163. T.H. Mokhothu, B.R. Guduri and A.S. Luyt, Kenaf fiber-reinforced copolyester biocomposites. *Polym Composites* 32, 2001-2009 (2011).
164. M.F.M. Alkbir, S.M. Sapuan, A.A. Nuraini and M.R. Ishak, Effect of geometry on crashworthiness parameters of natural kenaf fiber reinforced composite hexagonal tubes. *Mater Design* 60, 85-93 (2014).
165. M.N. Prabhakar, A.U.R. Shah, K.C. Rao and J.-I. Song, Mechanical and thermal properties of epoxy composites reinforced with waste peanut shell powder as a bio-filler. *Fibers Polym* 16, 1119-1124 (2015).
166. A.Q. Barbosa, L.F.M. da Silva, M.D. Banea and A. Öchsner, Methods to increase the toughness of structural adhesives with micro particles: an overview with focus on cork particles. *Mat.-wiss. u. Werkstofftech* 47, 307-325 (2016).
167. N. Benmansour, B. Agoudjil, A. Gherabli, A. Kareche and A.Boudenne, Thermal and mechanical performance of natural mortar reinforced with date palm fibers for use as insulating materials in building. *Energy Build* 8, 198-104 (2014).
168. V. Fiore, T. Scalici and A. Valenza, Characterization of a new natural fiber from *Arundo donax L.* as potential reinforcement of polymer composites. *Carbohydrate Polym* 106, 77-83 (2014).
169. V. Fiore, L. Botta, R. Scaffaro, A. Valenza and A. Pirrotta, PLA based biocomposites reinforced with *Arundo donax* fillers. *Composites Sci Technol* 105, 110-117 (2014).
170. M. A. Meyers, P.-Y. Chen, A. Y.-M. Lin and Y. Seki, Biological materials: Structure and mechanical properties. *Prog Mater Sci* 53, 1-206 (2008).
171. https://info2materia4pl.wordpress.com/2015/04/29/drewno-kilka-slow-o-naturalnym-kompozycie/ (2021)
172. F.G. Calvo-Flores and J.A. Dobado, Lignin as renewable raw material. *ChemSusChem* 3, 1227-1235 (2010).
173. T. Aro and P. Fatehi, Production and application of lignosulfonates and sulfonated lignin. *ChemSusChem* 10, 1861 (2017).
174. D. Kai, M.J. Tan, P.L. Chee, Y.K. Chua, Y.L.Yap and X.J. Loh, Towards lignin-based functional materials in a sustainable world. *Green Chem* 18, 1175-1200 (2016).
175. Ch. Dong and H. Takagi, Flexural properties of cellulose nanofiber reinforced green composites. *Composites B* 58, 418-421 (2014).
176. F.G. Calvo-Flores, J.A. Dobado, J. Isac-García and F. Martin-Martinez, *Lignin and Lignans as Renewable Raw Materials: Chemistry, Technology and Applications*. John Wiley & Sons, New York (2015).
177. N.M. Stark, L.M. Matuana and C.M. Clemons, Effect of processing method on surface and weathering characteristics of wood–flour/HDPE composites. *J Appl Polym Sci* 93, 1021-1030 (2004).
178. N.M. Stark, Effect of weathering cycle and manufacturing method on performance of wood flour and high-density polyethylene composites. *J Appl Polym Sci* 100, 3131-3140 (2006).

179. J.M. Pilarski and L.M. Matuana, Durability of wood flour-plastic composites exposed to accelerated freeze–thaw cycling. II. High density polyethylene matrix. *J Appl Polym Sci* 100, 35-39 (2006).
180. D. Ndiaye, E. Fanton, S. Morlat-Therias, L. Vidal, A. Tidjani and J.-L. Gardette, Durability of wood polymer composites: 1. Influence of wood on the photochemical properties. *Composites Sci Technol* 68, 2779-2784 (2010).
181. Y. Zhang, S.Y. Zhang and Y.H. Chui, Water vapor adsorption and volumetric swelling of melt-impregnated wood–polymer composites. *J Appl Polym Sci* 102, 2668-2676 (2006).
182. N. Pjanic, Research of wood-polymer compositions of DMTA methods. *Int J Sci Soc* 2, 19-24 (2020).
183. S. Mounir, Y. Maaloufa, A.b. Cherki and A. Khabbazi, Thermal properties of the composite material clay/granular cork. *Constr Build Mater* 70, 183-190 (2014).
184. A. Şen, J. Van den Bulcke, N. Defoirdt, J. Van Acker and H. Pereira, Thermal behaviour of cork and cork components. *Thermochim Acta* 582, 94-100 (2014).
185. R. Costa, A. Lourenço, V. Oliveira and H. Pereira, Chemical characterization of cork, phloem and wood from different *Quercus suber* provenances and trees. *Heliyon* 5, e02910 (2019).
186. A.Q. Barbosa, Improved toughness of adhesives filled with cork micro particles, Ph.D. thesis, University of Porto, Porto, Portugal (2017).
187. A.M.A. Pintor, C.I.A. Ferreira, J.C. Pereira, P. Correia, S.P. Silva, V.J.P. Vilar, C.M.S. Botelho and R..R. Boaventura, Use of cork powder and granules for the adsorption of pollutants: A review. *Water Res* 46, 3152-3166 (2012).
188. E. Conde, E. Cadahía, M.C. García-Vallejo and B.F. de Simón, Polyphenolic composition of *Quercus suber* cork from different Spanish provenances. *J Agric Food Chem* 46, 3166-3171 (1998).
189. L.M.C.C. Gil, Cork powder waste: An overview. *Biomass Bioenergy* 13, 59-61 (1997).
190. L.M.C.C. Gil, New cork powder particleboards with thermoplastic binding agents. *Wood Sci Technol* 27, 173-182 (1993).
191. E.M. Fernandes, V.M. Correlo, J.F. Mano and R.L. Reis, Cork–polymer biocomposites: Mechanical, structural and thermal properties. *Mater Design* 82, 282-289 (2015).
192. C.H. Vane, T.C. Drage and C.E. Snape, Bark decay by the white-rot fungus *Lentinula edodes*: Polysaccharide loss, lignin resistance and the unmasking of suberin. *Int Biodeter Biodegr* 57, 14-23 (2006).
193. K. Kemppainen, M. Siika-aho, S. Pattathil, S. Giovando and K. Kruus, Spruce bark as an industrial source of condensed tannins and non-cellulosic sugars. *Ind Crops Prod* 52, 158-168 (2014).
194. M. Stepczyńska, A. Pawłowska, K. Moraczewski, P. Rytlewski, A. Trafarski, D. Olkiewicz and M. Walczak, Evaluation of the mechanical and biocidal

properties of *Lapacho* from *Tabebuia* plant as a biocomposite material. *Materials* 14, 4241 (2021).
195. S. Cailol, B. Boutevin and J.-P. Pascault, Bio-sourced epoxy monomers and polymers, in *Handbook of Adhesive Technology*, third edition, A. Pizzi and K.L. Mittal (Eds.), pp. 443-470, CRC Press, Boca Raton, FL (2018).
196. A.E. Gerbase, C.L. Petzhold and A.P.O. Costa, Dynamic mechanical and thermal behavior of epoxy resins based on soybean oil. *J Amer Oil Chem Soc* 79, 797-802 (2002).
197. N. Boquillon and C. Fringant, Polymer networks derived from curing of epoxidised linseed oil: influence of different catalysts and anhydride hardeners. *Polymer* 41, 8603-8613 (2000).
198. N. Ghasemi Rad, Z. Karami, M.J. Zohuriaan-Mehr, A. Salimi and K. Kabiri, Linseed oil-based reactive diluents preparation to improve tetra-functional epoxy resin properties. *Polym Adv Technol* 30, 2361-2369 (2019).
199. M. Jaszek, M. Osińska-Jaroszuk, G. Janusz, A. Matuszewska, D. Stefaniuk, J. Sulej, J. Polak, M. Ruminowicz, K. Grzywnowicz and A. Jarosz-Wilkolazka, New bioactive fungal molecules with high antioxidant and antimicrobial capacity isolated from *Cerrena unicolor* idiophasic cultures. *BioMed Res Int* 497492 (2013).
200. A. Leonowicz and K. Grzywnowicz, Quantitative estimation of laccase forms in some white-rot fungi using syringaldazine as a substrate. *Enzyme Microbe Technol* 3, 55-58 (1981).
201. P. Widsten and A. Kandelbauer, Laccase applications in the forest products industry: A review. *Enzyme Microbe Technol* 42, 293-307 (2008).
202. Y. Zhang, D.-Q.; Yang, X.-M. Wang, M. Feng and G. He, Fungus-modified lignin sands its use in wood adhesives for wood composites. In: Proceedings of the International Conference of Wood Adhesives, Toronto, ON, Canada (2013).
203. https://zpe.gov.pl/a/wlokna-naturalne-sztuczne-i-syntetyczne/DGQKwcYV4 (2021).
204. https://www.ekologia.pl/wiedza/slowniki/leksykon-ekologii-i-ochrony-srodowiska/celuloza (2021).
205. https://www.sciencedirect.com/topics/earth-and-planetary-sciences/cellulose (2021).
206. https://www.intechopen.com/books/cellulose/ (2021).
207. http://www.scienceclarified.com/Ca-Ch/Cellulose.html (2021).
208. https://www.encyklopedialesna.pl/haslo/lignina/ (2021).
209. N.-E. El Mansouri, A. Pizzi and J. Salvado, Lignin-based polycondensation resin for wood adhesives. *J Appl Polym Sci* 103, 1690-1699 (2007).
210. V. Ibrahim, G. Mamo, P.-J. Gustafsson and R. Hatti-Kaul, Production and properties of adhesive formulated from laccase modified Kraft lignin. *Ind Crops Prod* 45, 343-348 (2013).
211. B.M. Wood, S.R. Coles and S. Maggs, Use of lignin as a compatibiliser in hemp/epoxy composites. *Composites Sci Technol* 71, 1804-1810 (2011).

212. G. Yamini, A. Shakeri, M.J. Zohuriaan-Mehr and K. Kabiri, Cyclocarbonated lignosulfonate as a bio-resourced reactive reinforcing agent for epoxy biocomposite: From natural waste to value-added bio-additive. *J CO_2 Utilization* 24, 50-58 (2018).
213. https://www.britannica.com/science/starch (2020).
214. V.K. Singh, Mechanical behavior of walnut (Juglans L.) shell particles reinforced bio-composite. *Sci Eng Composite Mater* 22, 383-390 (2015).
215. C. Guler, Y. Copur and C. Tascioglu, The manufacture of particleboards using mixture of peanut hull (*Arachis hypoqaea* L.) and European Black pine (*Pinus nigra* Arnold) wood chips. *Bioresource Technol* 99, 2893-2897 (2008).
216. N. Sutivisedsak, H.N. Cheng, Ch.S. Burks, J.A. Johnson, J. P. Siegel, E.L. Civerolo and A. Biswas, Use of nutshells as fillers in polymer composites. *J Polym Environ* 20, 305-314 (2012).
217. https://pph-rewa.pl/produkt/lupiny-orzecha/ (2020).
218. Y. He, H.-M. Wu, J.-B. Guo, W.-D. He and Y. Zhou, EPDM-G-GMA toughening of straw/polypropylene composites: mechanical properties, thermal stability and rheological properties. *Int Polym Process* 35, 50-57 (2020).
219. https://www.agropolska.pl/uprawa/rosliny-oleiste/wartosc-nawozowa-slomy-rzepakowej,150.html (2021).
220. H. Pińkowska and P. Wolak, The investigation of chemical composition of waste rapeseed biomass as a raw material for synthesis of useful chemical bioproducts under hydrothermal conditions: Part 1. Classical analytical methods, http://yadda.icm.edu.pl/yadda/element/bwmeta1.element.ekon-element-000165070244 (2020).
221. J.M.F. Abreu and A.M. Bruno-Soares, Characterization and utilization of rice, legume and rape straws. *Options Mediterraneennes. Serie B: Etudes et Recherches* 17, 39-51 (1999).
222. D. Paukszta, B. Doczekalska, A. Ostrowski and M. Bartkowiak, Modification of rapeseed straw with organic acid anhydrides. *J Composite Mater* 49, 1369-1378 (2014).
223. T. Xia, H. Huang, G. Wu, E. Sun, X. Jin and W. Tang, The characteristic changes of rice straw fibers in anaerobic digestion and its effect on rice straw-reinforced composites. *Ind Crops Prod* 121, 73-79 (2018).
224. P.R. Hornsby, E. Hinrichen and K. Tarverdi, Preparation and properties of polypropylene composites reinforced with wheat and flax straw fibers. Part II: Analysis of composite microstructure and mechanical properties. *J Mater Sci* 32, 1009-1015 (1997).
225. K.G. Ramawat and M.R. Ahuja: Fiber plants: An overview, in: *Fiber Plants. Biology, Biotechnology and Applications*, K.G. Ramawat and M.R. Ahuja (Eds.), pp. 3-15, Springer (2016).
226. I. Sanal and D. Verma, Construction materials reinforced with natural products, in: *Handbook of Ecomaterials*, L.M.T. Martínez, O.V. Kharissova and B.I. Kharisov, (Eds.), pp. 1-24, Springer, Switzerland (2019).

227. M. Zhan, R.P. Wool and J.Q. Xiao, Electrical properties of chicken feather fiber reinforced epoxy composites. *Composites A* 42, 229-233 (2011).
228. T. Sathishkumar, S. Satheeshkumar and J. Naveen, Glass fiber-reinforced polymer composites – a review. *J Reinforced Plast Composites* 33, 1258-1275 (2014).
229. G.H. Awan, L. Ali, K.M. Ghauri, E. Ramzan and E. Ehsan, Effect of various forms of glass fiber reinforcements on tensile properties of polyester matrix composite. *J. Faculty Eng.Technol.* 16, 33-39 (2009).
230. N. Gupta, B. S. Brar and E. Woldesenbet, Effect of filler addition on the compressive and impact properties of glass fiber reinforced epoxy. *Bull Mater Sci* 24, 219-223 (2001).
231. J.D. Bolt, D.P. Button and B.A. Yost, Ceramic-fiber-polymer composites for electronic substrates. *Mater Sci Eng A* 109, 207-211 (1989).
232. L.D. Maxim and M.J. Utell, Aluminosilicate fibers, in: *Encyclopedia of Toxicology*, third edition, P. Wexler (Ed.), pp. 156-160, Academic Press (2014).
233. H.N. Cheng, M.K. Dowd and A. Biswas, Biobased polymeric materials prepared from cotton byproducts. *ACS Symposium Series* 1105, 47-62 (2012).
234. Y.-Y. Li, B. Wang, B. Wang and M.-G. Ma, The enhancement performances of cotton stalk fiber/PVC composites by sequential two steps modification. *J Appl Polym Sci* 135, 46090 (2017).
235. https://hempiflax.com/wlokno/ (2021).
236. D.S. De Vasconcellos, F. Touchard and L. Chocinski-Arnault, Tension-tension fatigue behaviour of woven hemp fiber reinforced epoxy composite: A multi-instrumented damage analysis. *Int J Fatigue* 59, 159-169 (2014).
237. M.S. Islam, K.L. Pickering and N.J. Foreman, Influence of alkali fiber treatment and fiber processing on the mechanical properties of hemp/epoxy composites. *J Appl Polym Sci* 119, 3696-3707 (2011).
238. G. Lebrun, A. Couture and L. Laperrière, Tensile and impregnation behavior of unidirectional hemp/paper/epoxy and flax/paper/epoxy composites. *Composite Struct* 103, 151-160 (2013).
239. S. Liang, P.B. Gning and L. Guillaumat, Properties evolution of flax/epoxy composites under fatigue loading. *Int J Fatigue* 63, 36-45 (2014).
240. G. Coroller, A. Lefeuvre and A. Le Duigou, Effect of flax fibers individualisation on tensile failure of flax/epoxy unidirectional composite. *Composites A* 51, 62-70 (2013).
241. A. Triki, M. Karray and C. Poilane, Dielectric analysis for adhesion study of woven flax fibers reinforced epoxy composites. *J Electrostatics* 76, 67-72 (2015).
242. Z. Liu, S.Z. Erhan and D.E. Akin, 'Green' composites from renewable resources: Preparation of epoxidized soybean oil and flax fiber composites. *J Agric Food Chem* 54, 2134-2137 (2006).
243. U. Tayfun, M. Dogan and E. Bayramli, Investigations of the flax fiber/thermoplastic polyurethane eco-composites: Influence of isocyanate modification of flax fiber surface. *Polym Composites* 38, 2874-2880 (2015).

244. J. Militký and A. Jabbar, Comparative evaluation of fiber treatments on the creep behavior of jute/green epoxy composites. *Composites B* 80, 361-368 (2015).
245. T.T.L. Doan, H. Brodowsky and E. Mäder, Jute fiber/epoxy composites: Surface properties and interfacial adhesion. *Composites Sci Technol* 72, 1160-1166 (2012).
246. R. Masoodi and K.M. Pillai, A study on moisture absorption and swelling in bio-based jute-epoxy composites. *J Reinforced Plast Composites* 31, 285-294 (2012).
247. M. Jawaid, H.P.S. Abdul Khalil, O.Y. Alothman, M.T. Paridah and H.P.S. Abdul Khalil, Effect of oil palm and jute fiber treatment on mechanical performance of epoxy hybrid composites. *Int J Polym Anal Charact* 19, 62-69 (2014).
248. A. Gopinath, M. Senthil Kumar and A. Elayaperumal, Experimental investigations on mechanical properties of jute fiber reinforced composites with polyester and epoxy resin matrices. *Procedia Eng* 97, 2052-2063 (2014).
249. M. Boopalan, M. Niranjanaa and M.J. Umapathy, Study on the mechanical properties and thermal properties of jute and banana fiber reinforced epoxy hybrid composites. *Composites B* 51, 54-57 (2013).
250. https://pl.wikipedia.org/wiki/Sizal (2021).
251. S. Srisuwana, N. Prasoetsophaa and N. Suppakarn, The effects of alkalized and silanized woven sisal fibers on mechanical properties of natural rubber modified epoxy resin. *Energy Procedia* 56, 19-25 (2014).
252. M.Z. Rong, M.Q. Zhang and Y. Liu, The effect of fiber treatment on the mechanical properties of unidirectional sisal-reinforced epoxy composites. *Composites Sci Technol* 61, 1437-1447 (2001).
253. V.P. Arthanarieswaran, A. Kumaravel and M. Kathirselvam, Evaluation of mechanical properties of banana and sisal fiber reinforced epoxy composites: Influence of glass fiber hybridization. *Mater Design* 64, 194-202 (2014).
254. F.I. Romli, A.N. Alias, A.S.M. Rafie and D.L.A.A. Majid, Factorial study on the tensile strength of a coir fiber-reinforced epoxy composite. *AASRI Procedia* 3, 242-247 (2012).
255. M. Vaishnavi, S. Saarani, A. Varghese, D. Sathyan and K.M. Mini, Strength and workability characteristics of coir and nylon fiber reinforced self-compacting mortar. *Mater Today Proc* 46, 4696-4701 (2021).
256. N. Saba, F. Mohammad, M. Pervaiz, M. Jawaid, O.Y. Alothman and M. Sain, Mechanical, morphological and structural properties of cellulose nanofibers reinforced epoxy composites. *Int J Biol Macromol* 97, 190-200 (2017).
257. https://materialyinzynierskie.pl/wytwarzanie-ubran-z-celulozy-drzewnej/ (2021).
258. https://pl.wikipedia.org/wiki/W%C5%82%C3%B3kna_celulozowe (2021).
259. U. Nirmal, J. Hashim and K.O. Low, Adhesive wear and frictional performance of bamboo fibers reinforced epoxy composite. *Tribology Int* 47, 122-133 (2012).

260. A.V. Ratna Prasad and K. Mohana Rao, Mechanical properties of natural fiber reinforced polyester composites: Jowar, sisal and bamboo. *Mater Design* 32, 4658-4663 (2011).
261. K. Murali Mohan Rao, K. Mohana Rao and A.V. Ratna Prasad, Fabrication and testing of natural fiber composites: Vakka, sisal, bamboo and banana. *Mater Design* 31, 508-513 (2010).
262. H.M. Akil, M.F. Omar, A.A.M. Mazuki, S. Safiee, Z.A.M. Ishak and A. Abu Bakar, Kenaf fiber reinforced composites: A review. *Mater Design* 32 4107-4121 (2011).
263. A.H. Jamadi, N. Razali, M. Petrů, M.M. Taha, N. Muhammad and R.A. Ilyas, Effect of chemically treated kenaf fiber on mechanical and thermal properties of PLA composites prepared through fused deposition modeling (FDM). *Polymers* 13, 3299 (2021).
264. R. Mahjoub, J.M. Yatim, A.R.M. Sam and S.H. Hashemi, Tensile properties of kenaf fiber due to various conditions of chemical fiber surface modifications. *Constr Build Mater* 55, 103-113 (2014).
265. A. Manral and P.K. Bajpai, Analysis of properties on chemical treatment of kenaf fibers, *Mater Today Proc* 40, S35-S38 (2021).
266. R. Rahman and S.Z.F.S. Putra, Tensile properties of natural and synthetic fiber-reinforced polymer composites, in: *Mechanical and Physical Testing of Biocomposites, Fiber-Reinforced Composites and Hybrid Composites*, M. Jawaid, M. Thariq and N. Saba (Eds.), pp. 81-102, Woodhead Publishing (2019).
267. A.M. Fairuz, S.M. Sapuan, E.S. Zainudinand and C.N.A. Jaafar, Effect of filler loading on mechanical properties of pultruded kenaf fiber reinforced vinyl ester composites. *J Mech Eng Sci* 10, 1931-1942 (2016).
268. Ch.K. Yong, Ch.Y. Chee, Ch.H. Chuah and N.-S. Liou, Effect of fiber orientation on mechanical properties of kenaf-reinforced polymer composite. *BioResources* 10, 2597-2608 (2015).
269. Y.A. El-Shekeil, S.M. Sapuan K. Abdan and E.S. Zainudin, Influence of fiber content on the mechanical and thermal properties of kenaf fiber reinforced thermoplastic polyurethane composites. *Mater Design* 40, 299-303 (2012).
270. N. Chand and M. Fahim (Eds.), *Tribology of Natural Fiber Polymer Composites*, second edition. Woodhead Publishing (2021).
271. N. Chand and M. Fahim, Natural fibers and their composites, in: *Tribology of Natural Fiber Polymer Composites*, N. Chand and M. Fahim (Eds.), pp. 1-59, Woodhead Publishing (2021).
272. B.B. Kalita, N. Gogoi and S. Kalita, Properties of ramie and its blends. *Int J Eng Res General Sci* 1, 1-6 (2013).
273. S. Nam and A.N. Netravali, Green composites. I. physical properties of ramie fibers for environment-friendly green composites. *Fibers Polym* 7, 372-379 (2006).
274. S. Nam and A.N. Netravali, Green composites. II. Environment-friendly, biodegradable composites using ramie fibers and soy protein concentrate (SPC) resin. *Fibers Polym* 7, 380-388 (2006).

275. M.N. Cazaurang-Martinez, P.J. Herrera-Franco, P.I. Gonzalez-Chi and M. Aguilar-Vega, Physical and mechanical properties of henequen fibers. *J Appl Polym Sci* 43, 749-756 (1991).
276. M. Pathan Yasin, C. Venkata Ramana, K. Krishna Vamshi and A. Pradeep, A study of continuous henequen/epoxy composites. *Mater Today Proc* 18, 3798-3811 (2019).
277. K. Hintze, J.I. Tapia, E. Alvarado-Gómez and A. Encinas, Natural henequen fibers functionalized with a magnetic fatty acid mixture. *Mater Letters* 291, 129580 (2021).
278. D.C.O. Nascimento, A.S. Ferreira, S. N. Monteiro, R.C.M.P. Aquino and S.G. Kestur, Studies on the characterization of piassava fibers and their epoxy composites. *Composites A* 43, 353-362 (2012).
279. M. Ramesh, K. Palanikumar and K.H. Reddy, Mechanical property evaluation of sisal-jute-glass fiber reinforced polyester composites. *Composites B* 48, 1-9 (2013).
280. V.S. Srinivasan, S. Rajendra Boopathy and D. Sangeetha, Evaluation of mechanical and thermal properties of banana–flax based natural fiber composite. *Mater Design* 60, 620-627 (2014).
281. M. Jawaid, H.P.S. Abdul Khalil and A. Abu Bakar, Woven hybrid composites: Tensile and flexural properties of oil palm-woven jute fibers based epoxy composites. *Mater Sci Eng A* 528, 5190-5195 (2011).
282. R. Bhoopathi, M. Ramesh and C. Deepa, Fabrication and property evaluation of banana-hemp-glass fiber reinforced composites. *Procedia Eng* 97, 2032-2041 (2014).
283. R. Petrucci, C. Santulli and D. Puglia, Mechanical characterisation of hybrid composite laminates based on basalt fibers in combination with flax, hemp and glass fibers manufactured by vacuum infusion. *Mater Design* 49, 728-735 (2013).
284. V. Mohanavel, T. Raja, A. Yadav, M. Ravichandran and J. Winczek, Evaluation of mechanical and thermal properties of jute and ramie reinforced epoxy-based hybrid composites. *J Natural Fibers*, DOI: 10.1080/15440478.2021.1958432 (2021).
285. K. Hofmann and W. Glasser, Engineering plastics from lignin, 22. Cure of lignin-based epoxy resins. *J Adhesion* 40, 229-241 (1993).
286. E. Aracri, C. Diaz Blanco and T. Tzanov, An enzymatic approach to develop a lignin-based adhesive for wool floor coverings. *Green Chem* 16, 2597-2603 (2014).
287. A. Leonowicz and K. Grzywnowicz, Quantitative estimation of laccase forms in some white-rot fungi using syringaldazine as a substrate. *Enzyme Microbe Technol* 3, 55-58 (1981).
288. A. Pizzi, Recent developments in eco-efficient bio-based adhesives for wood bonding: Opportunities and issues. *J Adhesion Sci Technol* 20, 829-846 (2006).
289. W. Nieh and W.G. Glasser, Lignin derivatives with epoxy functionality. In: *Lignin: Properties and Materials*, W. Glasser and S. Sarkanen (Eds.), pp. 506-514, Amer. Chem. Soc., Washington, D.C. (1989).

290. K. Hofmann and W. Glasser, Engineering plastics from lignin, 23. Network formation of lignin-based epoxy resins. *Macromol Chem Phys* 195, 65-80 (1994).
291. W. Peng and B. Riedl, The chemorheology of phenol-formaldehyde thermoset resin and mixtures of the resin with lignin fillers. *Polymer* 35, 1280-1286 (1994).
292. M.A. Khan, S.M. Ashraft and V.P. Malhorta, Development and characterization of wood adhesive using bagasse lignin. *Int J Adhesion Adhesives* 24, 485-493 (2004).
293. M. Jaszek M. Osińska-Jaroszuk, J. Sulej, A. Matuszewska, D. Stefaniuk, K. Maciąg, J. Polak, Ł. Matuszewski and K. Grzywnowicz, Stimulation of the antioxidative and antimicrobial potential of the blood red bracket mushroom *Pycnoporus sanguineus* (higher *Basidiomycetes*). *Int J Medical Mushrooms* 17, 701-712 (2015).
294. T.G. Yashas Gowda, M.R. Sanjay, K. Subrahmanya Bhat, P. Madhu, P. Senthamaraikannan and B. Yogesha, Polymer matrix-natural fiber composites: An overview. *Cogent Eng* 5, 1446667 (2018).
295. A. Ashori, Wood-plastic composites as promising green-composites for automotive industries. *Bioresource Technol* 99, 4661-4667 (2008).
296. J.S. Renner, R.A. Mensah, L. Jiang, Q. Xu, O. Das and F. Berto, Fire behavior of wood-based composite materials. *Polymers* 13, 4352 (2021).
297. S. Yang, V. Chalivendra and Y. Kim, Electro-fracture studies of natural fiber composites. *J Natural Fibers* 18, 1044-1053 (2021).
298. A. Veerasimman, V. Shanmugam, S. Rajendran, D. J. Johnson, A. Subbiah, J. Koilpichai and U. Marimuthu, Thermal properties of natural fiber sisal based hybrid composites – a brief review. *J Natural Fibers*, DOI: 10.1080/15440478.2020.1870619 (2021).
299. I. Md Nazrul, F. Rahman, A.K. Das and S. Hiziroglu, An overview of different types and potential of bio-based adhesives used for wood products. *Int J Adhesion Adhesives* 112, 102992 (2022).
300. L.A. Heinrich, Future opportunities for bio-based adhesives - advantages beyond renewability. *Green Chem* 21, 1866-1888 (2019).
301. S. Sugiarto, Y. Leow, Ch.L. Tan, G. Wang and D. Kai, How far is lignin from being a biomedical material? *Bioactive Mater* 8, 71-94 (2022).
302. N. Sweeney, Biobased acrylic adhesives, in: *Handbook of Adhesive Technology*, third edition, A. Pizzi and K.L. Mittal (Eds.), pp. 385-403, CRC Press, Boca Raton, FL (2018).
303. I. Singh and Paramjot, Bioadhesives in drug delivery, in: *Handbook of Adhesive Technology*, third edition, A. Pizzi and K.L. Mittal (Eds.), pp. 575-590, CRC Press, Boca Raton, FL (2018).

3

Marble Dust Reinforced Epoxy Structural Adhesive Composites

Amar Patnaik[1]*, Pankaj Agarwal[1,2], Ankush Sharma[3], Deepika Shekhawat[1] and Tapan Kumar Patnaik[4]

[1]Department of Mechanical Engineering, Malaviya National Institute of Technology Jaipur, Jaipur, Rajasthan, India
[2]Department of Mechanical Engineering, Amity University Jaipur, Jaipur, Rajasthan, India
[3]Centre of Excellence for Composites, Ahmedabad Textile Industry's Research Association (ATIRA), Ahmedabad, India
[4]Department of Physics, GIET University, Gunupur, Odisha, India

Abstract

The mechanical properties of epoxy composites filled with non-functionalized oxide nanoparticles is influenced by a number of factors. The development of sophisticated materials also necessitates a direct relationship between a material and its qualities from an industrial standpoint. This work describes the fabrication and testing of marble dust nanoparticle reinforced epoxy structural adhesive composites. The behaviour of composite structure after reinforcing with marble dust has been highlighted in this study. The marble dust in different wt.% (i.e., 0, 1.5, 3 and 4.5 wt.%) is incorporated into LY 556 epoxy and composites fabricated via casting method. The primary focus of this study was to evaluate the effects produced on the physical, mechanical and tribological properties with varying filler (marble dust) content. The results indicated significant improvement in hardness and wear resistance of the composite, whereas tensile strength and flexural strength show decrement behaviour with increase in marble dust content. Tribological behaviour of adhesive composites was examined by assessing the wear tracks via FE-SEM. Results indicated presence of debris particles and mild abrasion caused by the hard marble dust particles in the matrix and could also be seen in SEM micrographs. Finally, TOPSIS is implemented to identify the best alterative from the developed epoxy adhesives based on physical, mechanical, thermal and tribological properties.

**Corresponding author*: patnaik.amar@gmail.com

K.L. Mittal and S.K. Panigrahi (eds.) Structural Adhesives: Properties, Characterization and Applications, (105–134) © 2023 Scrivener Publishing LLC

***Keywords*:** Polymers, marble dust (MD), structural adhesives, nanoparticles, mechanical properties, wear

3.1 Introduction

A material which is applied to the surfaces of two similar or dissimilar materials to permanently join them together is called an adhesive. Structural adhesive bonding and non-structural adhesive bonding are the two main kinds of adhesive bonding. Applications that may experience higher stresses up to their yield points require use of structural adhesives. For structural adhesive joints it is, therefore, essential to be able to transfer stress without compromising integrity, within the design constraints. These joints are expected to be strong enough throughout the service life. Whereas, non-structural adhesives are used to hold lightweight components in place rather than to sustain heavy loads, sometimes also called as holding adhesives which include pressure-sensitive tapes, and packaging adhesives. Epoxy resin based structural adhesives have high bond strength and superior performance. Tough modified epoxy polymers are used in structural adhesives. Hence, they are most suited for structural bonding in engineering applications involving materials, such as glass, metals, polymers, etc. Epoxy, acrylic, urethane, and other adhesives are among the current generations, and their usage is increasing day by day. Structural adhesives for permanent bonding of metals and other hard adherends are among the various types of liquid adhesives in commercial use. Adhesives come in the form of pastes, liquids, films, and supported films for structural applications. Based on origin, adhesives are categorized into two classes: natural and synthetic. The natural adhesives comprise animal glue, casein (milk)- and protein (fish glue) -based adhesives, vegetable-based glues, and natural rubber (latex) glue. Whereas, thermoplastics, thermosets, elastomers, and emulsions are synthetic adhesives. The term thermoplastic adhesive refers to an adhesive that requires no curing (i.e., no chemical reaction). By heating or adding a solvent to thermoplastics, they can be reprocessed several times. To name a few, polyacrylates, polyamides, poly(vinyl acetate) (PVAc) are entitled as thermoplastic adhesives. Thermosetting signifies adhesives that are cured or "set" on heating. Thermosetting involves polymer chains that are chemically crosslinked and cannot be reprocessed.

Adhesives are utilised in various sectors (refer to Figure 3.1(a)) such as biomedical (dentistry), construction, aerospace, automotive, packaging, etc. Adhesive formulations are primarily comprised of natural or synthetic polymers, as well as monomers and pre-polymers that can create polymers.

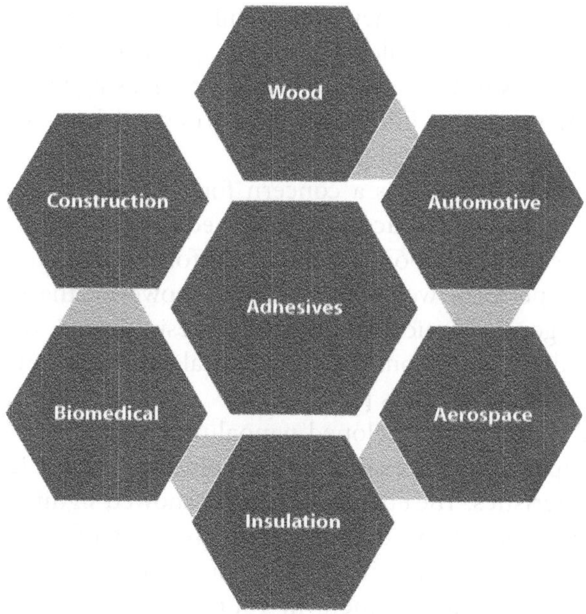

(a)

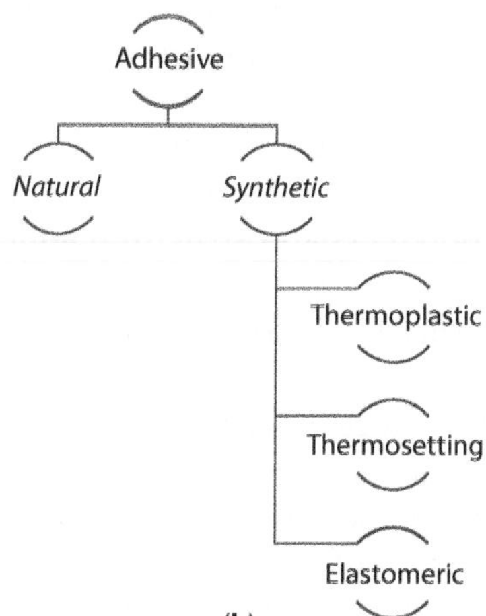

(b)

Figure 3.1 (a) Adhesives classification based on their intended use. (b) Classification of Adhesives [1].

Because adhesives shrink as they solidify, fillers can be employed to reduce shrinking. The selection of filler type and its quantity must be made wisely depending on the substrate and application. Increasing trend in adhesive properties with increasing filler percentage is observed and it tends to drop after reaching a maximum value.

However, the brittleness is a concern for automotive industry [2, 3]. Epoxies are the most widely accepted and used among various commercial structural adhesives [4]. Epoxy resins can perform well up to 100 to 125°C and some can perform well even at 200°C, however, the cost increases but they offer good chemical and corrosion resistance [5, 6]. The performance of bonded connections using structural adhesives can be improved through the addition of fillers [7, 8].

Fouly and Alkalla [9] developed nanoalumina filled epoxy nanocomposites and experimentally examined their physico-mechanical and tribological characteristics. The epoxy composites showed significant influence of low loading fraction on different characteristics of epoxy nanocomposites. Compared to neat epoxy, increasing Al_2O_3 nanoparticles from 0.1 to 0.4 wt.% showed increment in Young's modulus and compressive yield strength. Whereas tribological results indicated improvement in wear resistance at different loads and sliding times [9]. There are various kinds of adhesives and the classification of adhesives is presented in Figure 3.1(b).

Particulate-filled composites exhibit superior mechanical and tribological characteristics [10–14]. Marble dust nanoparticles are increasingly being used in polymers and structural adhesives due to their easy availability and low price. With further increment in nanoparticle loading, the epoxy's viscosity tends to increase, causing inadequate wetting of the particles by the resin and causing difficulty in achieving a uniform adhesive bond. Surface wettability, viscosity, and target properties of the adhesive joints are three important factors that must be taken into consideration while selecting proper nanoparticles. Aggregation and inadequate dispersion of nanoparticles are two obstacles that epoxy nanocomposite adhesion confront, and these prohibit epoxy adhesives from improving their final qualities. Partially cured epoxy is unable to increase the adhesive's final qualities, resulting in a loss in adhesion strength. Better dispersal of nanoparticles in epoxy matrix can be achieved by surface treatment of the nanoparticles with functional groups. Furthermore, functional reactive groups on the surface of filler particles can aid in curing procedure.

In order to improve the characteristics of epoxy adhesives, several types of nanoparticles with distinct organic or inorganic nature can be utilised.

Inorganic nanoparticles are the most prevalent form of nanoparticles employed to reinforce epoxy adhesives. Clay nanoplates, halloysite nanotubes, and metal oxides are all inorganic and mineral (marble dust) nanoparticles that are employed in epoxy adhesive compositions. Nanoparticles, depending on their origin, shape, and size, can help epoxy adhesives improve a variety of properties. Nanoparticles are gaining popularity owing to their potential to improve the mechanical strength of epoxy adhesives without compromising other important features such as heat stability and toughness. However, nanoparticles' shape and functionality prevent creation of epoxy crosslinks during curing reaction, lowering the crosslink density, ultimately reducing the glass-transition temperature and adhesion strength. The dispersal of nanoparticles having a diameter less than 100 nm offers a unique mix of physical, mechanical and wear properties. The nanoparticles present unique characteristic, high surface free energy, and high surface area resulting in improved interfacial area with the matrix material. As a result, dispersal of smaller percentage of nanoparticles in epoxy adhesive matrix can result in considerable improvements both in terms of fabrication and adhesively bonded joint performance such as cure reaction, thermal stability, enhanced tensile, and compressive strengths, and enhanced toughness. Nolte *et al.* [15] fabricated heavily loaded nanocomposites by reinforcing with alumina (up to 50 wt. % Al_2O_3) nanoparticulates into epoxy resin LY 556 along with HY 917 hardener. The results revealed particle agglomeration on increasing the particle concentration beyond 30 wt. % making this wt. % as the most optimal one_Muñoz *et al.* [16] modified epoxy resin by reinforcing with 20 nm sized titania (TiO_2) ceramic nanoparticles for multipurpose applications and results indicated improved mechanical and electrochemical properties. Reddy *et al.* [17] examined the changes occurring in the physical and mechanical properties of epoxy (LY556) polymer by reinforcing with 1, 2, 3, and 4 wt. % tungsten carbide (WC) nano-particulates for structural applications. Results revealed that up to 2 % WC particles, both tensile and flexural strengths enhanced, whereas beyond this loading mechanical properties decreased. Sharma and Gautam fabricated granite dust (5, 10, and 15 wt. % mass) reinforced epoxy resin polymer composites and utilization of granite resulted in_enhancement in the hardness, impact strength and wear resistance [18].

Non-biodegradable wastes have been determined to have the most negative influence on the environment, and all solid ceramic wastes fall into this category. In developing countries, the extraction and use of natural decorative stones such as marble and granite are at a higher level. During the processing, cutting, and sizing of these stones to make them usable, a large quantity of trash is created. These wastes are dumped in the open, and their poisonous nature has a harmful impact on both the environment and human health. In the past, several studies on the recycling of these wastes in the construction sector for the preparation of concrete as a replacement for cement have been done. Although this effort benefits society in terms of health and economy, it is insufficient because garbage continues to accumulate day by day.

In order to optimise the number of experiments there are a variety of mathematical approaches, which are chosen for minimising the complexity of the procedure which is selected on the basis of the problem nature and complexity level ascribed to the decision-making process. Multi-criteria decision making (MCDM) addresses the need for a numerical framework in the material selection procedure. MCDM helps in the overall evaluation by providing a platform for choosing, sorting, and ranking materials. When there are a number of materials with very comparable performances, taking into account these interdependencies might help lessen the chance of making the wrong choice [19].

The objective of the present work was utilization of marble dust-polymer composites by varying the marble dust content (0, 1.5, 3, and 4.5 wt.%) to make structural materials which delivered enhanced resultant properties in addition to lower costs. The performance of the fabricated marble dust filled epoxy composites was evaluated by conducting physical, mechanical and tribological tests. The relative coefficient friction of fabricated marble dust (MD) filled epoxy composites with respect to load and filler composition was examined. Also, the possible wear mechanism was characterized using Scanning Electron Microscopy analysis.

3.2 Materials and Methods

3.2.1 Procurement of Raw Materials

LY 556 epoxy (density = 1.14 g/cm^3; supplied by Savita Scientific & Plastic Products, Jaipur, India) along with the hardener HY 951 in the ratio of 10:1 was used as matrix material as per the manufacturer's specification. The LY 556 epoxy has long pot-life of approximately 40 minutes and is an anhydride-cured in addition to being low viscosity matrix. The waste marble slurry was collected from local marble cutting industry of Jaipur, Rajasthan, India. The slurry was dried in an oven at 100 °C for 24 hours in order to evaporate the moisture from the marble slurry. After that the marble dust (density = 2.68 g/cm^3) was ball milled at 350 rpm

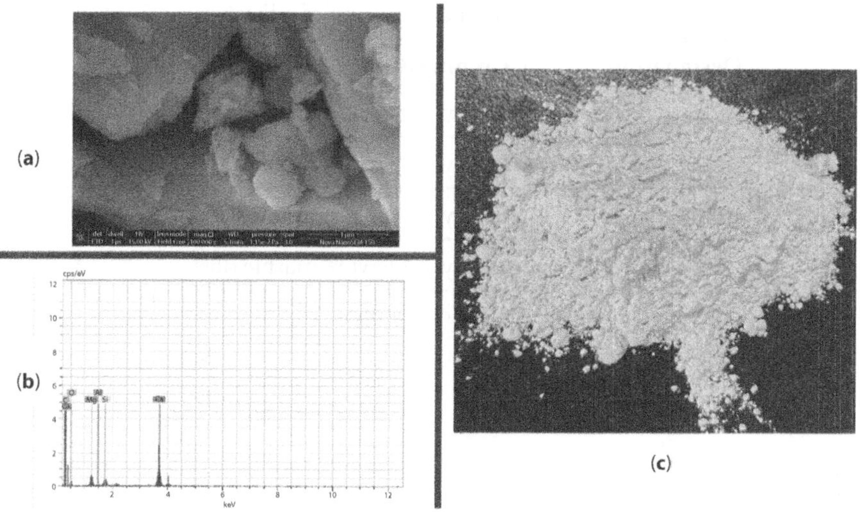

Figure 3.2 (a) SEM micrograph of marble dust, (b) Major constituents of marble dust, and (c) Photo of procured marble dust.

for 72 hr in toluene medium to reduce the size of waste marble dust to nanometer.

After ball milling of marble dust, the particles were obtained in the size range 100 nm to 550 nm and this marble powder was used as the reinforcement material in the composite. The composition of the marble dust is O (54 wt%), Ca (21 wt%), C (13 wt%), Mg (10.21 wt%). The SEM micrograph and major constituents of marble dust are illustrated in Figure 3.2.

3.2.2 Fabrication of Composites

The as-received marble dust was washed with distilled water to eliminate the undesirable dust particles followed by oven drying and crushing. The marble dust was weighed in accordance with the desirable wt.% (i.e., 0, 1.5, 3 and 4.5 wt.%) and was embedded into the resin-hardener mixture and mixed homogeneously. Table 3.1 gives the detailed composition and designation of the composites produced. Prior to pouring the prepared mixture into the mould a release agent, Chem-Trend (Chemlease 7000PMR EZ) was applied to the inner surface of the mould for preventing bonding between the mould and the resin. The composition was thoroughly mixed as illustrated in Figure 3.3 before pouring into wooden moulds measuring, 200 mm × 200 mm × 12 mm and cured for 24 hours at room temperatures. In order to avoid exothermic reaction,

Table 3.1 Composition and designation of adhesives.

Designation	Composition
EPM1	Epoxy + 0 wt.% marble dust
EPM2	Epoxy + 1.5 wt.% marble dust
EPM3	Epoxy + 3 wt.% marble dust
EPM4	Epoxy + 4.5 wt.% marble dust

*EPM: Marble dust filled epoxy composite.

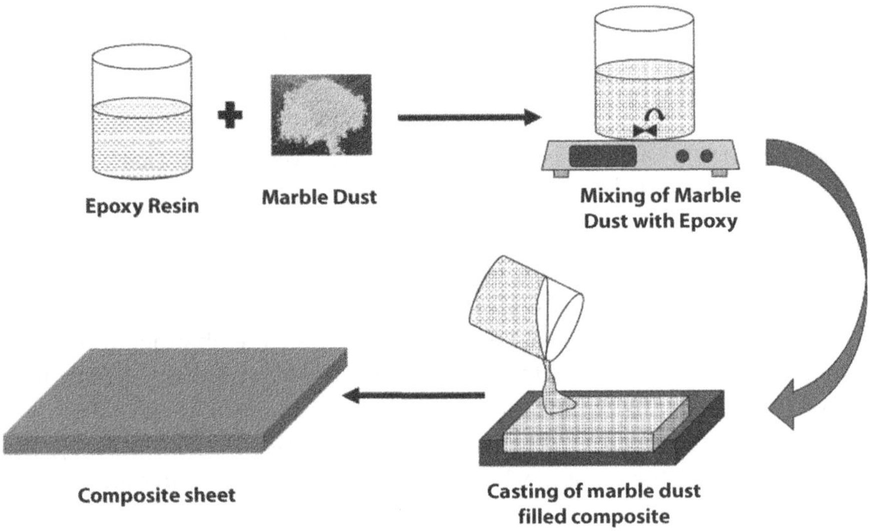

Figure 3.3 Fabrication methodology of a structural adhesive.

high temperature curing is generally not recommended in this case. The detailed methodology for composite fabrication is represented as a flow diagram in Figure 3.3 and after curing at room temperature the unfilled and marble dust filled epoxy composites are shown in Figure 3.4. (After the removal of cured composite from the mould the samples (1.5 wt.%, 3 wt.% and 4.5 wt.%) of marble dust filled epoxy composites as presented in Figure 3.3) were cut according to size and shape defined by ASTM standards as presented in Table 3.2 for the evaluation of physical, mechanical and tribological properties.

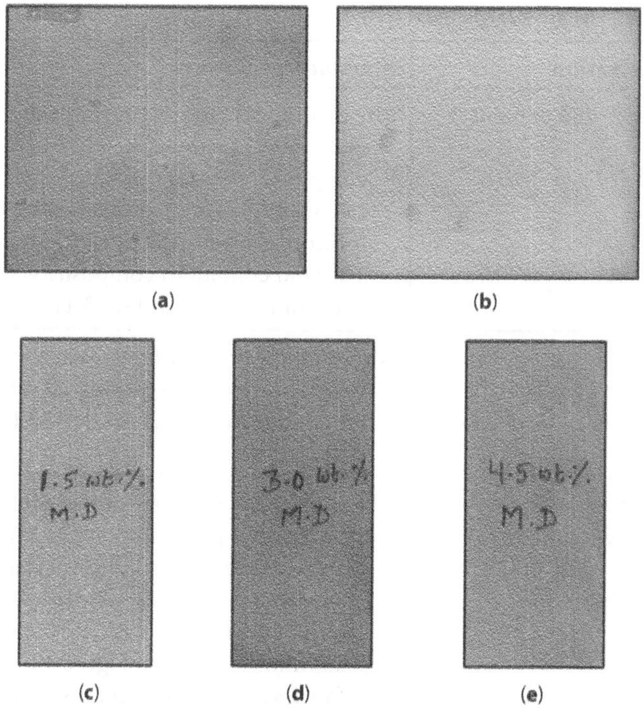

Figure 3.4 Fabricated Composites (a) Unfilled (neat epoxy composite), (b) Marble dust filled composite (c) 1.5 wt. % MD, (d) 3 wt.% MD and (e) 4.5 wt.% MD.

3.2.3 Physical and Mechanical Characterization

3.2.3.1 Density and Void Content

The experimental density of the sample was measured using Archimedes' Principle as per ASTM D792 standard using RADWAG density measurement equipment as illustrated in Figure 3.5. The theoretical density of the composite was computed using Rule of Mixtures (ROM) [20]. The void percentage in the composites was measured using the difference between theoretical and experimental densities as per Equation 3.1.

$$Void\ content(\%) = \frac{Density_{theor} - Density_{exper}}{Density_{theor}} \tag{3.1}$$

Table 3.2 ASTM standards used to test nanoparticle filled epoxy adhesive.

Characterization	Standard/procedure
Density and void content	The theoretical density of composite materials in terms of weight fraction was computed whereas the actual density was calculated by simple water immersion technique as per ASTM D792 standard. The void content in composite was determined through Eq. (3.1)
Hardness (Vickers Micro Hardness Tester)	Microhardness of the fabricated composites was measured by averaging 10 readings on a single sample as per ASTM E384 standard
Tensile test (Servo-hydraulic machine, HEICO [25kN], New Delhi, India)	The tensile tests were performed on flat straight side specimens with end tabs in displacement mode (2 mm/min) as per ASTM standard D3039-76. The size of tensile specimens was 150 × 10 × t (4.25 ± 0.25) mm as shown in Figure 3.7. Thickness (t) varies because of barium sulphate used in glass fiber reinforced polymer composites with 0, 10, 20, and 30 wt. %. Average values from three tests were reported
Flexural properties (Servo-hydraulic machine, HEICO [25kN], New Delhi, India)	Three-point flexural tests were performed under displacement mode at a loading rate 1 mm/min as per ASTM D7264 standard. Specimen dimensions were 80 mm × 10 mm × 4.25 mm
Impact test	The impact test on the fabricated epoxy adhesive composites was conducted using an impact tester machine as per ASTM D256 standard

3.2.3.2 Water Absorption

Water absorption was determined as per ASTM D5229 standard. The test coupons were immersed in distilled water at room temperature for 24 hours. The water absorption was computed by measuring the weight difference of test coupons before and after immersion in the water container. The samples were taken out from the water container and wiped with a

Figure 3.5 Density measuring equipment.

tissue paper before taking the final weight. The sample weights were measured at intervals of three hours.

3.2.3.3 Vickers Hardness

The hardness of marble filled composites was determined using Vickers hardness test equipment in accordance with ASTM E 384 standard as shown in Figure 3.6.

The sample surface to be tested was placed beneath Vickers pyramid indenter (having an angle of 136°) which was forced into the sample surface with the force applied for a fixed time of 10 s. After release of the force, the indentation image was viewed and a Vickers hardness number was generated on the screen.

3.2.3.4 Tensile Test

The tensile test was performed on a Universal Testing Machine (make: HEICO India, 25 kN capacity) with a crosshead speed of 1 mm/min as per ASTM D3039 standard. Three samples were tested, and average value of tensile strength was reported. The tensile modulus was computed from the slope of stress-strain curve obtained from the test. The pictorial view of the test set-up is shown in Figure 3.7.

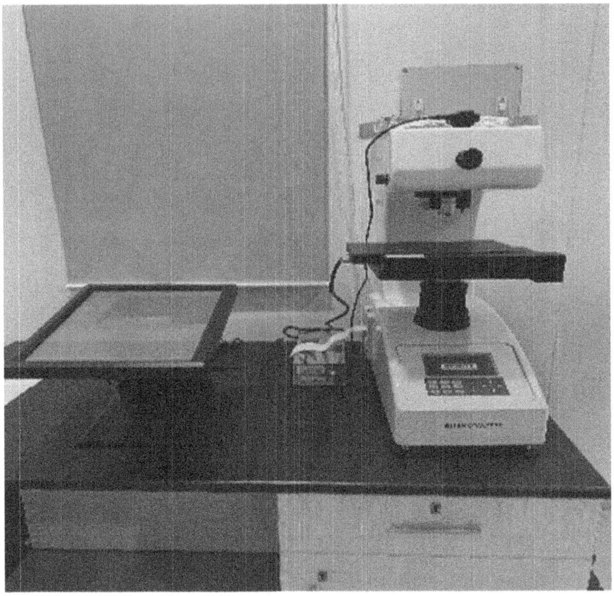

Figure 3.6 Vickers hardness tester.

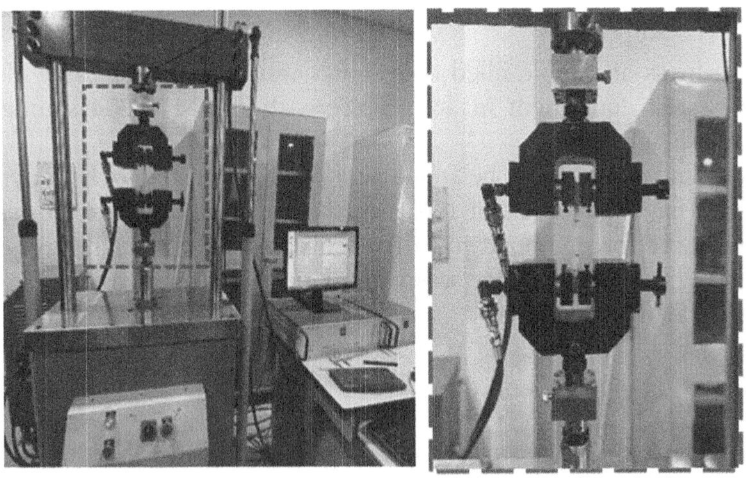

Figure 3.7 Universal testing machine (HEICO).

3.2.3.5 *Flexure Test*

The flexural test on the fabricated composites was performed in the three-point bending mode using the same Universal Testing Machine as used for tensile testing. The flexural test was conducted with a crosshead speed of 1.5 mm/min in accordance with ASTM D790 standard. For each composite, three samples were tested, and average value of flexural strength was presented. The flexural modulus was computed by considering the maximum deflection during the test and was calculated as:

$$Flexural\ Modulus = \frac{load * length^3}{4 * width * (thickness^2) * deflection} \tag{3.2}$$

3.2.3.6 *Impact Test*

The impact test on the fabricated epoxy adhesive composites was conducted using an impact tester machine as per ASTM D256 standard. All samples were prepared as per the standard. The impact energy was obtained by striking the hammer on the notched epoxy adhesive composites specimens. Three samples of each composite were tested and average value was reported.

3.2.3.7 *Thermal Conductivity*

The thermal conductivity of the developed epoxy adhesives composites was experimentally measured using thermal constants analyser. The thermal conductivity was evaluated using the transient plane source method. A Hot Disk probe was kept between two similar epoxy adhesive composite samples with a cross section of 20 mm×20 mm and all remaining sides of the test coupons were thermally insulated. The test was performed as per the procedure given by Sharma and Patnaik [21].

3.2.3.8 *Specific Wear Rate*

The specific wear rate is defined as the amount of material lost (in grammes) per unit of applied load per unit of travel distance. In general, parameters such as applied load, velocity or sliding speed, temperature buildup between surfaces, and period of operation, among others, have a significant impact on the specific wear rate. Surface wear occurs in three stages for almost every component in the system. During the primary

(a) (b)

Figure 3.8 (a) Pin-on-disc tribometer (b) test sample pins.

stage, surfaces get closer to each other, and the rate of wear fluctuates between high and low. The tribological investigation was carried out as per ASTM G 99 standard using DUCOM Pin-on-Disc Tribometer. Pins of diameter 5 mm and length 25 mm as illustrated in Figure 3.8, were slid for a distance of 188.5 meters against Inconel 31 steel disc at varied loads 5N, 10N, 15N, and 20N. The other parameters such as speed and time were kept constant at 3 m/s and 10 minutes, respectively. Samples and rotating disc were cleaned with acetone before and after tests. Resistance to adhesive wear was assessed through weight loss method. The weights were measured using RADWAG weighing machine. Figure 3.8 illustrates the experimental setup for pin-on-disc experiments. The specific wear rate (mm^3/N-m) were calculated as:

$$\textit{Specific Wear Rate} = \textit{mass loss} \div (\textit{density} \times \textit{load} \times \textit{sliding distance}) \tag{3.3}$$

where, mass loss in grams (g), density in g/cm^3, sliding distance in meters (m), load in Newton (N).

3.2.3.9 TOPSIS Approach

Technique for Order of Preference by Similarity to Ideal Solution (TOPSIS) was used to rank the alternatives on the basis of distance from positive ideal solution and negative ideal solution. The following steps shown in Figure 3.9 were followed to rank the alternatives. The weightage of each criterion is calculated using entropy method [22].

Where, D = Decision matrix, M = Decision matrix value, W = Weight normalized matrix, ω = Weight of the criterion, S^+ = Positive Ideal Solution, S^- = Negative Ideal Solution, $\mathbb{S}_i^+$ and $\mathbb{S}_i^-$ = alternative distance from positive and negative ideal solution, RC = Relative closeness

Step 1 Establishing the Decision Matrix

$$D = \{M_{ij}\} \text{ where } i = 1,2, \ldots m;\ j = 1,2 \ldots\ldots n$$

Step 2 Construction of Normal Desicion Matrix

$$M_{ij} = x_{ij} \Big/ \sqrt{(x_{ij}^2)} \quad where\ i = 1,2. \ldots m; j = 1,2 \ldots . n$$

Step 3 Construction of Weight Normalized Desicion Matrix

$$W_{ij} = \omega_j M_{ij}$$

Step 4 Determining Positive and Negative Ideal Solution

$$Positive\ ideal\ Solution\ : S^+ = \{W_1^+, \ldots\ldots, W_n^+\}$$
$$where\ W^+ = \{\max(W_{ij}) if\ j \in J; min(W_{ij}) if\ j \in J'\}$$
$$Negative\ ideal\ Solution\ : S^- = \{W_1^-, \ldots\ldots, W_n^-\}$$
$$where\ W^- = \{\min(W_{ij}) if\ j \in J; max(W_{ij}) if\ j \in J'\}$$

Step 5 Evaluating Separation Measures for Each Alternative

$$S_i^+ = \sqrt{\sum_j^n (W_{ij} - W_j^+)^2}\ i = 1, \ldots\ldots\ldots, m$$

$$S_i^- = \sqrt{\sum_j^n (W_{ij} - W_j^-)^2}\ i = 1, \ldots\ldots\ldots, m$$

Step 6 Calculation of Relative Closeness to Ideal Solution

$$RC_i = \frac{S_i^-}{S_i^+ + S_i^-}$$

Figure 3.9 Steps for TOPSIS.

3.3 Results and Discussion

3.3.1 Density and Void Content

The theoretical and experimental densities of marble dust filled epoxy composites are presented in Table 3.3. From the tabulated data the results reveal that the theoretical density is higher than the experimental density in all situations. The density of composite shows increase with addition of marble dust in the epoxy. The increase in density of the composite is attributed to the increasing weight percentage of filler material which possesses higher density compared to matrix material. The difference between theoretical and experimental densities is termed as void content in the composite and voids are intrinsic in all polymer composites and play a significant role in determining the physical and mechanical properties of

Table 3.3 Density and void content of marble dust filled epoxy composites.

Composite	Theoretical density (g/cm³)	Experimental density (g/cm³)	Void content (%)
EPM1	1.2108	1.1911	1.63
EPM2	1.2543	1.2309	1.87
EPM3	1.2997	1.2712	2.19
EPM4	1.3492	1.3114	2.80

the material. The data presented in Table 3.3 show that with increased percentage of marble dust the void content increases from 1.63 % to 2.80 %. This increase in void percentage may be because of air trapped during fabrication of the composites.

3.3.2 Water Absorption

Figure 3.10 shows moisture absorption percentage of marble dust filled epoxy composites with varying weight percentage of marble dust filler. The water absorption of marble powder reinforced composite material is found to increase with the increase in the percentage of filler, which is owing to poorer cohesion and porosity at higher filler concentrations.

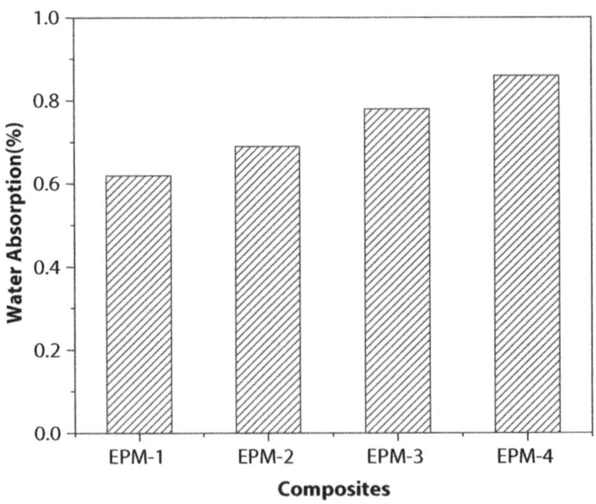

Figure 3.10 Water absorption of marble dust filled epoxy composites.

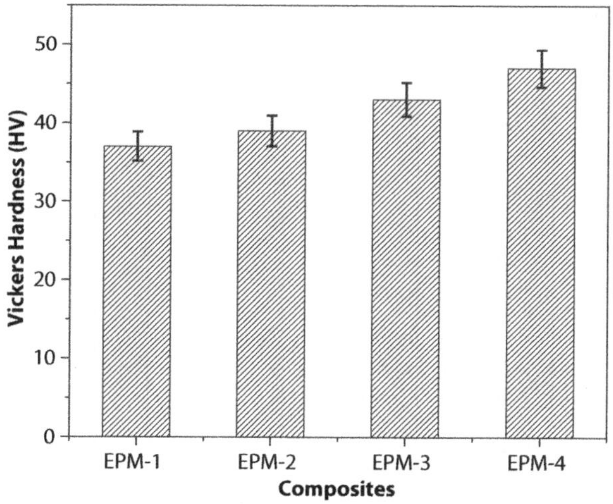

Figure 3.11 Hardness of fabricated composites.

The water absorption in epoxy composites can be due to multiple causes, including capillary effect at the interface between filler and adhesive and in micro-voids produced during the fabrication of composites. The increasing behaviour may also be due to higher filler concentrations resulting in reduced cohesion with higher porosity than at lower filler concentrations.

3.3.3 Hardness

Figure 3.11 represents the Vickers hardness of composites with increase in the weight percentage of marble dust. The results show that the hardness of neat epoxy composite is minimum at 37 HV in comparison to other three marble dust filled epoxy composites. With increase in the weight percentage of marble dust from 1.5 wt.% to 4.5 wt.%, the hardness increases from 39 HV to 47 HV. The enhancement in hardness is because of small particle size, toughness and rigidity of marble dust filler.

3.3.4 Tensile Strength and Tensile Modulus

The tensile strength and tensile modulus of marble dust filled epoxy composites with varying percentage of marble dust are presented in Figure 3.12. The tensile strength is maximum for neat epoxy in comparison to other three composites with a maximum of 65.2 MPa and decreases to 63.78 MPa, 59.53 MPa, and 57.97 MPa with increase in weight percentage (1.5%-4.5%) of marble dust filler. The reduction in tensile strength

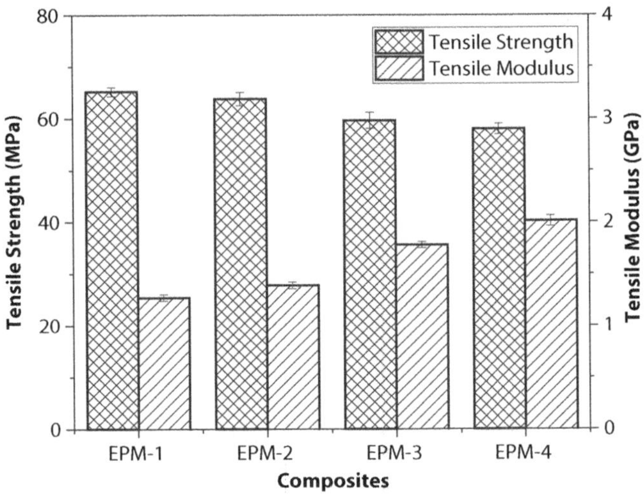

Figure 3.12 Tensile strength and tensile modulus of fabricated composites.

is due to increase in the percentage of voids that directly affects the tensile properties of composites. On the contrary, the tensile modulus of marble dust filled composites increases with increase in wt. % of marble dust filler. The EPM-4 shows the maximum tensile modulus of 2.01 GPa, because of high modulus of marble dust filler thus enhancing the modulus of composites.

3.3.5 Flexural Strength and Flexural Modulus

The flexural strength and flexural modulus of marble dust filled epoxy composites with varying weight percentage tested in a three-point bending mode on a universal testing machine are shown in Figure 3.13. The neat epoxy (EPM-1) shows the maximum flexural strength of 87.32 MPa in comparison to other three composites. The agglomeration of particles caused by heavy filler loading causes this decline. As the filler weight percentage increases, the contact area between the filler and matrix decreases because of insufficient wetting of the filler. As far as the flexural modulus is concerned, it has been discovered that the trend of variation is slightly different from the flexural strength.

It is observed that the flexural modulus of the composite increases with the addition of marble dust. This variation in the results was thought to have originated from the difference in orientation in fiber layers, filler

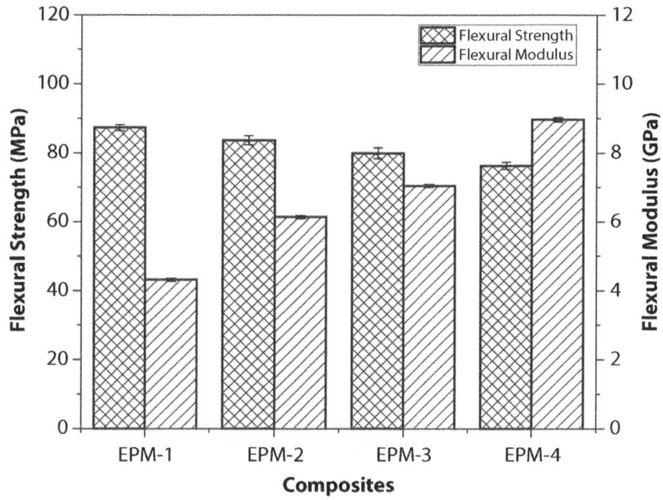

Figure 3.13 Flexural strength and flexural modulus of fabricated composites.

amount, polymer matrix, and coupling among fiber, filler, and matrix the prepared composite.

3.3.6 Impact Energy

The Izod impact test is one of the most widely used impact tests available to determine the energy absorbed by the material before failure when subjected to sudden loading. The Izod impact energy_of the epoxy adhesives was also examined, and the results are presented in Figure 3.14. It can be interpreted that impact energy of EPM4 is ~50% higher than that of EPM1. This may be attributed to the filling of the gaps of the epoxy matrix by nanoparticle marble dust which aids in the absorption of energy from impact.

3.3.7 Thermal Conductivity

The thermal conductivity results of epoxy adhesives at different weight percentages are presented in Figure 3.15. It shows that with increase in weight % of the nano-sized marble dust the experimental thermal conductivity increases. The experimental results are also compared with series model, parallel model, and Geometric Mean Model (GMM).

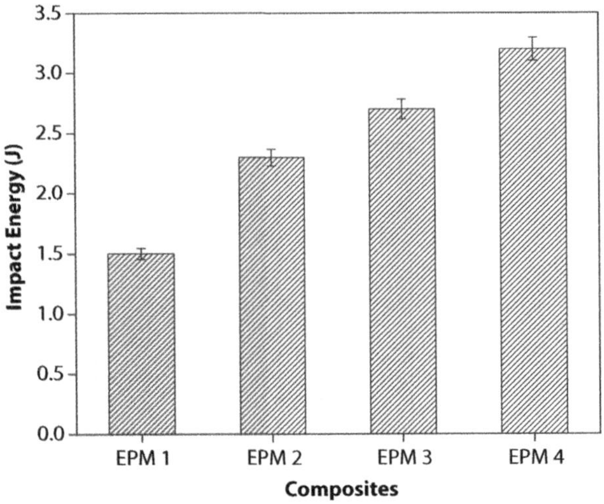

Figure 3.14 Impact energy of fabricated composites.

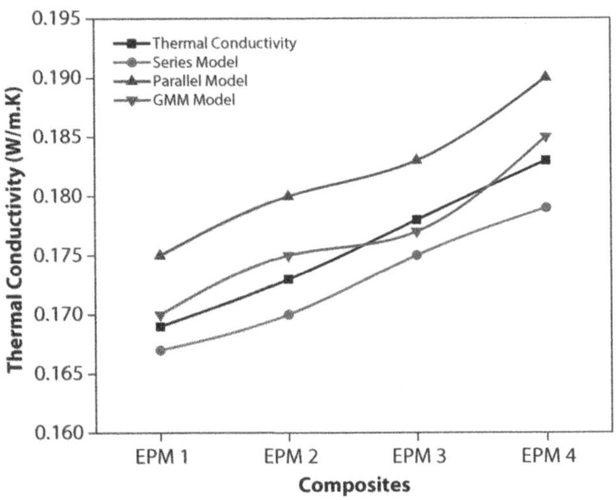

Figure 3.15 Thermal conductivity of fabricated composites.

The results demonstrate that the thermal conductivity of nano-marble dust filled epoxy adhesives is higher than the theoretically predicted values, which can be attributed to the elongated agglomerates of nano-sized marble dust in the adhesive matrix. It can also be anticipated that a variety of factors such as particle size and shape, surfactant, particle dispersion,

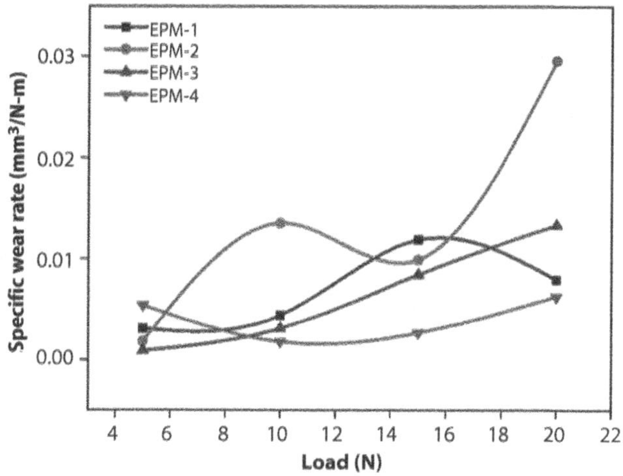

Figure 3.16 Specific wear rate of the fabricated composites at different loads.

thermal characteristics of nano- particles, and the interface between polymer and particles, all would have an impact on the thermal conductivity of nanocomposites. The order of thermal conductivity from Figure 3.15 is by Parallel Model > GMM model > Experimental > Series Model.

3.3.8 Specific Wear Rate

Specific wear rates of the fabricated samples obtained from pin-on-disc experiments are shown in Figure 3.16. There is an increasing trend in the specific wear rate of the samples with the increasing load. However, the addition of marble dust as a reinforcement has reduced the overall specific wear rate of the samples as witnessed in Figure 3.16.

Samples without reinforcement have the tendency to bend at higher loads during pin-on-disc experiments. Out of all the composites fabricated, EPM-3 shows highest resistance to sliding wear. The worn-out surfaces of epoxy samples were examined with SEM equipped with EDS. The SEM micrographs of worn-out surfaces are shown in Figure 3.17. The pure epoxy sample showed matrix stretching and debris on the sample surface. Pure epoxy sample showed the adhesive wear as the dominant mechanism. Figure 3.17 indicates the presence of wear debris along with a patch owing to mild abrasion caused by hard marble dust particles in the matrix. No voids were observed in 3 wt.% MD reinforced epoxy sample (EPM-3).

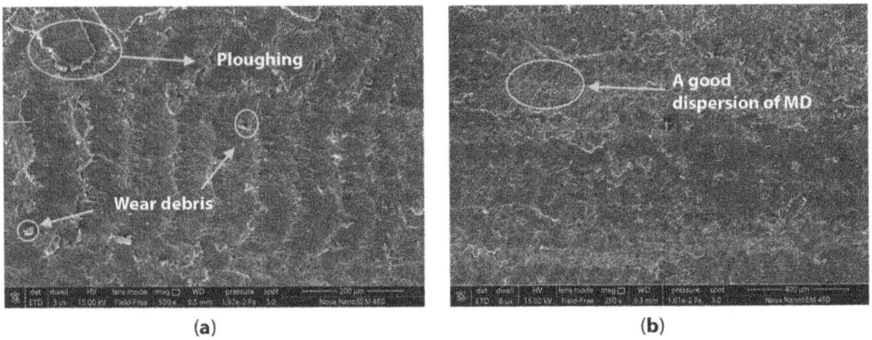

Figure 3.17 SEM micrographs of (a) worn-out surface under peak load (3 wt.% MD) at 500 magnification (b) worn-out surface under peak load (3 wt.% MD) at 250 magnification.

3.3.9 Ranking of Epoxy Adhesive Composites

The physical, mechanical, thermal conductivity and tribological properties were evaluated as per ASTM standards. For the purposes of ranking the composites, the results of these experiments are considered as criteria in evaluating their performance. The performance defining criteria are presented in Table 3.4.

A decision matrix was prepared based on the given alternatives as presented in Table 3.5. Because the decision matrix contains information obtained at different levels, it has been normalised and presented in tabular format as highlighted in Table 3.6 to make it more comparable. The weight for each criterion was calculated using the entropy method and multiplying the calculated weight in normalized decision matrix as shown in Table 3.7. Based on the analysis the epoxy adhesive composites were ranked as EPM 4 (Rank 1), EPM 3(Rank 2), EPM 2 (Rank 3) and presented in Table 3.8. According to the experimental results, the incorporation of nanosized marble dust has a considerable impact on the assessed properties of the epoxy adhesive composites.

Table 3.8 Ranking of the alternatives.

Composites	S_i^+	S_i^-	RC_i	Rank
EPM1	0.255377192	0.001436099	0.005592	4
EPM2	0.157178205	0.102845552	0.3955237	3
EPM3	0.03795534	0.217916214	0.8516625	2
EPM4	0.029252146	0.252594962	0.8962127	1

3.4 Summary and Conclusions

Marble dust reinforced epoxy composites were successfully fabricated via casting method. In the present work the influence of marble dust filled epoxy composites with varying the weight percentage on physical, mechanical, and wear properties was studied. With the addition of marble dust from 0 wt.% to 4.5 wt.% in the adhesive composites, increase in density was observed from 1.1911 g/cm^3 to 1.3114 g/cm^3. The void content was also found to increase from 1.63 % to 2.80 %. The water absorption percentage and hardness of the fabricated composites were also found to increase with the addition of marble dust. The tensile and flexural strengths were observed to decrease with increase in weight percentage of marble dust loading. In contrast, the tensile and flexural modulus are found to increase, indicating improvement in the stiffness of the composites. The specific wear rate of the composites indicates an increasing trend with the addition of marble dust reinforcement. EPM-3 shows higher resistance to sliding wear in comparison to the other composites considered in the present study. Based on the TOPSIS analysis the ranking of the epoxy adhesive is found in the order EPM4>EPM3>EPM2>EPM1.

Marble dust being easily available as a by-product from marble industry in the Indian state like Rajasthan, so it is economical to use marble dust as a reinforcement for structural adhesives. However, further research need to be conducted for optimum reinforcement quantity. From this study, the developed composites can be utilized in low-cost structural applications where high wear resistance is of prime importance. Lastly, this study also showed that structural material produced via low-cost methodology with good performances could be an area for future research for exploring different minerals as reinforcing materials that are economical and easily available. It has been discovered that using marble dust in a specified proportion reduces building costs and will improve the concrete qualities.

References

1. S. Ebnesajjad and A. H. Landrock, *Adhesives Technology Handbook*. William Andrew, Norwich, NY (2014).
2. M. Kothe, C. Kothe, and B. Weller, Epoxy resin adhesives for structural purposes–a new approach. in: Proc. Engineered Transparency Int. Conf., at Glasstee. pp. 383-392 Düsseldorf, Germany (2014).
3. G.-S. Chae, H.-W. Park, J.-H. Lee, and S. Shin, Comparative study on the impact wedge-peel performance of epoxy-based structural adhesives modified with different toughening agents, *Polymers*, 12, 1549 (2020).
4. C. D. Wright and J. M. Muggee, Epoxy structural adhesives, in: *Structural Adhesives: Chemistry and Technology*, S.R Hartson (Ed.) Plenum Press, NewYork (1986).
5. S.J. Park and M.K. Seo, Element and processing, in: *Interface Science and Technology*. vol. 18, pp. 431–499, Elsevier (2011).
6. P. Kumar, A. Patnaik, and S. Chaudhary, A review on application of structural adhesives in concrete and steel–concrete composite and factors influencing the performance of composite connections, *Int J Adhesion Adhesives* 77, 1–14 (2017).
7. H. Khakpour, M. R. Ayatollahi, A. Akhavan-Safar, and L. F. M. da Silva, Mechanical properties of structural adhesives enhanced with natural date palm tree fibers: effects of length, density and fiber type, *Composite Struct.* 237, 111950 (2020).
8. V. Tzatzadakis and K. Tserpes, Production of a novel bio-based structural adhesive and characterization of mechanical properties, *J Adhesion* 97, 936–951 (2021).
9. A. Fouly and M. G. Alkalla, Effect of low nanosized alumina loading fraction on the physicomechanical and tribological behavior of epoxy, *Tribol. International* 152, 106550 (2020).
10. J. Michels, R. Widmann, C. Czaderski, R. Allahvirdizadeh, and M. Motavalli, Glass transition evaluation of commercially available epoxy resins used for civil engineering applications, *Composites B.* 77, 484–493 (2015).
11. C. Grave, I. McEwan, and R. A. Pethrick, Influence of stoichiometric ratio on water absorption in epoxy resins, *J. Appl. Polym. Sci.* 69, 2369–2376 (1998).
12. M. Choudhary, A. Sharma, D. Shekhawat, V. R. Kiragi, R. Nigam, and A. Patnaik, Parametric optimization of erosion behavior of marble dust filled aramid/epoxy hybrid composite, in: Proc. International Conference on Sustainable Computing in Science, Technology & Management, Jaipur, India, 2484–2490 (2019).
13. D. Shekhawat, A. Singh, M. K. Banerjee, T. Singh, and A. Patnaik, Bioceramic composites for orthopaedic applications: A comprehensive review of mechanical, biological, and microstructural properties, *Ceramic Int.* 47, 3013–3030 (2021).

14. D. Shekhawat, A. Singh, A. Bhardwaj, and A. Patnaik, A Short Review on Polymer, Metal and Ceramic Based Implant Materials, *IOP Conf. Ser.: Mater. Sci. Eng.* 1017, 1–12 (2021).
15. H. Nolte, C. Schilde, and A. Kwade, Production of highly loaded nanocomposites by dispersing nanoparticles in epoxy resin, *Chem Eng Technol.* 33, 1447–1455 (2010).
16. B. K. Muñoz, A. Bosque, M Sanchez, V Utrilla, S. G. Prolongo, M.G. Prolongo, and A Urena, Epoxy resin systems modified with ionic liquids and ceramic nanoparticles as structural composites for multifunctional applications, *Polymer*, 214, 123233 (2021).
17. M. K. Reddy, V. S. Babu, K. V. S. Srinadh, and M. Bhargav, Mechanical properties of tungsten carbide nanoparticles filled epoxy polymer nano composites, *Mater. Today: Proc.* 26, 2711–2713 (2020).
18. A. Sharma and V. Gautam, Mechanical and wear characterization of epoxy resin-based functionally graded material for sustainable utilization of stone industry waste, *Advanced Manuf.* 303–318 (2021).
19. A. Fahmi, S. Abdullah, F. Amin, and A. Ali, Precursor selection for sol–gel synthesis of titanium carbide nanopowders by a new cubic fuzzy multi-attribute group decision-making model, *Int. J. Intelligent Systems,* 28, 699–720 (2019).
20. A. Sharma, M. Choudhary, P. Agarwal, S. K. Biswas, and A. Patnaik, Effect of micro-sized marble dust on mechanical and thermo-mechanical properties of needle-punched nonwoven jute fiber reinforced polymer composites, Polym. Composites, 42, 1–18 (2020).
21. A. Sharma and A. Patnaik, Experimental investigation on mechanical and thermal properties of marble dust particulate-filled needle-punched nonwoven jute fiber/epoxy composite, *J. Metals*, (JOM)70, 1284–1288 (2018).
22. M. Choudhary, T. Singh, A. Sharma, M. Dwivedi, and A. Patnaik, Evaluation of some mechanical characterization and optimization of waste marble dust filled glass fiber reinforced polymer composite, *Mater. Res. Express*, 6, 105702 (2019).

4

Characterization of Various Structural Adhesive Materials

Srujan Sapkal, Pooja Maske, S. K. Panigrahi and Himanshu S. Panda*

Defence Institute of Advanced Technology, Pune, India

Abstract

With the rapid development of new and upgraded versions of structural adhesives in the modern era, scientists have focused their attention on improving their chemical and physical properties. Structural adhesives have a wide range of applications in sectors such as transport, electronics, biomedical, and aviation because of their unique characteristics. This has motivated many researchers to study these unique characteristics of adhesives to further advance the development of improved versions of structural adhesives. There are several options in the category of structural adhesives, e.g., phenolic resins, epoxy resins, polyurethane resins, acrylic resins, cyanoacrylate resins, and silicone resins with unique properties and applications in several sectors. To study the characteristics of these adhesives, many advanced technical instrumental facilities are available today. For surface analysis, researchers have scanning electron microscopy (SEM) and contact angle measurement; for thermal analysis, techniques such as differential scanning calorimetry (DSC) and thermal gravimetric analysis (TGA) are available; and for chemical characterization, scientists use nuclear magnetic resonance (NMR) and Fourier transform infrared (FTIR) spectroscopy. For mechanical characterization, essential techniques such as tensile and peel tests are mentioned. With the intent of introducing characterization techniques for structural adhesives, this chapter covers the relevant studies briefly.

Keywords: Adhesives, structural adhesives, characterization techniques, adhesive applications

**Corresponding author*: himanshusp@diat.ac.in; hspanda3@gmail.com

K.L. Mittal and S.K. Panigrahi (eds.) Structural Adhesives: Properties, Characterization and Applications, (135–192) © 2023 Scrivener Publishing LLC

List of Abbreviations

AFM	Atomic force microscopy
AIBN	Azobis (isobutyronitrile)
APBA	3-(Acrylamido) phenylboronic acid
ASTM	American Society for Testing and Materials
ATR-FTIR	Attenuated total reflection-Fourier transform infrared spectroscopy
BSE	Back-scattered electron
BDE	1,4-Butanedioldiglycidylether
CNC	Cellulose nanocrystal
CNF	Cellulose nanofiber
DDM	4,4′-Diaminodiphenylmethane
DDS	4,4′-Diaminodiphenyl sulfone
DGEBA	Diglycidyl ether of bisphenol A
DMA	Dynamic mechanical analysis
DMPA	Dimethyl propionic acid
DOPA	3,4-Dihydroxyphenylalanine
DSC	Differential scanning calorimetry
DTA	Dynamic thermal analysis
EDX	Energy dispersive X-ray
GLC	Gas-liquid chromatography
HTPB	Hydroxyl-terminated polybutadiene
IR	Infrared
JKR	Johnson-Kendall-Roberts
LPP	Low pressure plasma
MDI	Diphenylmethane-4,4′-diisocyanate
4-Mpy	4 - Mercaptopyridine
NMR	Nuclear magnetic resonance
NP	Nanoparticle
OCNF-CA	Catechol- and aldehyde- modified cellulose nanofiber
PEEK	Poly (ether ether ketone)
PEI	Polyethylenimine

PES	Polyether sulphone
PU	Polyurethane
PyMA	1-Pyrenemethyl methacrylate
RF	Radiofrequency
SAXS	Small angle X-ray spectroscopy
SE	Secondary electron
SEM	Scanning electron microscopy
SERS	Surface enhanced Raman scattering
SFG	Sum frequency generation
SP	Soy protein
SPI	Soy protein isolate
TEM	Transmission electron microscopy
TGA	Thermogravimetric analysis
TGMDA	Tetraglycidyl methylene dianiline
U-SAXS	Ultra small angle X-ray spectroscopy
VUV	Vacuum UV
XPS	X-ray photoelectron spectroscopy

List of Symbols

$h\nu$	The energy of radiation with a frequency ν
BE	Binding energy
KE	Kinetic energy
ϕ	Work function
μ	Magnetic moment
γ	Gyromagnetic ratio
I	Magnetic spin
E	Energy
B_o	External magnetic field
θ	Angle
γ	Surface tension
σ_C	Critical strength

P_C	Critical load
A	Cross-sectional area
F	Force/Load

4.1 Introduction

Adhesives play an essential role in our daily lives, from small applications like joining electrical components to large applications like aircraft bonding. Adhesives allow to achieve the state of adhesion in order to join two adherends [1]. As per the ASTM standard, "an adhesive is a substance capable of holding materials together by surface attachment" [2]. A variety of adhesives are available in the market for a wide range of applications, including significant sectors like transport, manufacturing, electronics, and biomedical. Application-based selection of adhesives is usually performed, including key parameters like adhesive chemistry, type of adhesive cure, compatibility with adherends, pre-treatment of surfaces, joint design, demand, and testing [3]. Rapid development in resin research has replaced nuts and bolts with adhesives because of their low cost and beneficial properties for joining materials. Adhesives offer many functional characteristics for material joining, including load distribution over the whole joint area, good fatigue property, attenuation of mechanical vibrations, and reduced galvanic corrosion [4].

The use of adhesives has many advantages, including uniform distribution of stress, no restriction on the shape of materials, joining of different materials, corrosion resistance, better insulation and shock absorption [1].

Figure 4.1 shows the classification of adhesives from various aspects. It is possible to classify adhesives as structural and non-structural. Structural adhesives are primarily used for applications in which adherends will experience stresses up to their yield point, and hence better stress transmission property and high strength are expected from these adhesives. It is

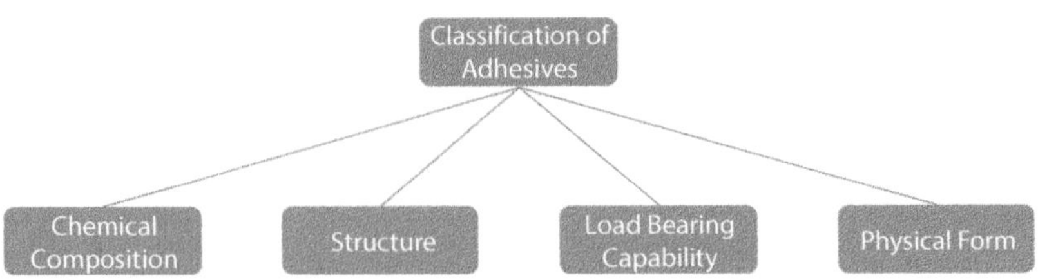

Figure 4.1 Classification of adhesives.

possible to classify adhesives differently, but here the classification of the structural adhesives is based on their functional groups. There are several options in structural adhesives as follows: (1) Phenolic resins, (2) Epoxy resins, (3) Polyurethane resins, (4) Acrylic resins, (5) Cyanoacrylate resins, and (6) Silicone resins. Each type has its unique advantages, which are discussed briefly in this chapter with their chemical properties. The research community has performed detailed characterization studies of these resins over the past decade using various techniques available. The characterization techniques used for adhesives are described in this chapter with suitable examples.

The structure of the chapter is as follows: First we discuss the classification of structural adhesives with their chemical and physical properties including their applications. Next, most of the chapter is devoted to the discussion of various advanced characterization techniques used for adhesives with relevant examples. Several advanced techniques to analyse the chemical composition of the adhesives are reviewed, such as Fourier transform infrared (FTIR) spectroscopy, nuclear magnetic resonance (NMR) and Raman spectroscopy are discussed at length with recent literature. Physical characterization of adhesives such as surface morphology, topography and microstructures is accomplished with techniques such as scanning electron microscopy (SEM), atomic force microscopy (AFM) and small-angle X-ray scattering (SAXS). To understand the thermal properties of the adhesives, such as glass transition temperature and other relevant properties variations with respect to temperature, techniques such as differential scanning calorimetry (DSC), dynamic thermal analysis (DTA) and thermo-gravimetric analysis (TGA) are considered. Finally, for mechanical testing, various ASTM standard techniques such as tensile and peel tests are discussed.

4.2 Various Structural Adhesives and their Properties

4.2.1 Phenolic Structural Adhesives

Phenolic resins result from the reaction of phenol and formaldehyde, and hence one can create a large variety of phenolic adhesives by varying the nature of phenol, molar ratio and pH of the medium [5]. Phenolic adhesives are primarily used in joining of metals and wood structures. Phenolic adhesives are based on novolacs and resoles, where the condensation reaction of phenol-formaldehyde under basic conditions leads to the formation of resoles, whereas acidic condensation reaction leads to novolacs [5, 6]. The structures for resole and novolac can be visualised from Figure 4.2.

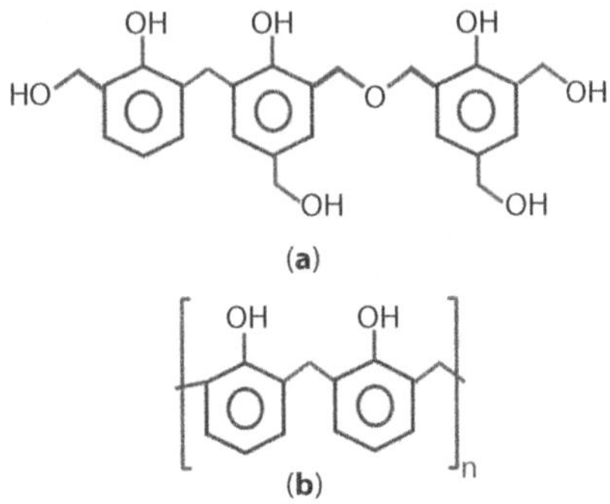

Figure 4.2 Structures of (a) resole and (b) novolac.

Resoles contain hydroxymethyl groups, which further condense at 130 - 200° C resulting in 3D network with the formation of methylene bridges and methylene ether. Thermoplastic novolacs require formaldehyde donor for crosslinking [6]. Based on the synthesis procedure, they can be classified as thermosetting phenolics and thermoplastic phenolics. Phenolic resins have several advantages, including strong mechanical properties, good processing performance, low cost, resistance to chemicals and high adhesion strength. Phenolic resins have found wide applications in electronics industries because of their high shear strength and heat resistance [5].

4.2.2 Epoxy Structural Adhesives

Epoxy resin was introduced in 1940 and is being used in one-component or two-component forms. Epoxy resins do need specific modifications using additives to increase their strength and toughness. Epoxy-based structural adhesives play a vital role in the adhesives market because of their wide range of applicability, capability to bond strongly to a wide range of materials, and chemical resistance. They have attracted much attention and have shown success in industrial areas such as aerospace, automotive, woodwork because of their excellent mechanical properties and ease of use [7–9]. Some essential properties of epoxies are low shrinkage, high strength, effective electrical insulation, low toxicity and low cost, excellent adhesion, low bond stresses, high creep resistance, dimensional stability, and high resistance to osmosis (prevents liquid transfer across the coating). One-component epoxy structural adhesives are available in paste and liquid forms, and heat is necessary for their curing because of its initiation at a higher temperature. Two-component epoxy usually

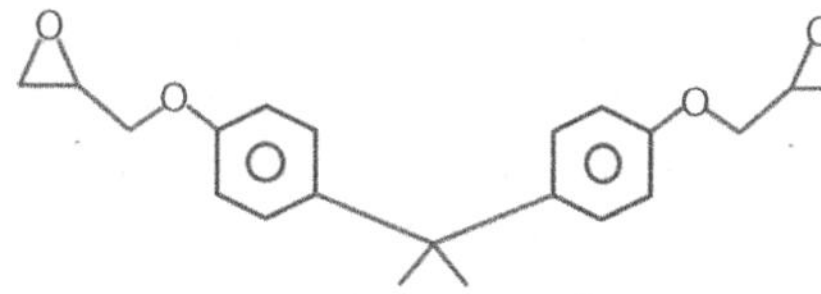

Figure 4.3 Structure of Diglycidyl Ether of Bisphenol-A (DGEBA).

contains epoxy resin, hardeners, fillers, accelerators, plasticisers, reactive diluents and toughening agents [4]. Also, epoxies have very good cohesive strength and very low creep because of their crosslinking and ability to participate in H-bonding [10]. Epoxies which are mostly used in common applications are based on "Diglycidyl Ether of Bisphenol-A" (DGEBA) [4], as shown in Figure 4.3.

It should be pointed out that the cured adhesive's glass transition temperature depends on the hardener's molecular structure. Recently, new hardeners have been introduced for epoxies for efficient curing and low odour [4]. Epoxies are significantly less sensitive to the addition of fillers to form composites, which is why it is possible to modify them with different additives [4].

4.2.3 Polyurethane (PU) Structural Adhesives

Otto Bayer and co-workers introduced polyurethanes in 1930. Polyurethane adhesives are derived from polyurethane which results from the reaction of polyol with isocyanate. The reaction between isocyanate group and the hydroxyl groups of polyol results in repeating urethane linkages in polyurethane [4, 11]. The structure of PU is shown in Figure 4.4.

Polyurethane adhesives are available as one-component as well as two-component systems where one-component system curing is facilitated by heat, and for two-component systems, thorough mixing is necessary before application. Polyurethane structural adhesives bond to many materials such as plastics, glass and metals. Usually, the joints formed with polyurethanes are impact-resistant and possess high strength at low

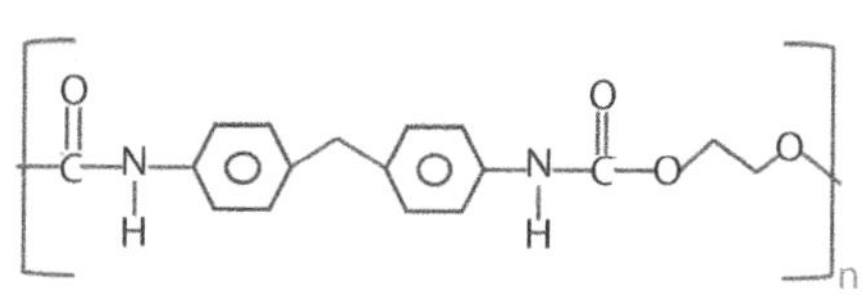

Figure 4.4 Structure of polyurethane.

temperatures. Polyurethane adhesives help in the construction of lightweight structures, and they have many applications as adhesives and sealants in construction industry and naval architecture [11, 12], laminated railway ties were developed using low-grade hardwoods with a two-component polyurethane structural adhesive for bonding [13]. Some of the notable mechanical properties of PU include good elongation ability (tensile strength=31.1 MPa [12], good cohesive strength [14]), good energy absorption, high resistance to harsh surroundings, thermal stability, chemical stability, ease of application and cost-effectiveness [11].

4.2.4 Acrylic Structural Adhesives

Structural acrylic adhesives are a small part of the large acrylic adhesives family [15]. A structural acrylic adhesive usually includes primary methacrylate monomer, secondary methacrylate monomer, a crosslinker, an initiator, a filler and an elastomer [15]. Acrylic adhesives are based on acrylate and methacrylate monomers, with irritating smell and are flammable in nature. They are two-component systems, polymerising by a free-radical mechanism similar to anaerobic adhesives [4]. They are different from other two-component systems because their cure is catalytic in nature and independent of stoichiometric mixing of catalyst. Applications of acrylics involve bonding of glass and metals and sporting equipment where fast curing without surface preparation is necessary. They have a few disadvantages like unpleasant odours and temperature resistance only up to 120°C [4]. Acrylics can be classified based on the odour of their monomers, like "High-odour" acrylics containing methyl or other lower alkyl methacrylates as the monomers as shown in Figure 4.5(a) and "Low-odour" acrylics use monomers like hydroxyethyl methacrylate, Figure 4.5(b) [2].

Acrylic structural adhesives have a wide range of applications in transportation, marine-based (bonding hull to a boat), and thermoplastic

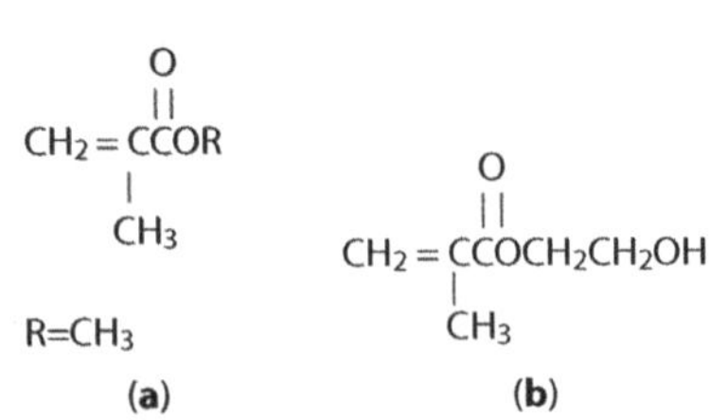

Figure 4.5 Structures of Acrylate Monomers (a) High-odour acrylics using methyl methacrylate as monomers (b) Low-odour acrylics using hydroxyethyl methacrylate.

bumpers. Aronovich and Boinovich [15] have discussed structural acrylic adhesives in great detail along with their functional comparison with other structural adhesives such as epoxies and polyurethanes. Since acrylic structural adhesives are higher in cost as compared to epoxies, they are not much explored in comparison with epoxies [15].

4.2.5 Cyanoacrylate Structural Adhesives

Cyanoacrylate adhesives were discovered by Eastman Kodak in 1958 and were commercialized in 1970. The cyanoacrylate adhesives are based on ethyl ester and are one-component adhesives. Structural Cyanoacrylate Adhesives are commonly known as Superglues because of their fast solidification. They are usually based on the ester of 2-cyanoacrylate, as shown in Figure 4.6 [4]. These adhesives require a surface adsorbed moisture layer (slightly alkaline) for the curing process to initiate. Earlier, cyanoacrylates were not considered as structural adhesives because of their thermoplastic nature, limited thermal resistance and brittleness, but these issues in cyanoacrylates have been addressed effectively in recent years with the help of elastomers [4].

Cyanoacrylate adhesives bond to a wide range of adherends including metals, plastics, and composites, except for highly acidic surfaces, and are widely used in various industrial applications [16].

4.2.6 Silicone Structural Adhesives

The name silicone came from an analogy with ketones in 1901 to describe new compounds with a general formula $R_2(SiO)_n$. Silicones can easily form three-dimensional structures via crosslinking which can be achieved through condensation, radical or addition reactions. Since the critical surface tension of wetting of solid silicones (24 mN/m) is higher than their liquid surface tension (20-23 mN/m) so this will assist in better film formation and surface coverage. Silicones are flexible in nature but have cohesive strength less than epoxies, making them cost-effective adhesives with broad applications in which flexibility of the joints is necessary.

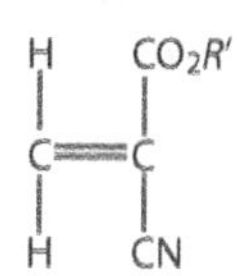

Figure 4.6 Structure of 2-cyanoacrylate.

They possess unique properties like low toxicity, easy wettability, high-temperature resistance, flexibility, high chemical resistance, and high optical clarity. Low-toxicity levels of silicones make them useful in biomedical applications. Silicones have a wide range of applications in sectors like automotive, aerospace, construction, domestic appliances, and electronics [17].

4.3 Characterization Techniques for Structural Adhesives

4.3.1 Chemical Characterization

Chemical characterization helps one to understand the chemical composition and chemical properties of the given material using sophisticated techniques developed by researchers over a period of decades. Various state-of-the-art techniques are available for the chemical characterization of structural adhesives, which will be discussed in this section.

4.3.1.1 Energy Dispersive X-ray (EDX)

EDX is necessary for the elemental analysis of various materials like ceramics, polymers, and metals. When an electron beam with sufficiently high energy impinges on the sample, the generation of characteristic X-rays takes place for various elements present in the sample. The detector captures these X-rays emitted from the material and reveals the presence of various elements present by performing analysis of the received signals. EDX is usually a part of SEM studies which are performed for the micro-scale and nano-scale morphological studies of materials. When an electron beam is incident on the material, electrons interact with atoms both inelastically and elastically. Elastic scattering is defined by the change in the electron's direction of motion without loss of energy, whereas inelastic collision is defined as loss of electron's energy without any noticeable change in the direction of motion. These collisions excite the atoms, and when they return to the ground state, they emit X-rays that are different and unique in wavelength for different atoms. The energy of X- rays will be proportional to the energy difference between the two energy levels from which the electron transition occurred. The detector system displays all mid-energy X-rays that are collected in any single analysis period [18] and a histogram plot of counts against energy is generated. Most of the scientific community is not fully aware of its possible applications. The spectrum of EDX microanalysis contains both semi-qualitative and semi-quantitative information. EDX technique is made useful in the study of drugs, such as in the study of drugs delivery in which the EDX is an important tool to detect nanoparticles (generally, used to improve the therapeutic performance of some chemotherapeutic agents. A detailed example showing the application of EDX and XPS for chemical analysis is presented in section 4.3.1.2.

4.3.1.2 X-ray Photoelectron Spectroscopy (XPS)

Albert Einstein discovered the photoelectric effect in 1905, and the XPS technique is primarily based on this phenomenon. The equation for photoelectrons can be described as

$$hv = BE + KE + \emptyset \tag{4.1}$$

Where *hv* is the energy of incident X-rays, BE is the binding energy of the electron (a measure of the energy of electron when it is bound to a certain orbital of an atom), KE is the kinetic energy of the emitted electron, and Ø is the work function of spectrometer. In XPS measurement, the binding energy is measured, which can be rewritten as

$$BE = hv - KE - \emptyset, \tag{4.2}$$

which is independent of the X-ray source used in the measurement [19, 20].

Analysis of a material surface is crucial because many phenomena such as corrosion, adhesion, charge transfer, adsorption, wettability, deposition, and catalysis occur only on the surface. XPS is considered the best analytical technique available for surface analysis; in this method, X-rays are incident on the surface, and then the detector measures the kinetic energies of the emitted electrons. Sample irradiation is performed with soft x-rays with energy less than 6 keV. XPS detects all elements from the periodic table except H and He with high surface sensitivity. It is also possible to detect the chemical states of the elements present on the sample surface [19, 20].

An interesting study was conducted by Horgnies *et al.* [21] on how the interfacial composition can affect the adhesion between aggregates and bitumen. This study was necessary to understand the interfacial dynamics between the aggregates and bitumen to enhance the lifetime of the roads. The authors used two aggregates (dolomite and granite), and the bitumen composition was comprised of asphaltenes micelles, and aromatic and saturated hydrocarbons. The chemical compositions of aggregates and bitumen were investigated with EDX and XPS, as given in Figure 4.7 and Table 4.1.

The interface between minerals and bitumen was characterized with advanced techniques like EDX and XPS before and after peeling. Using these techniques, it is possible to understand the role of various compounds formed at the interface on adhesion. Detailed analysis of the interface using EDX, XPS and mechanical tests revealed that alkali feldspars of granite were responsible for the weak interfacial bonding of aggregates with the bitumen.

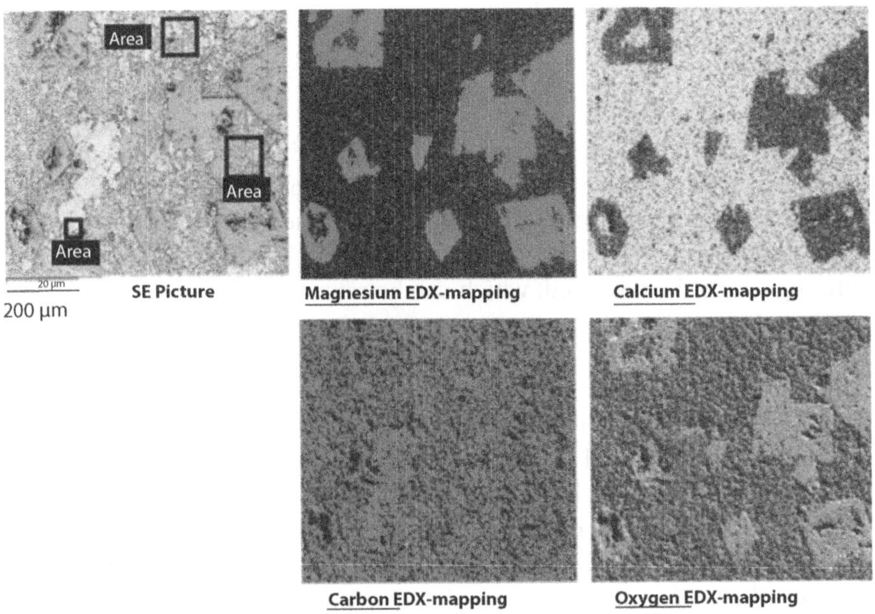

Figure 4.7 EDX maps and SEM micrographs of dolomite. Reprinted with permission from [21], Copyright 2011, Elsevier.

Table 4.1 Composition analysis of bitumen with EDX and XPS. Reproduced with permission from [21], Copyright 2011, Elsevier.

Atomic ratio (at. %)	Analysis by EDX	Analysis by XPS
Ca	95.5	96.5
O	1.8	1.9
S	2.7	1.6

4.3.1.3 Fourier Transform Infrared Spectroscopy (FTIR)

FTIR has many applications in examining polymer structures, analysis and inspection of biological tissues, cells, structural adhesives, and foods [22–24]. IR spectroscopy is important because it analyses the molecular vibrations present in the sample. In case of usual organic groups, they do display the characteristic IR bands because of fundamental molecular vibrations in the functional groups. A normal vibration mode absorbs the incident IR beam if a dipole moment change has occurred in the molecule during vibration [23]. It is important to note that when a molecule possesses centre of symmetry, all symmetric vibrations will be IR -inactive and will not be observed during the analysis. In contrast, the IR signals of polar groups are strong and intense in nature. Specific polymers like nylon, polyesters, polycarbonates can be readily spotted in FTIR because of their strong IR-response [25]. In the IR region (4000-1000 cm^{-1}), two kinds of vibrations are possible, which are (i) vibrations due to change in bond length, and (ii) vibrations due to change in bond angle. These vibration frequencies are very sensitive to H-bonding, the electronegativity of neighbours, the functional group present near the bonds, and the mass of the atoms involved in the vibration [23]. Based on the vast IR-spectra database collected for various functional groups over decades, it is now possible to analyse new materials using their IR–spectroscopy scans and computer-based tools. Other than IR spectroscopy, UV-spectroscopy is quite important and has wide range of applications in research as well as medical domains.

Recently, scientists have made it possible to determine the viscosity of adhesives based on the functional groups using FTIR technique [26, 27]. Also, after identification of the functional groups present in the compound, it is possible to determine the viscosity using the Orrick-Erbar and Letsou-Stiel equations.

Smart adhesives play a significant role in the biomedical and electronics fields [28, 29]. A fascinating study was conducted by Bhuiyan *et al.* [30] in which they successfully designed marine mussel-inspired (source of 3,4-dihydroxyphenylalanine (DOPA) having catechol side chain) smart adhesives which demonstrated reversible adhesion to wet surfaces. The exciting aspect is that the authors successfully implemented the FTIR characterization technique to verify the activation and deactivation of the adhesive coating with respect to adhesive property. The authors prepared adhesive polymers based on catechol (responsible for strong adhesion) with varying percentages of a conductive monomer, i.e., 1-pyrenemethyl methacrylate (PyMA). These materials were synthesized with AIBN-initiated free radical polymerization referred to as $D_xB_yPy_z$, and here D corresponds to DMA, B to APBA and Py to PyMA. The structures are given in Figure 4.8. JKR (Johnson-Kendall-Roberts) contact mechanics test was performed at pH values 3 and 9 to analyse variation in adhesion property w.r.t the pH value. For D10B10Py0, maximum adhesion strength was observed at pH 3 when it contacted the Ti surface. When the same surface was washed with a buffer of pH 9, a drastic decrease in the adhesion strength was observed. When placed under acidic condition of pH 3, the same adhesive successfully recovered its original adhesion strength. From the ATR-FTIR spectrum, it was observed that the reduced form of catechol contributed hugely to the bonding when in the presence of water under acidic conditions. To further support the conclusion, the authors successfully found a new peak at 1491 cm^{-1} in ATR-FTIR spectrum of pH-9 treated sample, corresponding to the formation of catechol-boronate complex. This result confirmed the pH sensitivity and reversibility of the adhesive coating adhesion.

Further investigations were performed to study the ability to deactivate adhesive coatings with the application of an electric field. It was observed that in the absence of an electric field, all untreated copolymer samples demonstrated average adhesion property. With an increase in the applied voltage, a decrease in the adhesion property for all the compositions was

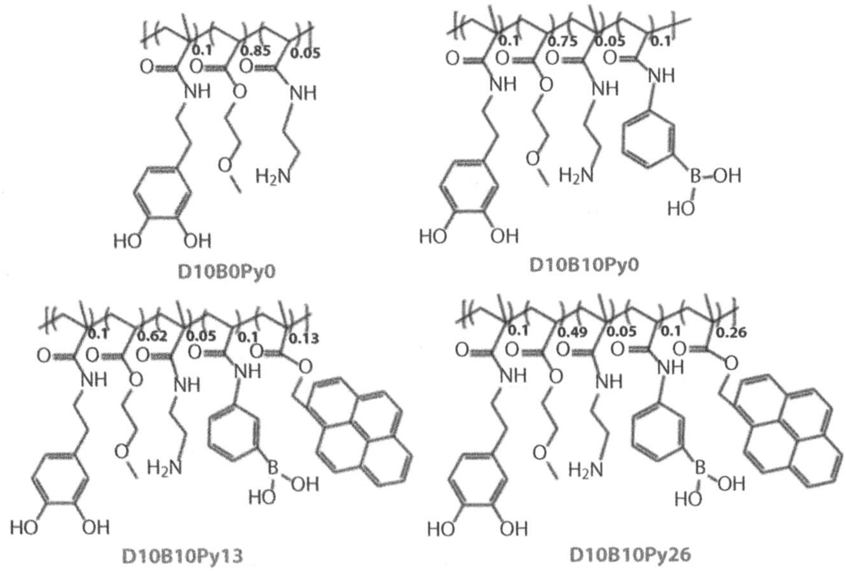

Figure 4.8 Synthesized smart adhesives to obtain reversible adhesion to wet surfaces. Reprinted with permission from [30], Copyright 2021, American Chemical Society.

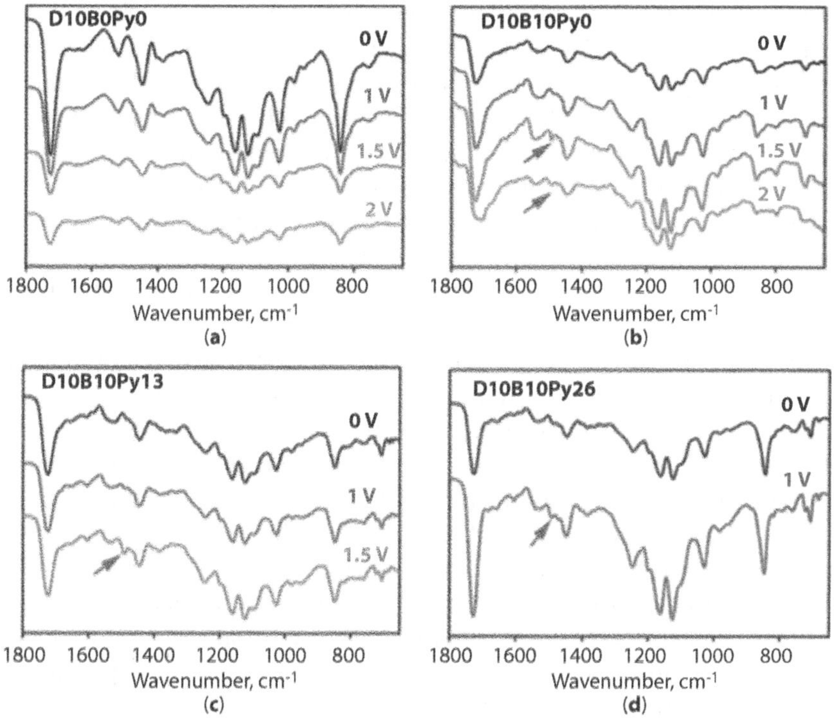

Figure 4.9 ATR-FTIR spectra indicating the formation of catechol-boronate complexation. Reprinted with permission from [30], Copyright 2021, American Chemical Society.

observed. Surprisingly in congruence with the hypothesis, the authors confirmed the formation of catechol-boronate complex using ATR-FTIR characterization as shown in Figure 4.9 when an external potential is applied.

Hygrothermal ageing is an essential phenomenon in adhesives and plays a vital role in the adhesive-substrate interface dynamics. Myers *et al.* [31] performed a crucial study highlighting the effects of hygrothermal ageing at buried interfaces of epoxies with different substrates. Epoxy adhesives are popular as an underfill material (for flip chip microelectronic packaging) and are widely used in the electronics industry [32]. Hygrothermal ageing causes undesirable effects like decreased bond strength and disruption of bonds at the interface. As pointed out by Myers *et al.* [31], for interfacial studies of adhesives earlier works in literature hugely relied on methods like FTIR, SEM and XPS, which are not entirely adequate to understand the dynamics occurring at the molecular level on the buried interfaces of adhesives. So authors used sophisticated methods like sum-frequency generation (SFG) vibrational spectroscopy [33] and ATR-FTIR spectroscopy to understand the molecular level dynamics at buried interfaces due to hygrothermal ageing. Here SFG is implemented primarily to analyse the response of epoxy interfaces to the externally created high temperature and humid conditions. The authors considered three different substrates for the analysis, namely SiO_2, deuterated polystyrene, and deuterated poly (ethylene terephthalate) with varying degrees of hydrophobicity. In the case of SiO_2, because of its hydrophilic nature, water remains at the interface after ageing treatment and reduces the interfacial bonding via hydrolysis reaction and replacement of hydrogen bonds between underfill and the substrate which causes loss of adhesion strength [34].

The analysis of SiO_2/epoxy interface was performed with sum frequency generation (SFG) spectroscopy and ATR-FTIR characterization techniques, as shown in Figure 4.10. For untreated SiO_2/epoxy interface, SFG signal at 2950 cm^{-1} was detected which corresponds to the vibration mode of the methyl group. This SFG result can validate the observed broad peak in the ATR-FTIR measurement in 2800-3000 cm^{-1} region which corresponds to the various stretching modes of the methyl group. Similar SFG and ATR-FTIR measurements were performed after 24 hr hygrothermal ageing of SiO_2/epoxy interface. SFG measurements showed peaks at 2875 and 2935 cm^{-1} and a broad peak in the range of 3000-3400 cm^{-1}, corresponding to epoxy-methyl stretching, epoxy-methyl resonance, and water, respectively. This indicated the changes that occurred in the molecular structure. Similar results were obtained for 24 hr treated samples in ATR-FTIR, where a broad peak from 2970 to 3700 cm^{-1} was obtained, which corresponds to

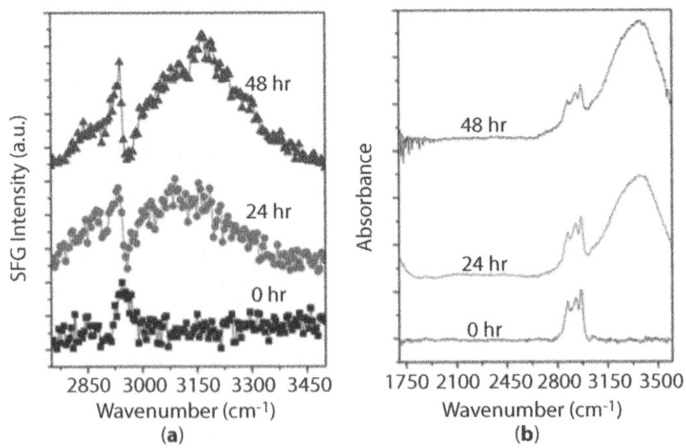

Figure 4.10 (a) SFG analysis of SiO_2/epoxy system (b) ATR-FTIR analysis of SiO_2/ epoxy system after different ageing time periods of 0 hr, 24 hr and 48 hr. Reprinted with permission from [31], Copyright 2014, American Chemical Society.

hydrogen-bonded water. Also, after 24 hr treatment, the lap shear strength of SiO_2/epoxy went down to zero. As one can see from Figure 4.10 (a), similar features were observed for 48 hr hygrothermal treatment.

The same analysis was performed for the remaining two substrates described in reference [31], comparing substrates with varying degrees of hydrophobicity. It can be interfered from their observations that the decrease in interfacial adhesion strength at poly (ethylene terephthalate)/ epoxy interface which is less than in SiO_2/epoxy situation but greater than in deuterated polystyrene/epoxy case.

Zhou and Cai [35] studied the hydroxyl-terminated polybutadiene (HTPB)-modified epoxy resin using methyl hexahydrophthalic anhydride as a curing agent and 2,4,6-tri(dimethylaminomethyl) phenol as a catalyst. The reaction between the epoxy resin and HTPB was monitored with FTIR in presence of 2,4,6-tri(dimethylaminomethyl phenol) (DMP-30) as catalyst, and mechanical tests confirmed the superior impact strength as compared to unmodified epoxy.

4.3.1.4 Gas-Liquid Chromatography (GLC)

Gas-liquid chromatography is a powerful tool for the chemical analysis of materials. In GLC analysis, the immobile phase is liquid, whereas the mobile phase is gas. The technique is primarily used for the chemical analysis of volatile components present in the solvent.

The sample of interest is usually dissolved in an appropriate solvent and then heated to evaporate the sample's volatile components. The mobile phase (usually a lightweight inert gas) carries the volatile components of the sample through a heated column. The GLC equipment comprises a rubber septum for injecting sample, a vaporization chamber, heaters, a column for carrier gas and detectors [36].

As shown in Figure 4.11, a micro-syringe is used to insert the sample into the vaporization chamber, maintained at a temperature of 55°C above the boiling point of the constituent with lowest boiling point identified among other chemical constituents present in sample. Subsequently, the volatile components of the sample get mixed with the carrier gas and travel through the temperature monitored column. The column can be operated in two modes, namely the isothermal program, and temperature program. In the isothermal condition of the column, the entire column is held at a fixed temperature very close to the middle point of the range on the boiling point scale comprised of boiling points of the constituents present in the sample. This method works best when the boiling point temperature range of the sample's constituents is very narrow. In the temperature program,

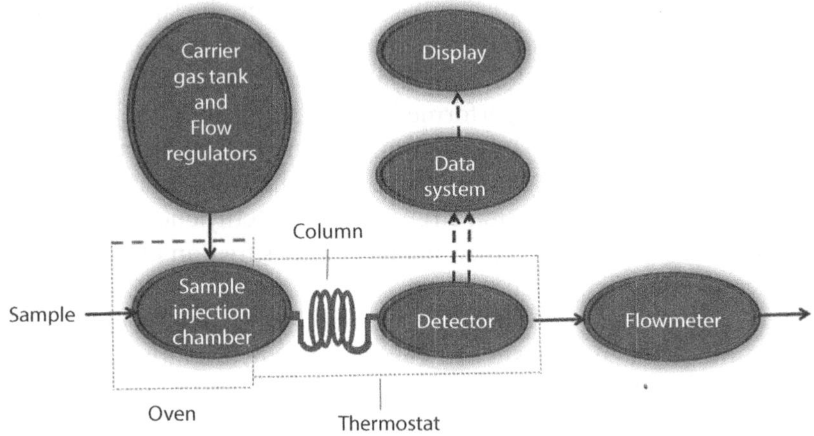

Figure 4.11 Schematic of equipment used for gas-liquid chromatography.

the temperature of the column is continuously increased in steps until all the components are separated, and sharp peaks of the signals at different temperatures for different components are observed [36]. Various types of detectors are already implemented in the Gas-liquid chromatographic analysis and some of them are mentioned below in Table 4.2.

Yan *et al.* [37] performed a safety evaluation of polyurethane adhesives which are used for food contact lamination films. They successfully identified six harmful migrant components (isocyanate residues) from additives that are found in the polyurethane using gas-chromatography, which exceeded the safety thresholds of daily intakes of KH-560 (epoxy-functional silane for improvement of thermal resistance of adhesive) and dimethyl propionic acid DMPA (additive for improvement of the adhesion strength).

Table 4.2 Various detectors used in GLC [36].

Type of detector	Applicable samples
Mass Spectrometer Detector (MSD)	All Samples
Flame Ionization Detector (FID)	Hydrocarbons, Alkali elements
Thermal Conductivity Detector (TCD)	Universal (all gases and vapours)
Electron Capture Detector (ECD)	Halogenated hydrocarbons
Atomic Emission Detector (AED)	Element-selective
Chemiluminescence Detector (CSD)	Oxidizing reagent
Photoionization Detector (PID)	Vapour and gaseous components

4.3.1.5 Nuclear Magnetic Resonance

W Pauli laid the foundation of NMR spectroscopy in 1924. Later in 1946, F Bloch and E Purcell experimentally demonstrated splitting energy levels of specific atomic nuclei because of an external applied magnetic field and won the Nobel Prize in physics [36].

The nucleus of an atom possesses properties like charge and spin, and hence it can behave like a magnet. Nuclei with an odd number of protons or neutrons will behave like a magnet. Neutrons and protons do possess spin, and the spin of the nucleus is defined by using I for the spin (discrete values of angular momentum) and m for the spin in the magnetic field. The nuclear shell model is implemented for the calculation of the spins of nuclei. The magnetic moment for a nucleus with non-zero spin is given as (μ: magnetic moment, γ: gyromagnetic ratio)

$$\mu = \gamma I \tag{4.3}$$

Also, it is observed that the molecular surroundings of a nucleus affect the absorption of RF radiation by the nucleus in the presence of a magnetic field which can be related to its molecular structure. Nuclei in different electronic environments absorb radiations with different frequencies to attain resonance. Hence, information about the nucleus' environment can be derived from its resonance frequency. When these nuclei are placed in a strong magnetic field, they try to align themselves along this external field, and the nuclei that are oriented along this direction are considered at lower energy or stable state.

In comparison, the nuclei magnets which are aligned opposite to the external magnetic field are considered at a higher energy state which are also statistically less in number. In the presence of an external magnetic field, based on the molecular environment of these small nuclei magnets, they orient themselves in various directions. When RF radiation is applied across these small magnets, they can absorb the energy from this RF radiation and get excited to a higher energy state. The amount of energy required for different magnets to reach the excited state will be different because of their different orientations and the surroundings, which enables the study of molecular structure using NMR spectroscopy [36, 38].

The energy provided by RF radiation is

$$\mathrm{E} = \mathrm{h}\nu \tag{4.4}$$

where ν is the NMR operating frequency given as

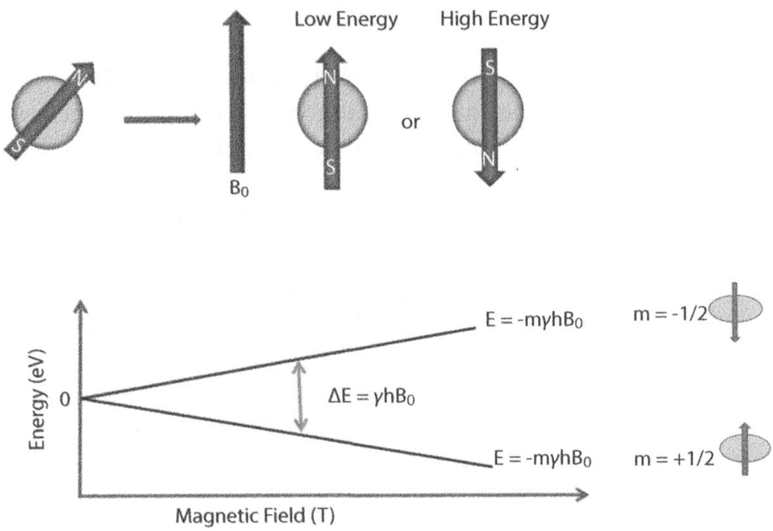

Figure 4.12 The phenomenon of spin alignment and energy level splitting in presence of external magnetic field. ΔE: Energy difference, B_o: Magnetic field, γ: Gyromagnetic ratio, E: Energy, m: spin.

$$\nu = \frac{\gamma}{2\pi} B_o \tag{4.5}$$

where γ is gyromagnetic ratio, different for different nuclei and B_o is the external applied magnetic field. The visuals of magnetic spins alignment and the splitting of energy levels in presence of external magnetic field are provided in Figure 4.12. In the NMR study, four nuclei that are of greatest use are ^{1}H, ^{13}C, ^{19}F, ^{31}P and the spin quantum number for these nuclei is 1/2. Hence, each nucleus will have two spin states which are I = +1/2 and I = -1/2.

Instead of conventional adhesives based on petroleum resources, biomass-derived adhesives with low prices and good performance dominate the market. Although they have remarkable properties like easy accessibility and value-added functionality, they lack good mechanical properties like low structural bond strength. Wang *et al.* [39] found the versatility of mussel-mimic chemistry in combination with bio-derived polymers to achieve high-strength adhesive bonding.

As shown in Figure 4.13, bio-inspired soy protein (SP) adhesive was produced by involving mussel mimicking motifs OCNF-CA (catechol- and aldehyde-modified cellulose nanofibers) and linking protein regions

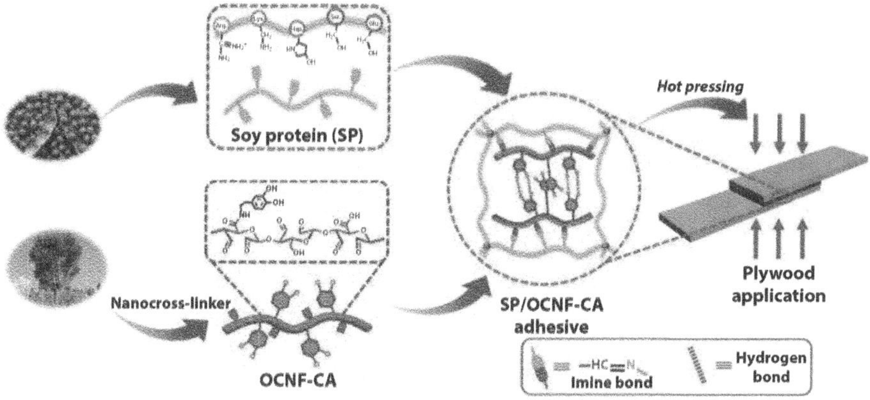

Figure 4.13 Scheme depicting the thermoset adhesive formed with a dual crosslinked network of OCNF-CA and SP. Reprinted with permission from [39], Copyright 2020, American Chemical Society.

connected with interfacial H-bonds and Schiff link between OCNF-CA and SP. This covalent bonding between OCNF-CA and SP provided cross-linking of polymers, enhancing cohesion strength. ^{13}C - NMR spectroscopy was used to investigate the formation of OCNF-CA/SP adhesive. NMR studies revealed the peaks for given SP composites and are shown in Figure 4.14, confirming the microchemical structure of OCNF-CA/SP. Details of the NMR signals obtained are also provided in Table 4.3.

^{13}C - NMR study revealed imine linkage between SP and aldehydes through D-glucose resonance peak at 92.8 ppm. Also, peak at 157.8 ppm

Table 4.3 Signals obtained in C-13 NMR spectrum of OCNF-CA/SP adhesive for investigation of microchemical structure [39]. Reproduced with permission from [39], Copyright 2020, American Chemical Society.

^{13}C - NMR spectroscopy signal	Chemical shift (ppm)
Amino acid carbonyls	155–175
Aromatic Carbons	115–130
α-Carbon of amino acids	45–65
β-Carbon of amino acids	25–45
Methyl groups	15–25
D-Glucose resonance	92.8

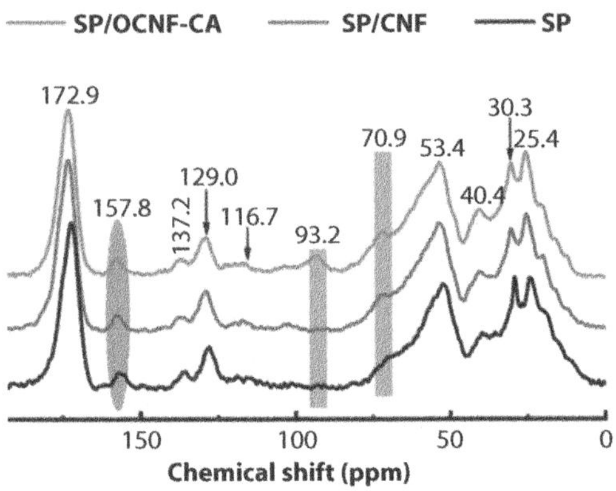

Figure 4.14 Signals observed in ^{13}C - NMR study of SP composites. Reproduced with permission from [39], Copyright 2020, American Chemical Society.

indicated a good miscibility between OCNF-CA and SPI because of covalent bonds.

In order to make use of tannin, Liu *et al.* [40] studied the depolymerisation of condensed tannin performed in NaOH plus urea medium. The changes in the molecular structure of the depolymerized tannin were observed with ^{13}C-NMR and IR spectroscopies. Results demonstrated that the decrease in polymerisation degree from 7 to 2 mainly occurred because of C-4 and C-8 bonds breaking. Tannin-modified phenolic resin was prepared by copolycondensation of phenol, tannin depolymerization products and formaldehyde in a basic medium. The study showed that tannin-modified phenolic resin had improved thermal stability and bonding strength than untreated tannin-modified phenolic resins.

4.3.1.6 Raman Spectroscopy

C V Raman discovered the paradigm-shifting phenomenon in which he observed that when light is incident on the sample from a high-intensity source, a tiny fraction of the incident light gets scattered with wavelength shifted in the visible region of electromagnetic spectra, representing the chemical structure of the sample. He was awarded the Nobel Prize for this discovery in 1931 [36].

Raman spectroscopy provides a detailed analysis of phase, chemical structure and crystallinity of the material. In the Raman spectrum, each peak corresponds to a specific bond vibration present in the molecule, and Raman spectrum will be a chemical fingerprint for any material.

Figure 4.15 and Figure 4.16 represent the phenomenon of Raman scattering and the block diagram of Raman spectroscopy equipment, respectively.

The principle of Raman spectroscopy is based on the inelastic scattering of photons when a laser beam in visible or IR frequency range impinges on the sample. The photons interact with molecular vibrations of the sample, which results in the scattered photons with reduced or increased energy. These shifts in the energies of the scattered photons give information about the vibrational modes present in the system [41].

Raman spectroscopy is crucial because it provides unique fingerprints for different materials, and the intensity of the peaks in the spectrum is

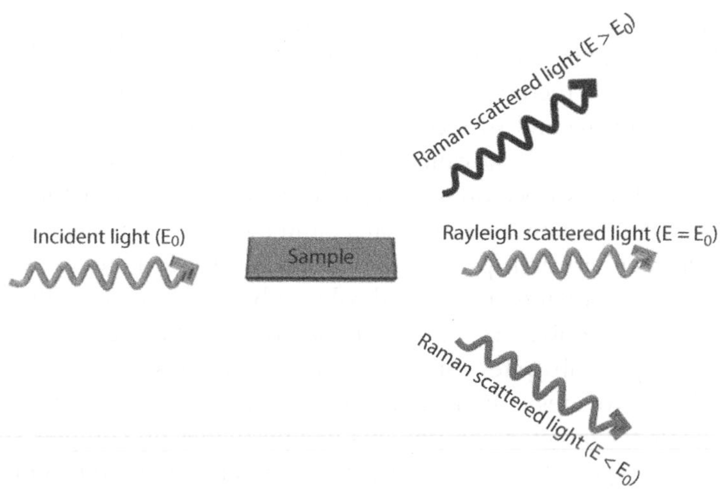

Figure 4.15 Scattering of light from a sample with different energies.

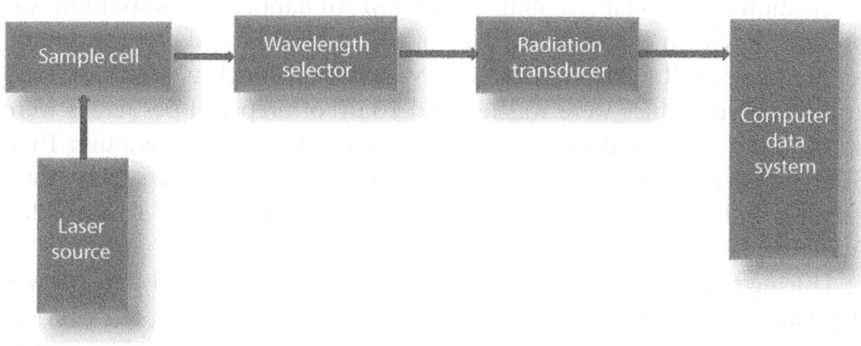

Figure 4.16 Block diagram for Raman spectroscopy equipment.

directly proportional to the concentration. Besides metals and alloys, Raman spectroscopy is suitable for a wide range of materials, including organic, inorganic and biological materials [36].

As an instrument, a Raman spectroscopy equipment contains many components like laser source, sample cell, spectrometer, and radiation transducer.

In modern days, adhesive tapes are effectively used for extraction of target components from surfaces with complex geometrical shapes. Efficient and qualitative extraction of chemical components from a given material surface is essential for technological advancement in environmental pollutant monitoring, analysis of explosive traces and pesticides on fruits and vegetables. Worldwide pesticides are used in cultivation of fruits and vegetables and these pesticides are harmul and cause several problems related to health and environment. For determination of these pesticides, there are already several methods discussed such as enzyme-linked immunosorbent assay (ELISA) and surface enhanced Raman scattering (SERS) [42]. SERS technique is more advanced and produces enhanced Raman signal from molecules adsorbed on the prepared surface which have found many applications in domains like food safety and biotechnology engineering. It is possible to enhance the properties of SERS substrate for better extraction of targeted components in which flexible substrates are available such as paper,cotton and polymers. Chen *et al.* [42] developed a flexible and adhesive SERS tape (substrate) for detection of pesticides on the surfaces of fruits which provides enhanced absorbption as well as Raman spectroscopy signals for targeted components like pesticides on complex surfaces of fruits and vegetables. For analysis, five different commercial tapes were used, namely 3M 500*1/2 transperant adhesive tape, 18mm *14m Wingtai transperant adhesive tape, 3M 200C double-sided adhesive tape, 3M black electric tape, and HM-3 guangbo double-sided foam adhesive tape. For fabrication of SERS tape, a 50μL of 0.005 M Au nanoparticles solution was poured slowly on the sticky surface of the tape and dried at atmospheric temperature for Raman analysis. Now, various fruits such as apples, oranges and green vegetables were washed and 10 μL of prepared pesticides were spread onto the peeled square shape and dried at room tmeperature. Then SERS tape was pressed onto the dried surfaces of fruits and vegetables for 3 seconds and then peeled off slowly. The visuals of the process are provided in Figure 4.17.

During the preparation of SERS tape, Au nanoparticles dispersion was deposited onto the adhesive tape. Due to presence of water, the acrylate component became swellable and then process of de-swelling for acrylate component takes place during the dry-evaporation.

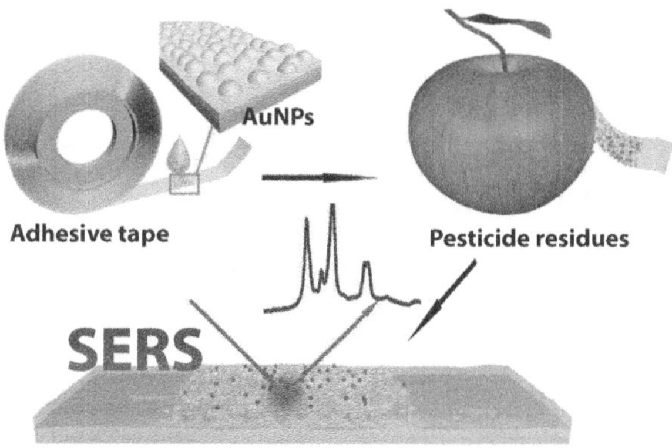

Figure 4.17 Schematics of SERS tape fabrication along with the target extraction process from fruit surface for SERS analysis. Reprinted with permission from [42], Copyright 2016, American Chemical Society.

In agriculture, sulfur and pyridine pesticides are mostly used because of their insecticidal activity. Hence, among the pyridine family 4-Mpy is used as a model probe to evaluate the performance of different SERS substrates for detection of sulfur and pyridine based pesticides. Among the five commercial tapes, transparent adhesive tape shows distinct peaks of 4-Mercaptopyridine which was used as Raman model probe for testing of SERS tapes as shown in Figure 4.18. Also, it shows low background noise and stable baseline and hence 3M transparent adhesive tape composed of polypropylene film and acrylate adhesive was chosen as substrate of SERS tape.

For further analysis and comparison of the selected substrate with different modifications, the four samples were prepared which are, transparent adhesive tape, 4-Mpy deposited adhesive tape, Au nanoparticles coated adhesive tape, and 4-Mpy deposited and Au nanoparticles coated SERS tape. The Raman spectrum of SERS tape showed Raman peaks at 715 and 1093 cm^{-1} corresponding to spectral features of 4-Mpy with significant increase in Raman peak intensities as shown in Figure 4.19. It is observed that SERS tape exhibits excellent SERS activity and also stability with adequate stickiness required for the target extraction from the complex surfaces. Surfaces of various substances such as fruits and vegetables were tested and good results were obtained with excellent SERS activity as shown in Figure 4.20.

Tatar *et al.* [43] performed an extensive analysis of the interphase between epoxy and concrete, which showed that the adhesive bonding between the epoxy and concrete primarily depends on the interlocking

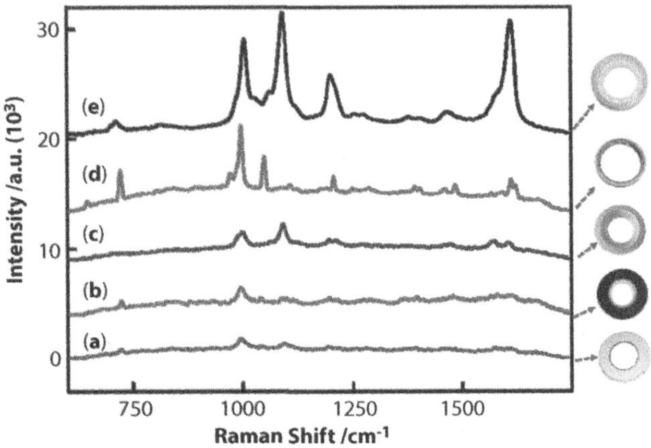

Figure 4.18 Raman spectra of 4-Mercaptopyridine from Au modified adhesive tape (a) double -sided adhesive tape; (b) electrical tape; (c) Wingtai transparent adhesive tape; (d) double- sided adhesive foam tape; (e) 3M adhesive tape (Exposure time: 10s; laser wavelength: 633 nm; laser power: 1.7 mW). Reprinted with permission from [42], Copyright 2016, American Chemical Society.

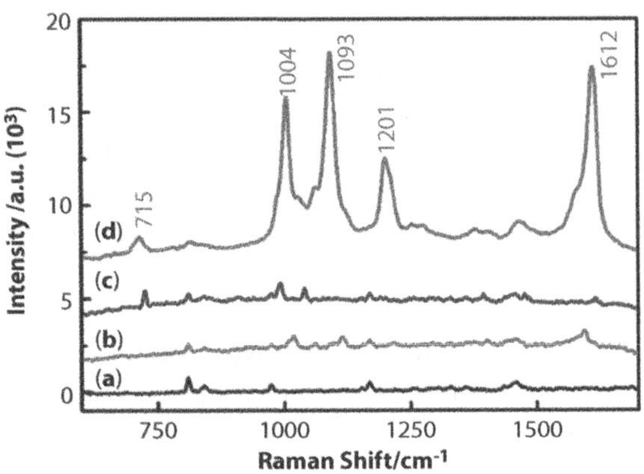

Figure 4.19 Comparative analysis of SERS signals from (a) adhesive tape; (b) adhesive tape coated with Au nanoparticles; (c) 4-Mpy on adhesive tape; (d) 4-Mpy on adhesive tape coated with Au nanoparticles. Reprinted with permission from [42], Copyright 2016, American Chemical Society.

between the adhesive and the concrete. The ability of the adhesive to penetrate the concrete substrate and wettability are the main contributing factors. Their interphase region was extensively studied with Raman spectroscopy, and a positive correlation was obtained between the pore size and depth of epoxy penetration into the substrate.

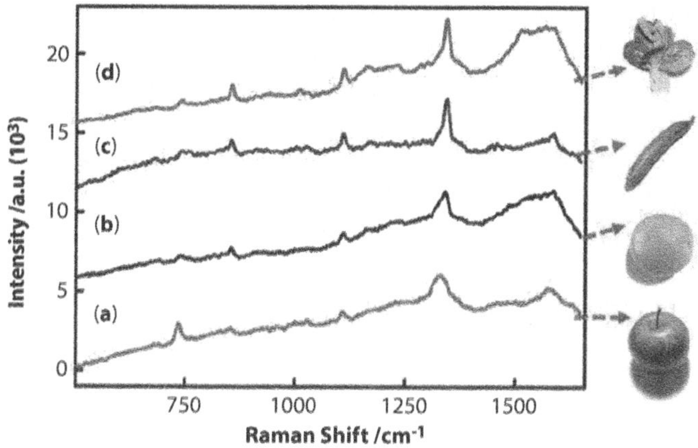

Figure 4.20 Raman spectra of parathion methyl extracted from peel of (a) apples; (b) oranges; (c) cucumber; (d) vegetables using SERS tape. Reprinted with permission from [42], Copyright 2016, American Chemical Society.

4.3.2 Physical Characterization

Physical characterization techniques help to analyse the physical attributes of the sample, which might include the size of particles, grain boundaries, grain size distribution, morphology and interface dynamics. Here the commonly used techniques for physical characterization are discussed.

4.3.2.1 Contact Angle Measurement

Measurement of contact angle provides a measure of the wettability of a solid by a liquid. The angle subtended by the liquid at the interface with the solid surface is called contact angle. The contact angle primarily depends on the interaction between the liquid phase and the solid phase. The balance between the cohesive and adhesion forces determines the value of the contact angle at the liquid-solid interface. When the value of the contact angle is small, the magnitude of adhesion force is much larger than the cohesive force, so the liquid will have greater wetting tendency. Mittal [44] also showed that joint strengths are very closely related to the surface energetics and dynamics; also interfacial adhesion can be improved by modification of the surface which can react with an adherate [45]. An adherate signifies any material (e.g., coating, adhesive, metal layer (metallized plastic)) that goes on top of an adherend. The term "adherate" is analogous to "adsorbate" in the field of adsorption. Moreover, when the contact angle value is large, the magnitude of cohesive force surpasses the magnitude of adhesion force. Figure 4.21 below depicts the measurement of contact angle and different cases of wetting [46]. While considering wetting phenomenon theoretically, three boundaries plays the central role, namely, vapour-liquid, liquid-solid and solid-vapour, as shown in Figure 4.22.

Here, three arrows indicate the interfcial tensions acting at three interfaces. When the droplet is in equilibrium, these forces balance each other, and one obtains the relation (Young Equation)

$$cos\theta = \frac{\gamma_{sv} - \gamma_{ls}}{\gamma_{lv}} \tag{4.6}$$

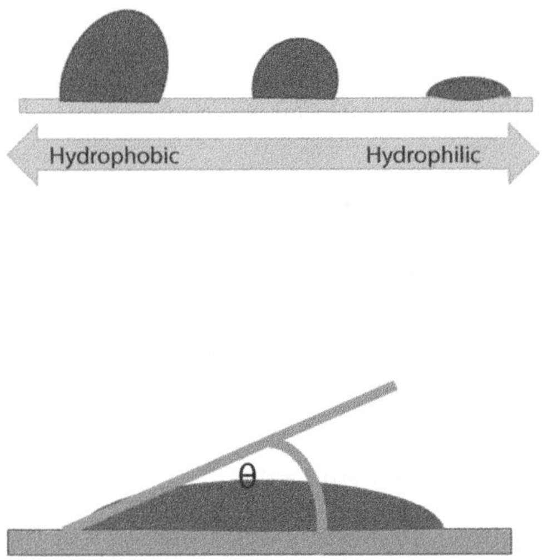

Figure 4.21 Contact angle and variation of contact angle: smaller angle corresponds to hydrophilic surface and larger angle corresponds to hydrophobic surface.

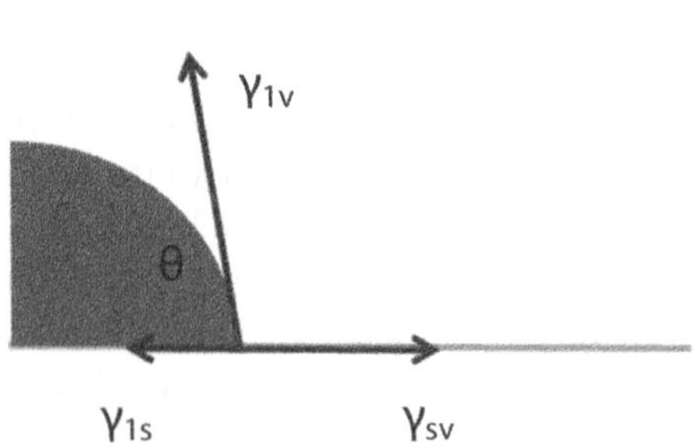

Figure 4.22 Various forces at the liquid-substrate interface. γ_{lv}, γ_{ls}, γ_{sv}: interfacial tensions at liquid-vapor interface, liquid-solid interface, and solid-vapor interface.

It is essential to consider Cassie's law for an effective contact angle measurement on chemically heterogeneous surfaces. For two chemical materials, the effective contact angle (θ_C) is given by the Cassie-Baxter equation,

$$cos\theta_C = \sigma_1\, cos\theta_1 + \sigma_2\, cos\theta_2 \tag{4.7}$$

Where, θ_1 and θ_2 are contact angles for respective chemical components with their fractional surface areas σ_1 and σ_2 in the composite substrate [47, 48].

Performing contact angle measurements is straightforward and requires simple assembly consisting of a light source, a camera and a tilting stage. The procedure for measurement is manifested in Figure 4.23.

Wen *et al.* [49] synthesized larch bark modified phenolic resins to prepare strawboards. Dynamic wettability of the synthesized resin was measured on the rice straw and the variation of contact angle with respect to time was studied. The results showed a better performance of bark-modified phenol-formaldehyde resin on all surfaces regardless of whether the rice straw was treated or untreated with alkaline solution (with the lowest equilibrium contact angle).

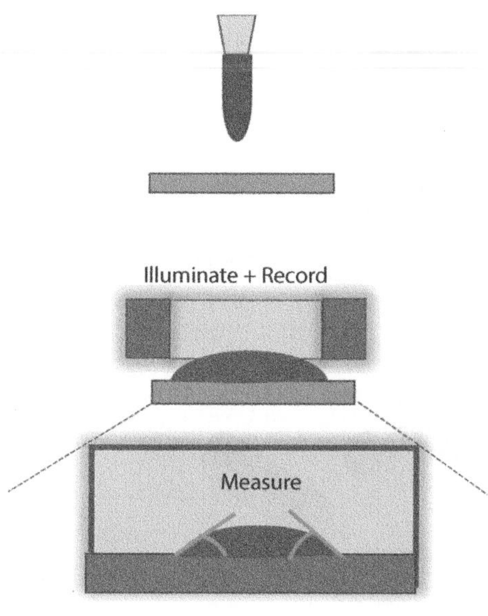

Figure 4.23 Schematic of contact angle measurement instrument.

4.3.2.2 Scanning Electron Microscopy (SEM)

A scanning electron microscope (SEM) is an instrument that uses an electron beam to produce microscopic images of various material surfaces. SEM is a frequently used instrument in research because of its high resolution as compared to optical microscopes. When a focused electron beam impinges on the sample and interacts with the atoms present on the surface, it produces different signals (Figure 4.24), which provide helpful information about the surface morphology and elemental composition. For different kinds of signals, different detectors are placed in the sample chamber for detection purposes. The whole process of imaging the sample is carried out in vacuum conditions to avoid electron scattering, noise in the signal and to maintain the high quality of imaging. The primary signals in SEM are secondary electrons (SEs) and backscattered electrons (BSEs) for imaging purposes, while characteristic X-ray signals are for compositional analysis of the material. BSEs emerge from elastic collisions, whereas SEs originate from inelastic collisions of electrons. Here BSEs provide information about deeper regions below the surface, whereas SEs provide information exclusively about the sample surface. The complete working of SEM with its various components can be visualised from Figure 4.24 [50].

Ebewele *et al.* [51] studied the morphology and performance of a phenolic adhesive based on different kinds of fillers added using SEM, DSC and

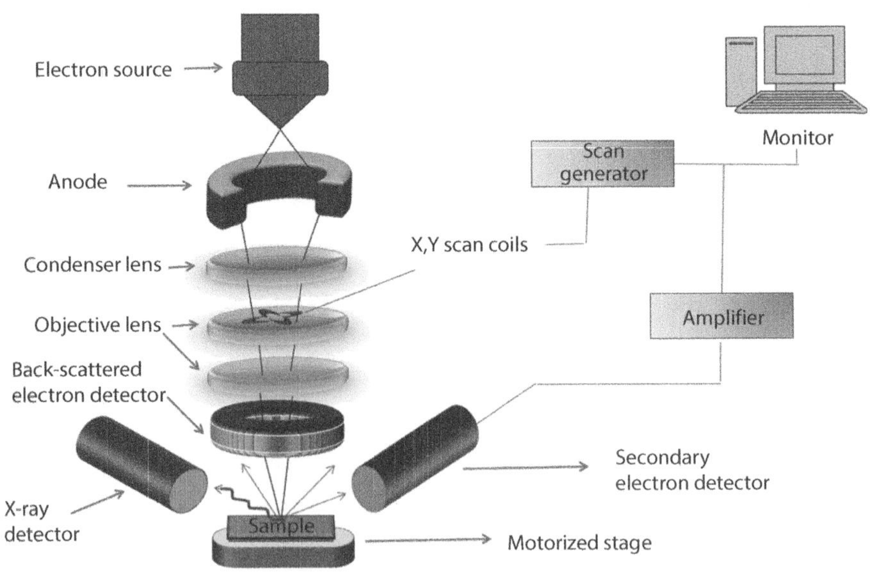

Figure 4.24 Detailed working principle of SEM.

IR spectroscopy. It was observed that the filler does not affect the curing dynamics, but it has a profound effect on the morphology and topography of adhesive-wood interphase as well as on bulk properties of the adhesive.

4.3.2.3 Gelation Time

Gelation is defined as the transition of a polymer from viscous liquid to elastic solid when the curing process is going on. Gelation time is a crucial parameter in determining the shelf-life of resins and is used for quality control purposes [52].

Gelation refers to the formation of continuous covalent bonds in the system, which is rather a slow process. The methods to determine the gelation time are diverse and can be classified as manual tests, automatic measurements (gel timers), and tests with research instruments [52]. Some of the essential methods [52] are mentioned in Table 4.4. The method-specific ASTM standards are provided in the table as reference.

Table 4.4 Methods for determination of gelation time [52].

Method	Gel time determination principle
DMA (Dynamic Mechanical Analysis) [ASTM D4065, D4440, D5279]	Point of intersection of the elastic modulus and the loss modulus
DMA (Dynamic Mechanical Analysis) [ASTM D4065, D4440, D5279]	Extrapolated onset of elastic modulus growth
RAS (Resonance Acoustic Spectroscopy) [ASTM E2001 – 18]	The onset of the elastic modulus growth
US (Ultrasonic)	The extrapolated onset of increase in sound velocity
US (Ultrasonic)	Point of curvature change in the ultrasonic pulse velocity curve
DEA (Dielectric Analysis) [ASTM D7416 – 09]	The inflection point (point of curvature change) in the log plot of ionic conductivity
Luminescence	Change in the slope of luminescence intensity vs time plot

Huang *et al.* [53] synthesized silicone–epoxy resins by hydrosilylation of 1, 2-epoxy-4-vinyl-cyclohexane with 1,3,5,7-tetramethylcyclotetrasiloxane. They found high reactivity in silicone–epoxy resins when aluminium complex/silanol compound was used as a catalyst and good UV resistance in cured silicone–epoxy resins. From the measurement of gelation time, it is concluded that the reactivity of silicone-epoxy increases with the Si-H bond in the molecule. The presence of Si-H bonds present in the silicone-epoxy adhesive was confirmed via IR spectroscopy and ^{1}H-NMR. Detailed procedure is provided in Huang *et al.* [53] regarding the synthesis of silicone-epoxy resins.

Sarkar and Adhikari [54] studied the properties such as adhesion strength and thermal stability of lignin modified phenolic resin. It was reported by the authors that modification of phenolic resin with lignin is more thermally stable than pure phenolic resin. The authors also reported the gelation time of resin which was measured by the ASTM standard method (D 2471-88). In this method, 415 ml of lignin modified phenolic resin was placed in an open container with test temperature maintained at 70° C. For every 15 second, authors inserted a wooden probe vertically into the reacting mass and calculated the gelation time when the reacting mass did not adhere to the probe. Effects of formaldehyde to phenol molar ratio, sodium to phenol molar ratio and phenol to lignin weight ratio on gelation time for phenolic resin are determined. It was observed that with increasing formaldehyde to phenol molar ratio (1.7 to 3.5), gelation time decreased from 140 min to 25 min.

4.3.2.4 *Small Angle X-ray Scattering (SAXS)*

SAXS is an important technique for structural analysis of polymers, nano-powders, nanocomposites, and bio-macromolecules. The working principle of SAXS is straightforward in which a solution of the sample is prepared in a quartz tube, and the tube is irradiated with a monochromatic X-ray beam and the scattered rays are detected with a detector. Next, measurement is performed for the pure solvent to subtract it from the original pattern of the solution to obtain all peaks from the sample of interest. The resulting intensity plots of the peaks provide information like shape and size of the particles dispersed in the solvent. Other characteristics are also obtained from the SAXS about the sample such as molecular weight, maximum dimension, and excluded particle volume [55]. Also, the principle of measurement is based on the elastic scattering of X-rays while travelling through the material and analysis of their scattering through a detector at small angles (0.1 – 10°). SAXS is capable to extract structural information at a scale of 1-100 nm and Ultra-SAXS can probe even smaller dimensions of the material [56].

Pruksawan *et al.* [57] conducted an excellent study of the toughening effect in epoxy by incorporating cellulose nanocrystals. They performed a very nice study using Ultra-SAXS and SAXS. A discussion of Pruksawan *et al.* [57] work is provided in the "mechanical characterization" section 4.3.4 of this chapter.

4.3.2.5 Atomic Force Microscopy (AFM)

AFM is a high-resolution scanning microscopic technique used to study the surface properties of materials, including morphology and physical properties. Other properties can also be measured with AFM, such as nanoparticle size and shape in 3D, dispersion studies of nanofillers in matrices and degree of coverage of a substrate with given nanoparticles morphology [58]. AFM instrument is a scanning probe microscope with a small tip (made of silicon) to scan the sample surface. When this small tip interacts with the sample, it causes the attached cantilever to deflect, which is directly proportional to the force experienced by the tip. This allows to form a picture of the surface which can be visualized in three dimensions [59]. The working principle of AFM is depicted in Figure 4.25.

Arikan *et al.* [60] investigated polymer surface functionalization to investigate the significance of surface modification to increase the adhesion to polymers. Interestingly, they investigated the effects of different pre-treatment methods, such as low-pressure plasma (LPP) and

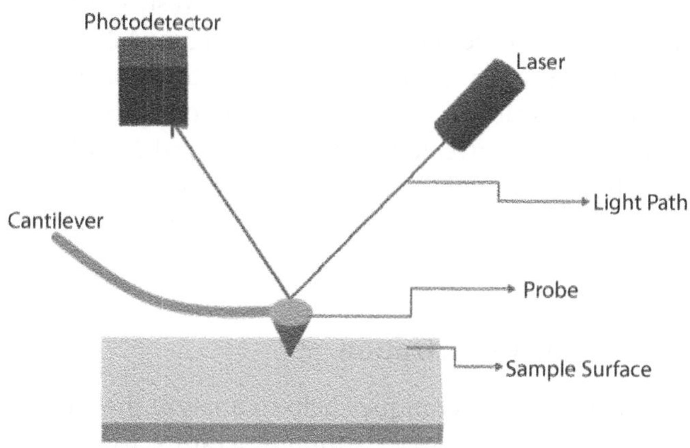

Figure 4.25 Schematic and working principle of AFM.

Table 4.5 O/C atomic ratio from XPS measurements for different polymer materials [60]. Reproduced with permission from [60], Copyright 2019, Elsevier.

Material	Untreated	VUV	VUV + ethanol wash	LPP	LPP + ethanol wash
PEEK amorphous	0.14	0.37	0.16	0.33	0.19
PEI	0.14	0.26	0.16	0.38	0.18
PES	0.19	0.4	0.19	0.5	0.2
RTM6	0.12	0.24	0.15	0.24	0.18

vacuum-UV (VUV), on four polymers, namely poly (ether ether ketone) (PEEK), polyetherimide (PEI), polyethersulfone (PES) and thermoset epoxy resin RTM6 (Hexel Corp.), for evaluation of the fact that surface treatment is essential for better adhesion. Ethanol and water are used in this research as polar solvents to wash the treated surface to measure the difference in adhesion after washing. AFM analysis was performed for surface characterization of various polymers with different pre-treatment methods. The AFM study performed for the amorphous PEEK polymer is shown in Figure 4.26, and similar AFM studies were performed for other polymers. XPS was used to perform chemical analysis of pre-treated surfaces of polymers. All polymers exhibited chemical activation, as shown in Table 4.5. In short, it is established from the XPS study that if fragments of polymers are highly oxidized with no bonding to bulk, these can be easily removed from the surface with polar liquid washing. Authors performed contact angle measurements for pre-treated and washed polymer samples with DI water and diiodomethane, as shown in Figure 4.27. Measured contact angle values of proper test liquids are used to determine the surface free energy of polymeric substrates using different procedures [61, 62]. Surface free energy for untreated samples is around 50 mN/m, and after VUV and LPP treatments, it increased up to 75 mN/m. Surface free energy primarily increased because of the formation of oxygen-rich groups and thus increase in the polar nature.

4.3.3 Thermal Characterization

Thermal characterization techniques help to analyze the variations in the properties of a material with respect to the temperature, which include methods such as differential scanning calorimetry (DSC), thermogravimetric

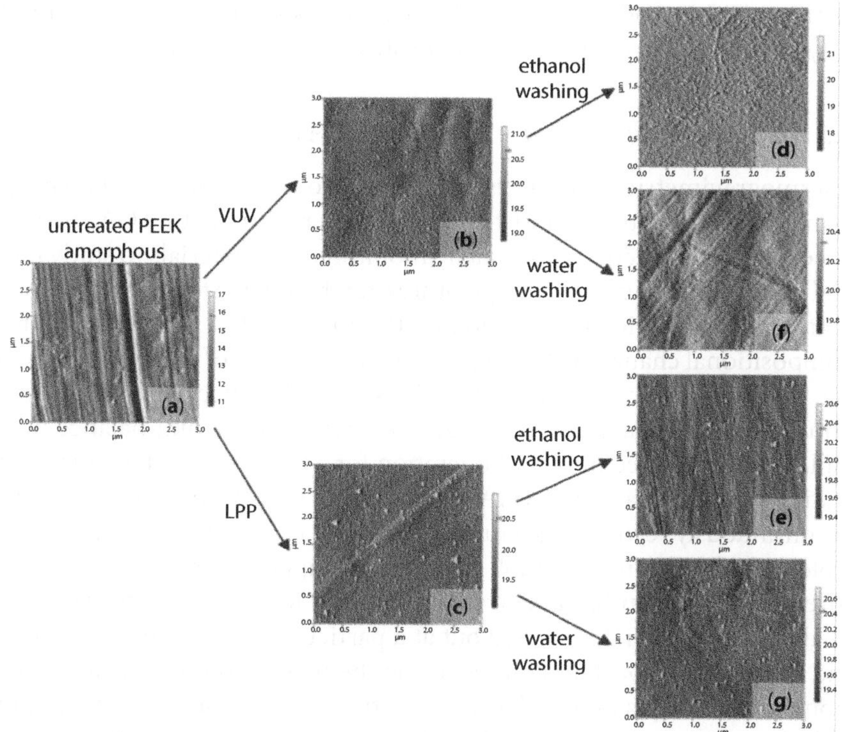

Figure 4.26 AFM images of amorphous poly (ether ether ketone) (PEEK) polymer. Reprinted with permission from [60], Copyright 2019, Elsevier.

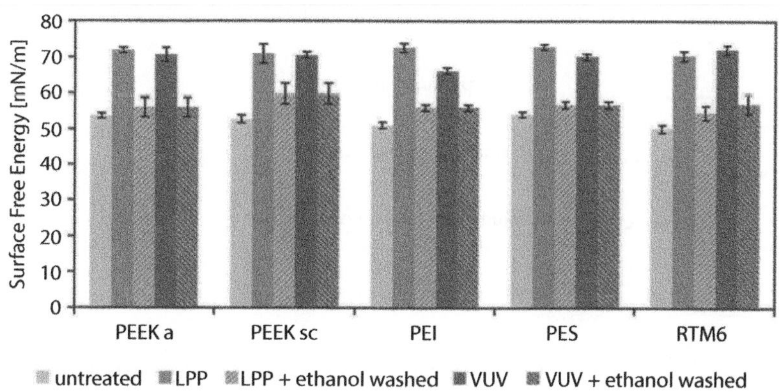

Figure 4.27 Surface free energy for vacuum-UV (VUV) and low-pressure plasma (LPP) treated samples and washed samples after treatments [60]; poly(ether ether ketone) (PEEK), polyetherimide (PEI), polyethersulfone (PES) and thermoset epoxy resin (RTM6). (a: amorphous, sc: semi-crystalline). Reprinted with permission from [60], Copyright 2019, Elsevier.

analysis (TGA) and differential thermal analysis (DTA). These methods play crucial role in characterization of structural adhesives.

4.3.3.1 *Thermogravimetric Analysis (TGA)*

Thermogravimetric analysis is performed to study the weight change of materials as a function of time and temperature. With TGA, thermal decomposition reactions are usually monitored for materials to study their ability to withstand high temperatures for better performance. TGA of materials can provide information on thermal stability, oxidative stability, compositional changes with temperature, the lifetime of reaction products, the kinetics of decomposition reactions, and the effect of the corrosive atmosphere on materials. TGA balance operates on a null-balance principle.There are three modes of operation for TGA, namely dynamic TGA, static TGA and quasistatic TGA. In dynamic TGA, the mass of the sample is continuously recorded as a function of increasing temperature. It is possible to deduce from the graph of mass vs temperature at which temperature decomposition has occurred and gas has formed. In static TGA, the analysis of a material is carried out at a particular temperature by holding the temperature constant and recording the mass of the sample as a function of time. In quasistatic TGA, the heating of the sample is done in multiple temperature intervals and held at these intervals until stabilization.

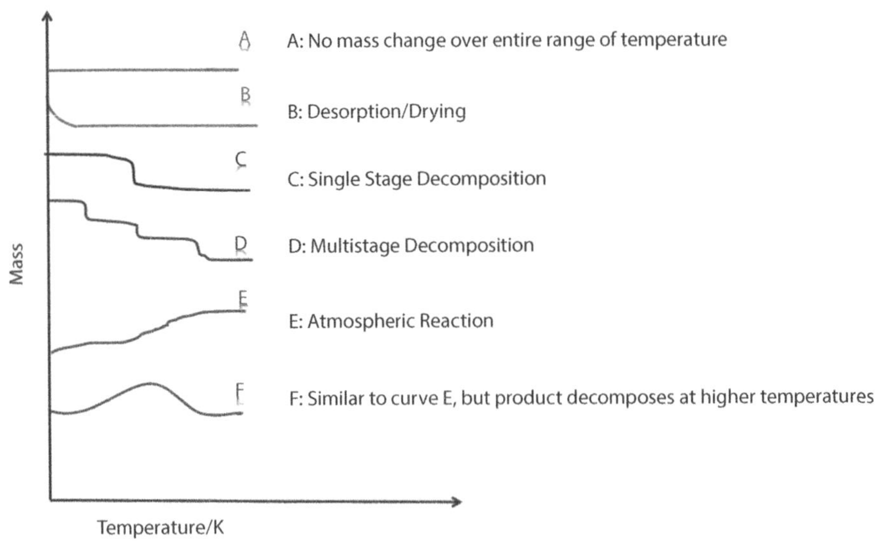

Figure 4.28 Different possible TGA plots [63, 64].

This is useful in characterization of materials at different temperatures [63]. Different kinds of TGA curves obtained during measurements are shown in Figure 4.28.

4.3.3.2 *Differential Thermal Analysis (DTA)*

In DTA, the difference between the temperature of the sample and the reference material is plotted as a function of increasing temperature, under a controlled temperature program so that the increment of sample temperature is linear in nature. The working principle of DTA can be visualised from Figure 4.29. In this diagram, R is the inert reference, whereas S is the sample. The "Es" output voltage (converted to temperature as T_s) from the sample thermocouple provide feedback and helps the system control current input to the furnace, so that sample temperature increases linearly at a given rate. The output across reference and sample thermocouples ΔE is converted to ΔT and the thermogram is plotted [36].

In general, to understand the thermogram, consider the thermogram of a polymer in Figure 4.30. The first decrease in ΔT is due to the glass transition phenomenon in polymer. This does not involve any enthalpy change of polymer. The following two maxima in the graph correspond to crystallization and oxidation processes which are exothermic. Oxidation occurs only when heating is done in the air. Furthermore, the minima represent

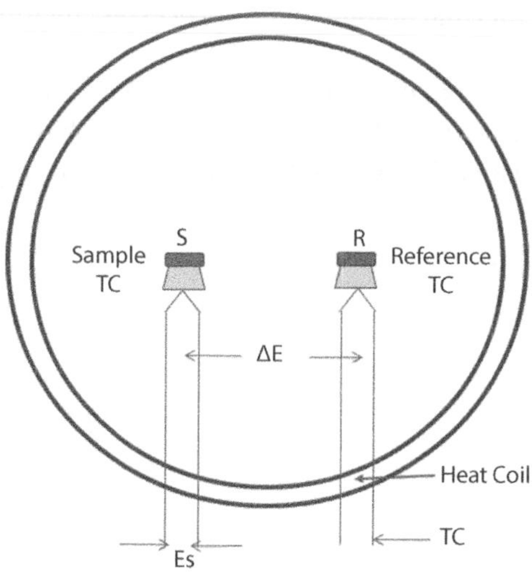

Figure 4.29 Schematic of DTA instrument. TC: thermocouple, ΔE: difference across sample TC and reference TC, E_S: output voltage from sample thermocouple.

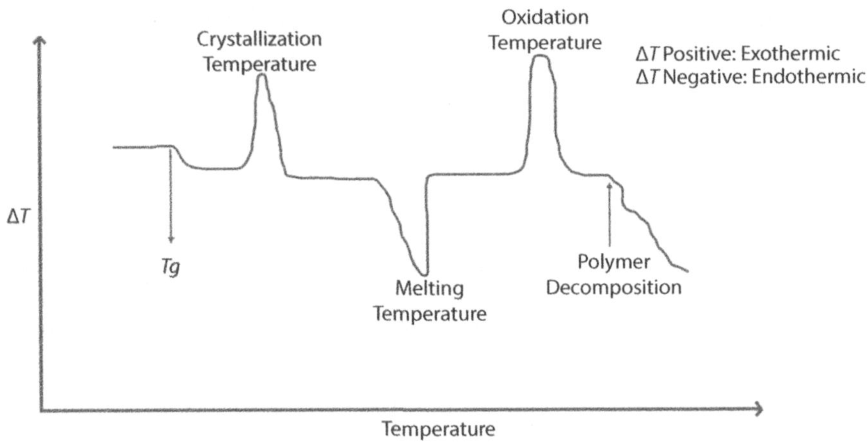

Figure 4.30 Differential thermogram for a polymeric material.

the melting of the polymer, which is endothermic. Finally, the polymer decomposition starts, which results in negative ΔT in the thermogram.

4.3.3.3 Differential Scanning Calorimetry (DSC)

DSC is primarily a thermal analysis procedure that monitors the heat flow ΔQ across the sample as a function of time or temperature, as shown in Figure 4.31. DSC is beneficial in determining sample properties such as heat capacity, T_g, thermal stability, cure processes, melting and oxidation behaviour. The three different types of DSC instruments are power-compensated DSC, heat-flux DSC, and modulated DSC. In a DSC experiment, heat is supplied to both sample and the reference such that the temperature of both should rise identically with time. The heat flow required (in or out)

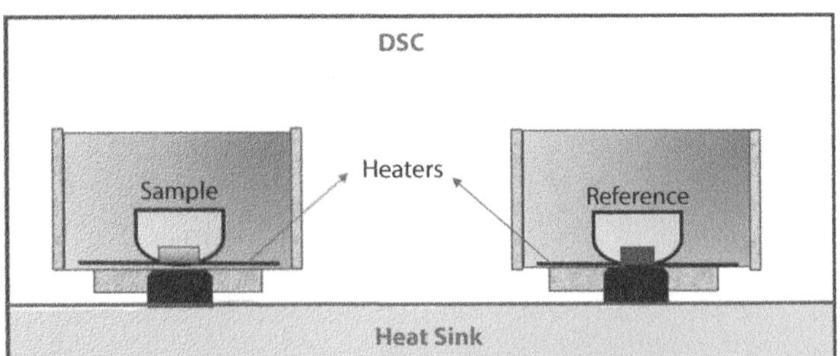

Figure 4.31 Power-compensated DSC.

to match the sample and the reference temperatures will be the amount of heat released or absorbed by the sample [36].

Ghiyasi *et al.* [65] performed a study in which improved adhesion (enhanced crosslinking) was achieved for epoxy adhesives using polyethylenimine modified silica. It was hypothesized that using nano-silica one can decrease the percolation threshold to a smaller percentage for the filler network to form. With this motivation, the authors prepared SiO_2 nanoparticles to be surface modified with polyethylenimine which plays a significant role because of its interaction with hydroxyl groups of SiO_2 through hydrogen bonds; also, it can react with epoxides. Moreover, the bulky structure of the polymer provides dispersion stability to SiO_2 nanoparticles. They are potential candidates in the application of structural adhesives because of their enhanced crosslinking property. Next, the compatibility of unmodified SiO_2 with the epoxy adhesive was examined which concluded that both PEI-modified SiO_2/epoxy and SiO_2/epoxy are potential candidates for structural adhesive applications.

TGA analysis of SiO_2 and SiO_2 - PEI nanoparticles was performed and is shown in Figure 4.32. Up to 100°C, the weight loss occurred because of the removal of moisture present in the sample. Next, removing physically adsorbed water molecules causes a step-like decrease in the thermogram between 100° C and 250° C. The decrease in the weight continues because of the condensation reaction of silanols. In the case of SiO_2 - PEI, the behaviour is similar below 100° C as for SiO_2, but with an increase in temperature, the rate of decrease of mass increases rapidly. The primary

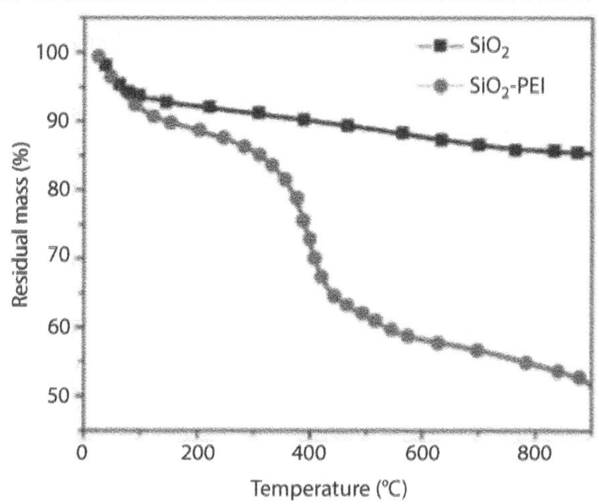

Figure 4.32 TGA thermograms of SiO_2 - PEI and SiO_2 nanoparticles. Reprinted with permission from [65], Copyright 2018, Elsevier.

decomposition reaction of PEI on the surface of SiO_2 occurs in the range of 280-430° C. The final residue of the SiO_2 - PEI is only 52% at the end of TGA analysis which is very low as compared to unmodified SiO_2 (85%). This supports the FTIR conclusion that the PEI is successfully adsorbed on the surface of silica nanoparticles.

DSC measurements were performed on the epoxy/SiO_2 - PEI and epoxy/SiO_2 at four different temperature increment rates of 2.5, 5, 7.5 and 10°C/min, as shown in Figure 4.33. In all cases, unimodal peaks were obtained, which confirmed the single-step kinetics of the reaction throughout the measurement.

Aran Ais *et al.* [66] performed an experiment in which they added diphenylmethane-4,4′-diisocyanate (MDI) and different mixtures of rosin acid in the synthesis process of thermoplastic polyurethane as an internal tackifier (added during the synthesis of TPU) in order to improve the adhesion to poly(vinyl chloride) (PVC). The details related to the synthesis are provided in Table 4.6. The equivalent units in Table 4.6 are

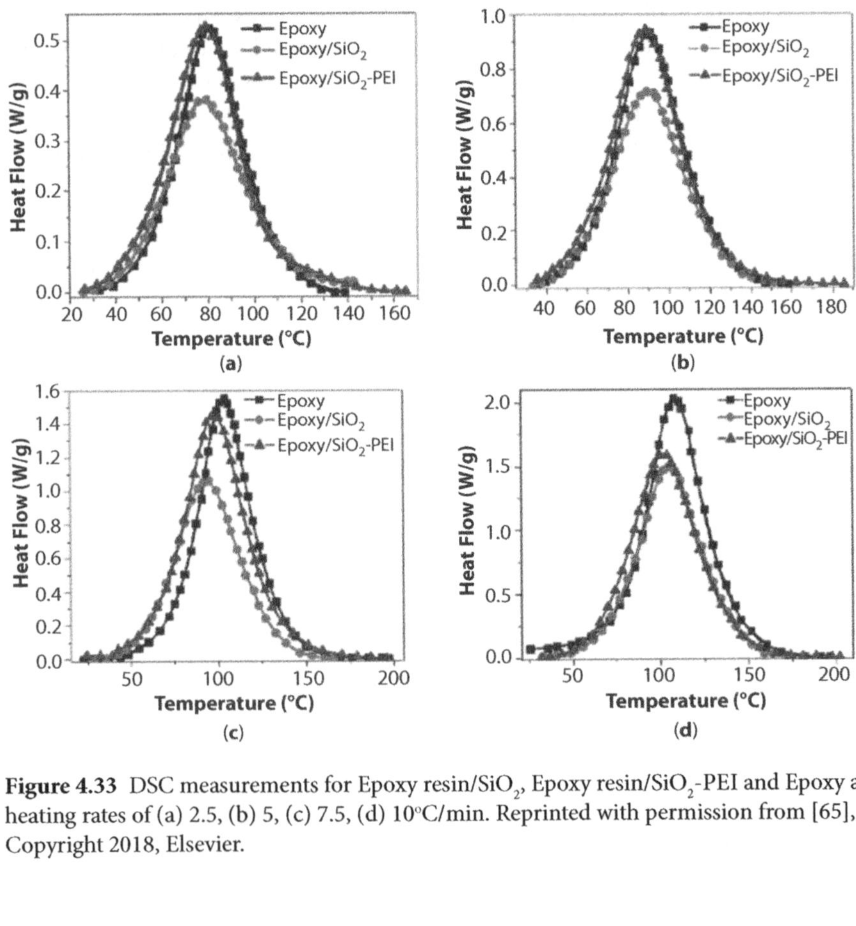

Figure 4.33 DSC measurements for Epoxy resin/SiO_2, Epoxy resin/SiO_2-PEI and Epoxy at heating rates of (a) 2.5, (b) 5, (c) 7.5, (d) 10°C/min. Reprinted with permission from [65], Copyright 2018, Elsevier.

Table 4.6 Nomenclature and hard segment content of thermoplastic PUs. Reproduced with permission from [66], Copyright 2005, Elsevier.

PU	MDI (equivalent)	Polyadipate of 1,4 butanediol (equivalent)	Rosin acid (equivalent)	1,4 butanediol (equivalent)	Hard segments (wt%)
M0	0.135	0.123	-	0.012	10.4
M5R	0.135	0.116	0.007	0.012	12.4
M10R	0.135	0.111	0.012	0.012	14.3
M20R	0.135	0.098	0.025	0.012	19.2

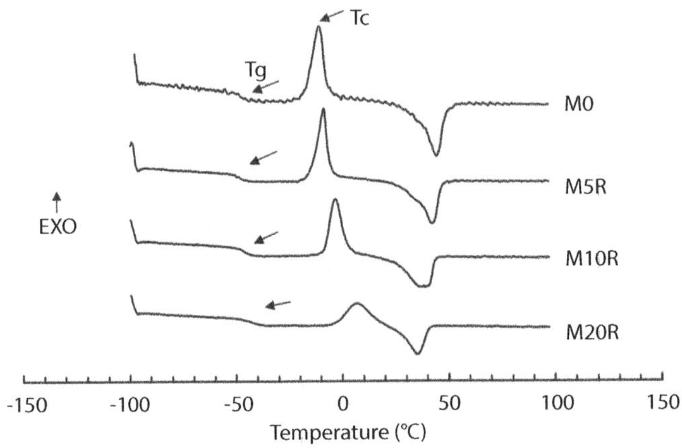

Figure 4.34 DSC curves of rosin acid-modified polyurethanes (crystallization of PUs); refer to Table 4.6 for nomenclature. Reprinted with permission from [66], Copyright 2005, Elsevier.

taken with respect to equivalents of OH groups present in macroglycol. Different amounts of rosin acid were incorporated during the thermoplastic polyurethane synthesis process, and nomenclature is provided in Table 4.6.

The DSC analysis of prepared thermoplastic polyurethanes is given in Figure 4.34. An increasing trend was observed in the crystallization temperature of the polymers with an increasing amount of rosin acid.

4.3.4 Mechanical Characterization

In the construction of large structures, joining one or more materials together is essential and structural adhesives play a significant role in this as they have found applications in automotive, construction, and electronic industries. The application of adhesives in joining processes helps eliminate mechanical fasteners, which are used to fabricate more complex shapes.

In order to ensure the optimal performance of adhesives, performing proper analysis and testing of mechanical performance of adhesives are crucial. This provides deep insights into the mechanical properties of the structural adhesives based on the mode of load application such as shear, impact, tensile or combinations thereof.

In industries, different functional testings are performed to evaluate properties of adhesives like impact testing, tack and green strength testing, and environmental testing. Usually, impact testing (ASTM D 950) data show that with the increase in the thickness of the adhesive layer, apparent strength also increases. Tack is the adhesive's crucial ability, which quantifies the adhesive's instantaneous bonding with the substrate under light pressure. Environmental testing usually includes the exposure of the adhesively-bonded areas to atmospheric variables like heat and moisture to estimate their performance for a longer duration of thousands of hours.

4.3.4.1 Tensile Test

Tensile loads come into play in adhesive design due to the bending of substrates or loading applied to the joint. Usually, an adhesive design is considered under tensile load when the stress experienced is perpendicular to the adhesive layer applied at the joint. The tensile strength is defined as the maximum tensile load at fracture divided by the cross-sectional area (Figure 4.35). With the determination of maximum possible tensile load P_C until fracture, one can calculate the tensile strength (σ_C) using (A is cross-sectional area):

$$\sigma_C = \frac{P_C}{A} \tag{4.7}$$

π-tensile test is very popular among researchers to study the bonding at wood-wood and metal-metal interfaces as explained in ASTM D897-78 [2].

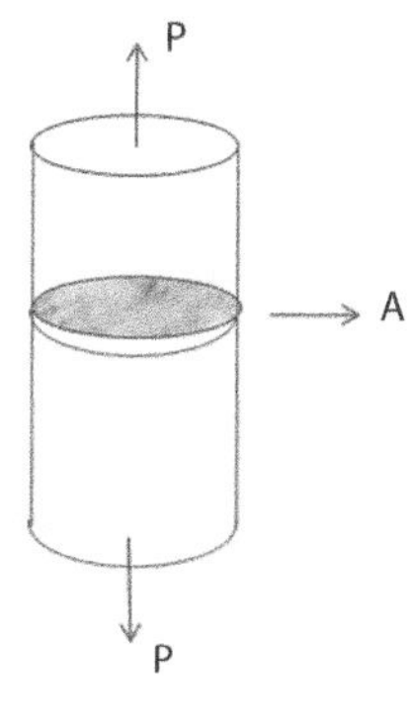

Figure 4.35 Tensile stress test. P: applied stress, A: cross-sectional area.

Yang *et al.* [67] studied diglycidyl ether of bisphenol-A (DGEBA) epoxy resin to improve mechanical properties at liquid nitrogen temperature (77 K). Significant improvement was observed in tensile strength, impact strength and ductility by appropriate addition of H30 (hydroxyl functionalized hyperbranched polymer) at 77 K. They found that tensile strength increased up to 17.7% and impact strength up to 26.3% by addition of 10 wt% H30.

4.3.4.2 *Lap Shear Test*

Adhesives are very efficient in handling the shear load as compared to tensile load or peel load. Under the shear load, the applied load acts along the same plane as the adhesive layer. The lap shear test assesses the strength of the adhesive joint to resist the in-plane forces and stresses caused by the external stimuli.

In the lap shear test, two adherends are overlapped with each other up to a certain length in such a way that the adhesive forms a layer in the overlapped region of the two adherends. This layer bonds the two specimens together once the adhesive is cured as per instructions. Then the specimen obtained is placed in the grip of a lap shear testing machine and pulled apart at different rates until failure. The test provides the minimum and maximum values of the failure load and the type of failure. Figure 4.36 shows the lap shear test.

For metallic adherends, ASTM D1002-72 is the popular lap shear test in which two standard metallic adherends with 0.064-inch thickness are used with an overlap length of 0.5 inches [2].

Other shear tests are available, namely the compressive and torsional shear tests, which are frequently used for mechanical analysis.

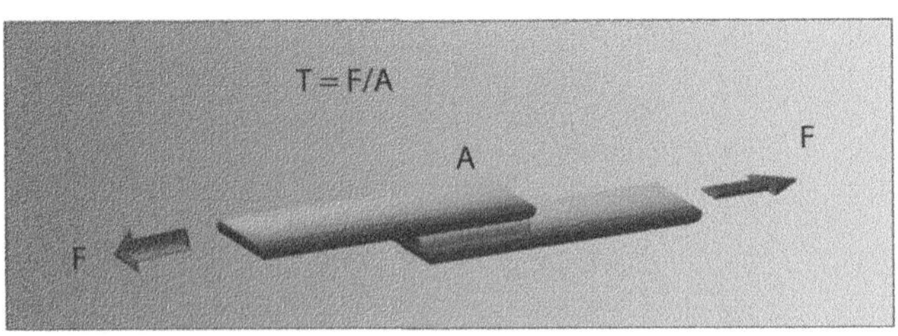

Figure 4.36 Single lap shear test; F: max load, A: area of joint and T: lap shear strength.

4.3.4.3 *Dynamic Mechanical Analysis (DMA)*

DMA is a technique that is used for the measurements of viscoelastic properties of resins and polymers. DMA primarily records the response of the material to a given periodic oscillating force at given environmental conditions (e.g., temperature variation). DMA measurements enable one to measure and analyse the viscoelastic properties of polymers as a function of temperature and frequency.

DMA primarily deals with the stress and strain generated in the material and analyses the kinetic properties. Also, creep recovery and dynamic testing are essential features of DMA [68].

As shown in Figure 4.37, a periodic force is applied to the sample, and deformation is recorded as a function of applied force. Stress and strain generated in the sample as a function of temperature can also be determined using a temperature-controlled furnace. Various deformation modes are available in DMA measurements such as tension, compression, shear, dual cantilever bending, and 3-point bending. The properties which can be measured with DMA are storage modulus, loss modulus, loss tangent and their behaviours as a function of frequency and temperature.

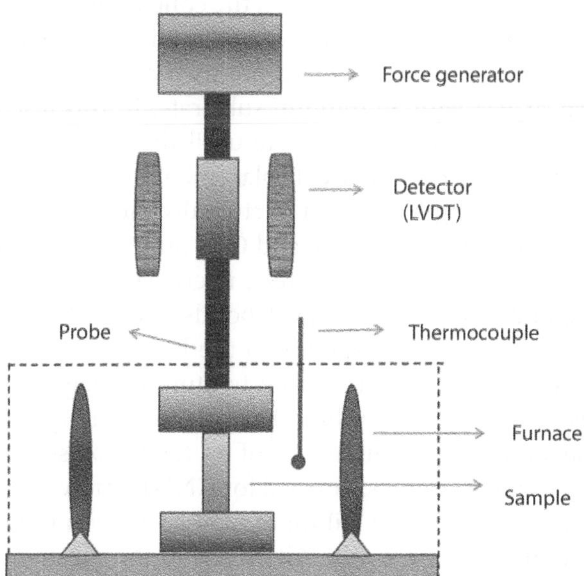

Figure 4.37 Various components of DMA equipment.

Trovati *et al.* [69] synthesized two different types of polyurethane structures using polycondensation, namely rigid PU resin and semi-rigid PU resin with mass proportions of 1:1 and 1:1.5, respectively. DMA analysis showed higher storage modulus in rigid PU and higher dissipation energy in semi-rigid PU.

Mechanical characterization techniques play an essential role in the characterization of structural adhesives. In real-life applications, to decide appropriate adhesives, properties of adhesives such as tensile strength, creep and fatigue resistance, hardness, curing requirements, and flexibility are very important. Structural adhesives, e.g., epoxies have applications in automotive industry, and good mechanical performance is expected from epoxy resins. Pruksawan *et al.* [57] found that incorporating cellulose nanocrystal (CNC) aggregates drastically improved the mechanical properties such as adhesion strength and fracture toughness of epoxy resin. Epoxy adhesives containing CNC aggregates demonstrated fracture toughness of 389 J/m^2 and adhesion strength of 29 MPa, and both values are higher compared to the virgin epoxy adhesive. Here authors implemented two different methods for dispersion of CNCs in epoxy adhesive, namely ball milling and solvent exchange method. The grinding process inside the ball milling machine was carried out for 15 and 90 minutes. The nomenclature for samples was followed according to the type of cellulose (CNC or carbon nano-fibre (CNF)), the content of cellulose (10, 20 or 40 phr), dispersion method and time of ball milling process. DGEBA and 4,4′-diaminodiphenylmethane (DDM) are used as models for analysis in a single lap shear test with aluminium substrate. In this test, shear stress was applied to the aluminium substrate until the joint eventually broke, and adhesion strength and fracture displacement of joints were measured. Values of adhesion strength (σ_{ad}) and fracture displacement (ΔL_f) for epoxy adhesive turned out to be 13.5 MPa and 0.16 mm, respectively when failure mode was analysed visually. Here it is crucial to note that cohesive failure corresponds to the breaking of bonds within the adhesive layer, whereas adhesion failure corresponds to the bond breaking at the interface of adhesive and the substrate. The fracture propagated through the bulk for epoxy adhesive, concluding that adhesion strength was stronger than cohesion strength. Measurements of σ_{ad} (adhesion strength) and ΔL_f (fracture displacement) were performed for CNF-reinforced epoxy and are shown in Figure 4.38 (a). The failure of CNF20-SE was visible and indicated a weak bonding between the adhesive and the substrate. In the case of CNC aggregates, a uniform distribution within epoxy was observed, and both CNC20-B0 and CNC20-B90 showed higher values of σ_{ad} and ΔL_f as shown in Figure 4.38 (b). Among all the prepared samples, CNC20-B90

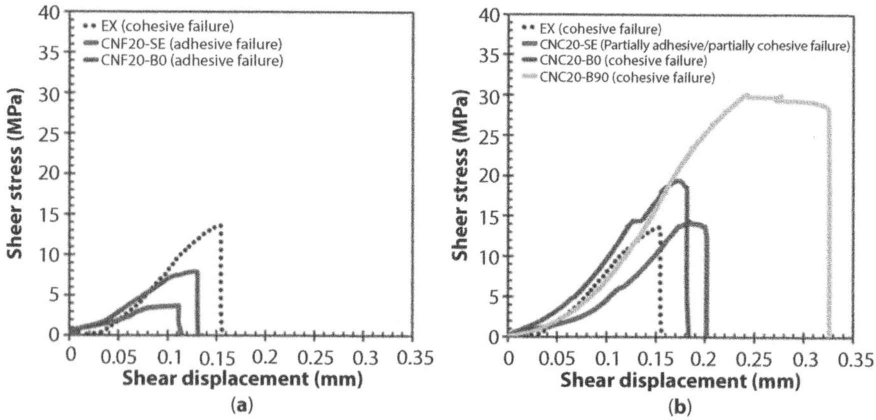

Figure 4.38 Single lap shear test plots for (a) CNF-epoxy adhesives (b) CNC-epoxy adhesives. EX adhesive: epoxy adhesive with DGEBA and DDM. CNF20-SE and CNF20-B0: CNF reinforced epoxy adhesive. CNC20-SE, CNC20-B0/B90: CNC reinforced epoxy adhesive. Reprinted with permission from [57], Copyright 2020, American Chemical Society.

showed the highest value of σ_{ad} (29.2MPa). Coincidentally, the viscosity of CNC-epoxy was low, which helped to attain adequate adhesion with the substrate.

For further analysis, a tensile test and DMA were performed. During the tensile test, fracture strain, tensile strength and tensile modulus were calculated for CNC-epoxy adhesives, as shown in Figure 4.39 (a). Evidently, CNC has high values of E (tensile modulus) and σ (tensile strength), and hence the addition of CNCs in epoxy increased the values of E and σ by 29-41% and 28-35%, respectively. DMA analysis showed that storage modulus increased by 24-35% when CNC was incorporated into epoxy, as shown in Figure 4.39 (b), which is consistent with the tensile test results. A study of linear expansion coefficient was carried out using TMA to understand the dimension variation as a function of temperature, as shown in Figure 4.39 (c). It is necessary to understand the behaviour of the linear expansion coefficient because only a slight difference is desired between the coefficients of adhesive and substrate for proper bonding at the interface. It is also observed that on addition of CNC in epoxy, a decrease in linear coefficient of about 14% was observed, which is expected because of the low value of α (linear expansion coefficient) for CNCs. Similarly, DSC analysis was performed to calculate the glass transition temperature in Figure 4.39 (d) and high T_g values indicate thermal stability of the adhesives.

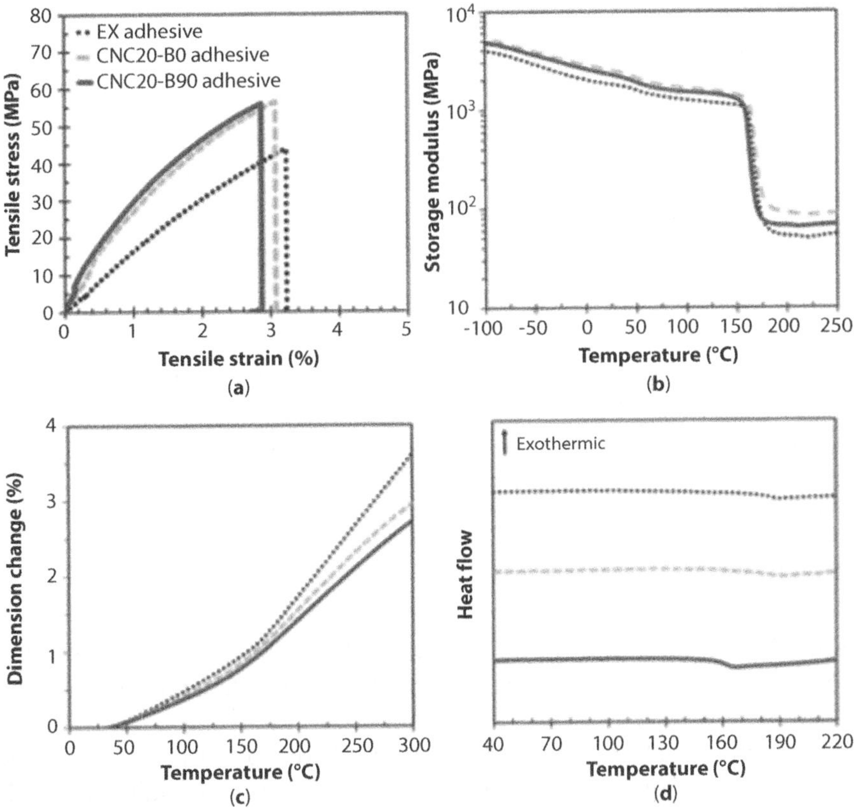

Figure 4.39 (a) Plots of tensile stress vs strain for adhesives, (b) Storage modulus profile of adhesives calculated using DMA, (c) Dimension variations with respect to the temperature, (d) DSC analysis of adhesives [57]; EX adhesive: epoxy adhesive with DGEBA and DDM (no nano cellulose) CNC B0 and CNC-B90: cellulose reinforced adhesives. Reprinted with permission from [57], Copyright 2020, American Chemical Society.

Now, to summarize our chapter on characterization techniques, Table 4.7 contains information about properties and applications of various structural adhesives discussed. Table 4.8 contains all the characterization techniques discussed along with their specific applications.

Table 4.7 Different structural adhesives with their properties and applications [2].

Structural adhesive	Properties	Applications	Ref(s)
Phenolic	Resistance against harsh environment, strong mechanical properties, good processing performance, low cost, resistance to chemicals, high adhesion strength	Electronics industries	[2]
Epoxy	Low shrinkage, high strength, electrical insulation, low toxicity and low cost, excellent adhesion to the substrate, superior mechanical properties, high resistance to osmosis	Fibreglass repairs, carpentry and woodworking, wood and metal fillers, reinforced bolts, jewellery making	[2, 4, 8]
Polyurethane	Good bond durability, substantial adhesion to a variety of substrates, good chemical resistance, high peel strength, formulation flexibility, high impact resistance	Bonding of structural plastics, manufacturing of particle boards, automotive industry	[2, 70]
Acrylic	Room temperature cure, free radical cure, low oxygen sensitivity, high shear strength, high chemical resistance, high peel strength, purported to be tolerant of various types of substrate surface contaminants	Plastic toys, electronic components bonding, magnets bonding jewellery, golf clubs	[2, 3]

(Continued)

Table 4.7 Different structural adhesives with their properties and applications [2]. (*Continued*)

Structural adhesive	Properties	Applications	Ref(s)
Cyanoacrylate	Rapid curing at room temperature, high bond strength, capable of joining a variety of substrates, lower cost as compared to other structural adhesives, high resistance to organic liquids having low and medium polarity, 100% reactive, storage-stable	Aircraft interior parts, bumper pads on automobiles, optical lenses, appliance nameplates, camera components, rubber stamps, vacuum cleaners, pens	[2, 71]
Silicone	Having a lower surface tension value makes silicones capable of wetting themselves which helps in better film formation, low glass transition temperature, high percentage of free volume compared to hydrocarbons	Used as elastomers for insulation purposes because of their thermal ageing resistance, as coating materials because of their chemical and weathering resistance, and as structural adhesives because of their good adhesion property	[17]

Table 4.8 Characterization techniques with their applications.

Characterization technique	Applications
EDX	Elemental analysis
FTIR	Identification and analysis of chemical composition of material
XPS	Elemental analysis and oxidation state
GLC	Identification of volatile components in the sample
NMR	Identification of polymers and characterization of constitutional, conformational and structure of polymers
Raman Spectroscopy	Determination of phase, chemical structure, and crystallinity
Contact Angle	Determination of surface free energy, wettability
SEM	Surface morphology
Gelation Time	Study of curing behaviour
SAXS	Particle size measurement
AFM	Study of surface morphology in 3D
TGA	Analysis of mass change w.r.t. temperature
DSC	Determination of glass transition temperature
Mechanical Tests	Measurement of tensile strength, toughness

4.4 Summary

Structural adhesives play an essential role in this technological era having numerous applications in major sectors like biomedical, electronics, construction, aviation and automotive. In the last few decades, researchers have invented efficient and more advanced structural adhesives possessing enhanced mechanical, thermal, chemical, and electrical properties. Characterization of previously developed adhesives with advanced techniques has helped immensely to gain insights into the adhesion mechanism and enabled the scientific community to invent improved versions of structural adhesives. This chapter provided a detailed outline with examples of many advanced characterization techniques used for the analysis of structural adhesives. For surface analysis, discussion of techniques like SEM and AFM; for thermal analysis, one can employ techniques like DSC and TGA; for chemical characterization, one can use NMR and FTIR, which finally has led to improvement in the adhesive properties and eventually in the application of adhesives in industries. Finally, for mechanical characterization, techniques like tensile test and peel test are mentioned with ASTM standards. These characterization techniques help in an in-depth analysis of the sample with all the important details, and hence they have a wide range of applications in all technological sectors.

Acknowledgements

The authors are thankful to Dr. C. P. Ramanarayanan, Vice-Chancellor of DIAT (DU), Girinagar, Pune, and Department of Metallurgical and Materials Engineering and DIAT (DU) for the support and the encouragement provided.

References

1. S. Ebnesajjad and A. H. Landrock, Joint design, in: *Adhesive Technology Handbook*, S. Ebnesajjad and A. H. Landrock (Eds.), second edition, pp.159-181, William Andrew, Norwich, NY (2009).
2. S. R. Hartshorn, *Structural Adhesives: Chemistry and Technology*, Springer US, Boston (MA)(1986).
3. E. J. C. Kellar, Key issues in selecting the right adhesive, in: *Advances in Structural Adhesive Bonding*, D. Dillard (Ed.), pp. 3-19, Woodhead Publishing, Cambridge (UK) (2010).
4. D.J. Dunn, *Engineering and Structural Adhesives*, Rapra Review Reports 15 (2004).
5. Y. Xu, L. Guo, H. Zhang, H. Zhai, and H. Ren, Research status, industrial application demand and prospects of phenolic resin. RSC Advances 9, 28924–28935 (2019).
6. H.J. Fabris and W.G. Knauss, Synthetic polymer adhesives, in: *Comprehensive Polymer Science and Supplements*, Vol. 7, G. Allen and J. Bevington (Eds.), pp. 131-177, Elsevier (1989).
7. P. Zuo and A.P. Vassilopoulos, Review of fatigue of bulk structural adhesives and thick adhesive joints. Intl. Materials Reviews 66, 313–338 (2021).
8. G.S. Baş and E. Sancaktar, Mechanical behavior of toughened epoxy structural adhesives for impact applications. Chem Engineering 4, 1-43 (2020).

9. K. Hodd, Epoxy resins, in: *Comprehensive Polymer Science and Supplements*, Vol. 5, G. Allen and J. Bevington (Eds.), pp. 667-699, Elsevier (1989).
10. D. Montarnal, F. Tournilhac, M. Hidalgo, and L. Leibler, Epoxy-based networks combining chemical and supramolecular hydrogen-bonding cross-links. J. Polym. Sci. Part A: Polym. Chem 48, 1133–1141 (2010).
11. H.M.C.C. Somarathna, S.N. Raman, D. Mohotti, A.A. Mutalib, and K.H. Badri, The use of polyurethane for structural and infrastructural engineering applications: A state-of-the-art review. Constr. Build. Mater. **190**, 995–1014 (2018).
12. T.E. Youssef, H. Al-Turaif, and A.A. Wazzan, Investigations on the structural and mechanical properties of polyurethane resins based on Cu(II) phthalocyanines. Int. J. Polym. Sci., 1-10 (2015).
13. M. Gong, S. Delahunty, Y.H. Chui, and L. Li, Use of low grade hardwoods for fabricating laminated railway ties. Constr. Build. Mater. 41, 73-78 (2013).
14. E. Sudoł, and E. Kozikowska, Mechanical properties of polyurethane adhesive bonds in a mineral wool-based external thermal insulation composite system for timber frame buildings. Materials 14, 1-19 (2021).
15. D.A. Aronovich, and L.B. Boinovich, Structural acrylic adhesives: A critical review, Rev. Adhesion Adhesives 9, 65-122 (2021).
16. B. Burns, Cyanoacrylates: Towards high temperature resistant instant adhesives. A critical review. Rev. Adhesion Adhesives 5, 361-390 (2017).
17. F. de Buyl, Silicone sealants and structural adhesives. Int. J. Adhesion Adhesives 21, 411–422 (2001).
18. M. Scimeca, S. Bischetti, H.K. Lamsira, R. Bonfiglio, and E. Bonanno, Energy dispersive X-ray (EDX) microanalysis: A powerful tool in biomedical research and diagnosis. European J. Histochemistry 62, 1-10 (2018).
19. J.D. Andrade, X-ray photoelectron spectroscopy (XPS), in: *Surface and Interfacial Aspects of Biomedical Polymers*, J.D. Andrade (Ed.), pp. 105-195, Springer US, Boston (MA) (1985).
20. F.A. Stevie, and C.L. Donley, Introduction to x-ray photoelectron spectroscopy. J. Vac. Sci. Technol. A 38, 063204 (2020).
21. M. Horgnies, E. Darque-Ceretti, H. Fezai, and E. Felder, Influence of the interfacial composition on the adhesion between aggregates and bitumen: Investigations by EDX, XPS and peel tests. Int. J. Adhesion Adhesives 31, 238–247 (2011).
22. S.G. Kazarian and K.L.A. Chan, ATR-FTIR spectroscopic imaging: Recent advances and applications to biological systems. Analyst 138, 1940-1951 (2013).
23. C. Berthomieu and R. Hienerwadel, Fourier transform infrared (FTIR) spectroscopy. Photosynthesis Res. 101, 157–170 (2009).
24. R. Valand, S. Tanna, G. Lawson, and L. Bengtström, A review of Fourier transform infrared (FTIR) spectroscopy used in food adulteration and authenticity investigations. Food Additives Contaminants: Part A 37, 19–38 (2020).

25. M.K. Bin Bakri, M.R. Rahman, P.L.N. Khui, E. Jayamani, and A. Khan, Use of sustainable polymers to make green composites, in: *Advances in Sustainable Polymer Composites*, M. R. Rahman (Ed.), pp. 109-129, Elsevier (2021).
26. M.A. Al-Ghouti, Y.S. Al-Degs, and M. Amer, Application of chemometrics and FTIR for determination of viscosity index and base number of motor oils. Talanta 81, 1096–1101 (2010).
27. M. Mohammadi, M. Khanmohammadi Khorrami, H. Ghasemzadeh, P. Noor, and S. Zandbaaf, Artificial neural network for quantitative and qualitative determination of the viscosity of nanofluids by ATR-FTIR spectrometry. Infrared Phys. Technol. 118, 103900 (2021).
28. J. Park, M. Kim, Y. Lee, H.S. Lee, and H. Ko, Nanomaterials: Fingertip skin-inspired microstructured ferroelectric skins discriminate static/dynamic pressure and temperature stimuli. Science Advances 1, 1-13 (2015).
29. M.Q. dos Reis, M.D. Banea, L.F.M. da Silva, and R.J.C. Carbas, Mechanical characterization of a modern epoxy adhesive for automotive industry. J. Brazilian Soc. Mech Sci. Eng. 41, 340 (2019).
30. M.S.A. Bhuiyan, B. Liu, J. Manuel, B. Zhao, and B.P. Lee, Effect of conductivity on in situ deactivation of catechol–boronate complexation-based reversible smart adhesive. Biomacromolecules 22, 4004-4015 (2021).
31. J.N. Myers, C. Zhang, K.W. Lee, J. Williamson, and Z. Chen, Hygrothermal aging effects on buried molecular structures at epoxy interfaces. Langmuir 30, 165–171 (2014).
32. R. Gomatam and K.L. Mittal (Eds.) *Electrically Conductive Adhesives*, CRC Press, Boca Raton, FL (2008).
33. P.T. Wilson, L.J. Richter, W.E. Wallace, K.A. Briggman, and J.C. Stephenson, Correlation of molecular orientation with adhesion at polystyrene/solid interfaces. Chem. Phys. Letters 363, 161–168 (2002).
34. S. Luo and C.P. Wong, Influence of temperature and humidity on adhesion of underfills for flip chip packaging, in: *2001 Proceedings. 51st Electronic Components Technology Conference (Cat. No.01CH37220)*, pp. 155–162, IEEE (2001).
35. W. Zhou and J. Cai, Mechanical and dielectric properties of epoxy resin modified using reactive liquid rubber (HTPB). J. Appl. Polym. Sci. 124, 4346–4351 (2012).
36. D. A. Skoog, F. J. Holler and S. R. Crouch, *Principles of Instrumental Analysis*, seventh edition, Cengage Learning, Boston (2016).
37. J.W. Yan, C. Hu, L.H. Tong, Z.X. Lei, and Q.B. Lin, Migration test and safety assessment of polyurethane adhesives used for food-contact laminated films. Food Packaging Shelf Life 23, 100449 (2020).
38. Atta-Ur-Rahman, *Nuclear Magnetic Resonance: Basic Principles*, first edition, Springer-Verlag, New York (1986).
39. Z. Wang, H. Kang, H. Liu, S. Zhang, C. Xia, Z. Wang, and J. Li, Dual-network nanocross-linking strategy to improve bulk mechanical and water-resistant

adhesion properties of bio-based wood adhesives. ACS Sustainable Chem. Eng. 8, 16430–16440 (2020).
40. J. Liu, L. Wang, J. Li, C. Li, S. Zhang, Q. Gao, W. Zhang, and J. Li, Degradation mechanism of Acacia mangium tannin in NaOH/urea aqueous solution and application of degradation products in phenolic adhesives. Int. J. Adhesion Adhesives 98, 102556 (2020).
41. S.P. Mulvaney and C.D. Keating, Raman spectroscopy. Anal Chem 72, 145–158 (2000).
42. J. Chen, Y. Huang, P. Kannan, L. Zhang, Z. Lin, J. Zhang, T. Chen, and L. Guo, Flexible and adhesive surface enhance Raman scattering active tape for rapid detection of pesticide residues in fruits and vegetables. Anal Chem 88, 2149–2155 (2016).
43. J. Tatar, N.R. Brenkus, G. Subhash, C.R. Taylor, and H.R. Hamilton, Characterization of adhesive interphase between epoxy and cement paste via Raman spectroscopy and mercury intrusion porosimetry. Cement Concrete Composites 88, 187–199 (2018).
44. K.L. Mittal, The role of the interface in adhesion phenomena. Polym. Eng. Sci. 17, 467–473 (1977).
45. K.L. Mittal, Adherate: A new and general term, J. Adhesion 6, 377–378 (1974).
46. A W. Adamson and A P. Gast, *Physical Chemistry of Surfaces*, sixth edition, Wiley (1997).
47. A. West, Experimental methods to investigate self-assembly at interfaces, in: *Interface Science and Technology*, Vol. 21, V.Ball (Ed.), pp. 131–241, Elsevier (2018).
48. H.Y. Erbil and C.E. Cansoy, Range of applicability of the Wenzel and Cassie–Baxter equations for superhydrophobic surfaces. Langmuir 25, 14135–14145 (2009).
49. M.Y. Wen, J.Y. Shi, and H.J. Park, Dynamic wettability and curing characteristics of liquefied bark-modified phenol formaldehyde resin (BPF) on rice straw surfaces. J. Wood Sci. 59, 262–268 (2013).
50. T. Klein, E. Buhr, and C. Georg Frase, TSEM: A review of scanning electron microscopy in transmission mode and its applications, in: *Advances in Imaging and Electron Physics*, Vol. 171, P. Hawkes (Ed.), pp. 297–356, Elsevier (2012).
51. R.O. Ebewele, B.H. River, and J.A. Koutsky, Relationship between phenolic adhesive chemistry and adhesive joint performance: Effect of filler type on fraction energy. J. Appl. Polym. Sci. 31, 2275–2302 (1986).
52. A.A. Shimkin, Methods for the determination of the gel time of polymer resins and prepregs. Russian J. General Chem. 86, 1488-1493 (2016).
53. W. Huang, Y. Zhang, Y. Yu, and Y. Yuan, Studies on UV-stable silicone–epoxy resins. J. Appl. Polym. Sci. 104, 3954–3959 (2007).
54. S. Sarkar and B. Adhikari, Lignin-modified phenolic resin: synthesis optimization, adhesive strength, and thermal stability. J. Adhesion Sci. Technol. 14, 1179–1193 (2000).

55. A.G. Kikhney and D.I. Svergun, A practical guide to small angle X-ray scattering (SAXS) of flexible and intrinsically disordered proteins. FEBS Letters 589, 2570–2577 (2015).
56. G. Fagherazzi, Small angle X-ray scattering, Acta Cryst. Sec. A 39, 500 (1983).
57. S. Pruksawan, S. Samitsu, Y. Fujii, N. Torikai, and M. Naito, Toughening effect of rodlike cellulose nanocrystals in epoxy adhesive. ACS Appl. Polym. Mater. 2, 1234–1243 (2020).
58. S. Mourdikoudis, R.M. Pallares, and N.T.K. Thanh, Characterization techniques for nanoparticles: comparison and complementarity upon studying nanoparticle properties. Nanoscale 10, 12871–12934 (2018).
59. D. Johnson, and N. Hilal, Characterisation and quantification of membrane surface properties using atomic force microscopy: A comprehensive review. Desalination 356, 149–164 (2015).
60. E. Arikan, J. Holtmannspötter, F. Zimmer, T. Hofmann, and H.J. Gudladt, The role of chemical surface modification for structural adhesive bonding on polymers - Washability of chemical functionalization without reducing adhesion. Int. J. Adhesion Adhesives 95, 102409 (2019).
61. F.M. Etzler, Determination of the surface free energy of solids. Rev. Adhesion Adhesives 1, 3–45 (2013).
62. F.M. Etzler, Characterization of surface free energies and surface chemistry of solids, in: *Contact Angle, Wettability and Adhesion*,Vol. 3, K.L. Mittal (Ed.), pp. 219–264, CRC Press, Boca Raton, FL (2003).
63. L. F. G. Herrera and H. Price, Thermogravimetric analysis (TGA). Chemistry Libre Texts. https://chem.libretexts.org/@go/page/214241 (2020).
64. M.E. Brown, *Introduction to Thermal Analysis*, second edition, Kluwer Academic Publishers, Dordrecht (2004).
65. S. Ghiyasi, M.G. Sari, M. Shabanian, M. Hajibeygi, P. Zarrintaj, M. Rallini, L. Torre, D. Puglia, H. Vahabi, M. Jouyandeh, F. Laoutid, S.M.R. Paran, and M.R. Saeb, Hyperbranched poly(ethyleneimine) physically attached to silica nanoparticles to facilitate curing of epoxy nanocomposite coatings. Prog. Organic Coatings 120, 100–109 (2018).
66. F. Arán Aís, A.M. Torró-Palau, A. César Orgilés-Barceló, and J.M. Martín-Martínez, Addition of rosin acid during thermoplastic polyurethane synthesis to improve its immediate adhesion to PVC. Int. J. Adhesion Adhesives 25, 31–38 (2005).
67. J.P. Yang, Z.K. Chen, G. Yang, S.Y. Fu, and L. Ye, Simultaneous improvements in the cryogenic tensile strength, ductility and impact strength of epoxy resins by a hyperbranched polymer. Polymer 49, 3168–3175 (2008).
68. P. Raja, and A. Barron, Dynamic Mechanical Analysis. (2021). https://chem.libretexts.org/@go/page/55857
69. G. Trovati, E.A. Sanches, S.M. de Souza, A.L. dos Santos, S.C. Neto, Y.P. Mascarenhas, and G.O. Chierice, Rigid and semi rigid polyurethane resins: A structural investigation using DMA, SAXS and Le Bail method. J. Molecular Structure 1075, 589–593 (2014).

70. B. Burchardt, Advances in polyurethane structural adhesives, in: *Advances in Structural Adhesive Bonding*, D.A. Dillard (Ed.), pp. 35-65, Elsevier (2010).
71. K.L. Shantha, S. Thennarasu, and N. Krishnamurti, Developments and applications of cyanoacrylate adhesives. J. Adhesion Sci. Technol. 3, 237–260 (1989).

5

The Effects of Shear and Peel Stress Distributions on the Behavior of Structural Adhesives in Tubular Composite Joints

Mohammad Shishesaz

Engineering School, Department of Mechanical Engineering, The Shahid Chamran University of Ahvaz, Ahvaz, Iran

Abstract

This chapter is concerned with the study of peel and shear stress distributions in adhesively bonded composites under external loads. It includes the design aspects of an adhesively bonded joint including the effects of mechanical and geometric properties of the adherends, type of adhesively bonded joint, and the shapes of a few adhesively bonded joints on stress distribution in the adhesive layer of tubular joints. Furthermore, the application of torsional and axial loads on the induced stresses is examined to introduce their effects on the adhesive stresses. Additionally, the effect of local damages in terms of voids or debonds in the adhesive layer as well as delamination in the composite adherends are discussed to determine their adverse effect in terms of adhesive shear and peel stresses on the joint performance. Moreover, the use of hybrid joints is surveyed to explore the benefits of combined mechanical and bonded joints on the adhesive layer performance. In contrast to the linear analysis, the nonlinear analysis is also introduced to seek the effect of adhesive nonlinear behavior on the induced stresses, affecting the overall joint performance. Furthermore, a short survey is presented on design aspects of tubular joints to better understand the effect of geometric and mechanical properties of the adhesive layer and adherends on the overall performance of the tubular joints.

Keywords**:** Composite pipe, tubular joint, adhesive layer, stress distribution, design aspects, adhesion failure

Email: mshishehsaz@scu.ac.ir

K.L. Mittal and S.K. Panigrahi (eds.) Structural Adhesives: Properties, Characterization and Applications, (193–254) © 2023 Scrivener Publishing LLC

5.1 Introduction

5.1.1 A Brief Review of Loads (Stresses) and Failure of Adhesively Bonded Tubular Composite Joints

Adhesives have been widely used in a variety of ways for joining parts together [1, 2]. One major concern in using adhesively bonded joints is the magnitude of peak stresses that develop in the adhesive layer.

In general, adhesively bonded tubular composite joints can be classified as follows:

1. Adhesively bonded metal to metal pipes (tubes) with different properties.
2. Adhesively bonded metal to non-metal pipes (tubes) (or non-metal to non-metal pipes) with different properties).
3. Adhesively bonded metal to laminated composite pipes (tubes).
4. Adhesively bonded two laminated composite pipes (tubes).

In these cases, the two pipes (tubes) can be joined by a layer of adhesive forming a lap joint (see Figure 5.1) or a socket joint of any shape as shown in Figure 5.2. However, other shapes may be used to transfer the mechanical and thermal loads via the joint from one adherend to the other (see Figure 5.3). These joints may experience axial or torsional loads, as well as bending moments (or their combinations). Depending on the type of applied loads, typical stresses are developed in the adhesive layer that, if not controlled, may result in different types of adhesion failures [1–3]. These failures can be caused by progression of stresses beyond the adhesion strength values, due to a separation at the adherend-adhesive interface in the coupling region [4], presence of an initial defect or a crack in the adhesive layer, or a delamination in the adherend. In such cases, different approaches are adopted to predict the bond strength, as well as how to strengthen the joint and control the advancement of stresses in the adhesive layer [5, 6]. Although in most cases the adhesive behavior is assumed to be linear elastic, yet in some cases a nonlinear behavior must be considered for its response to the applied loads due to its polymeric nature [7, 8].

One of the major factors influencing stress distribution in the adhesive layer of a composite joint with laminated adherends (tubes) is the layer arrangement in each adherend. A few studies have focused on how a delamination between these layers can influence the overall stress distribution in the adhesive layer [9, 10] that eventually may cause the joint failure. Furthermore,

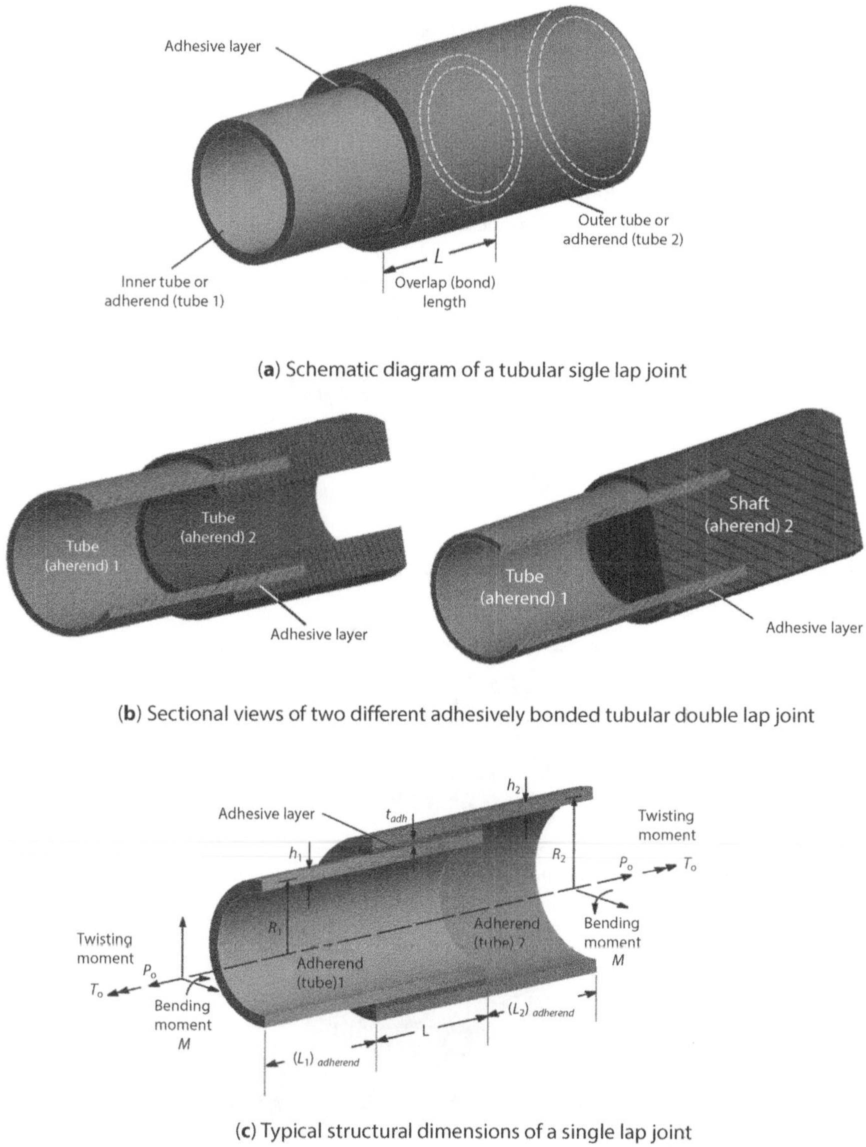

Figure 5.1 Schematic diagrams of two different adhesively bonded tubular lap joints with typical structural dimensions for a single lap joint.

for pipes in torsion, the improper stacking sequence of the adherends layers bonded by a layer of defective adhesive, can increase the torsional stresses in the joint, and hence leading to its failure. This issue has been addressed in [11] where the adhesive layer suffers from a circular void or debond.

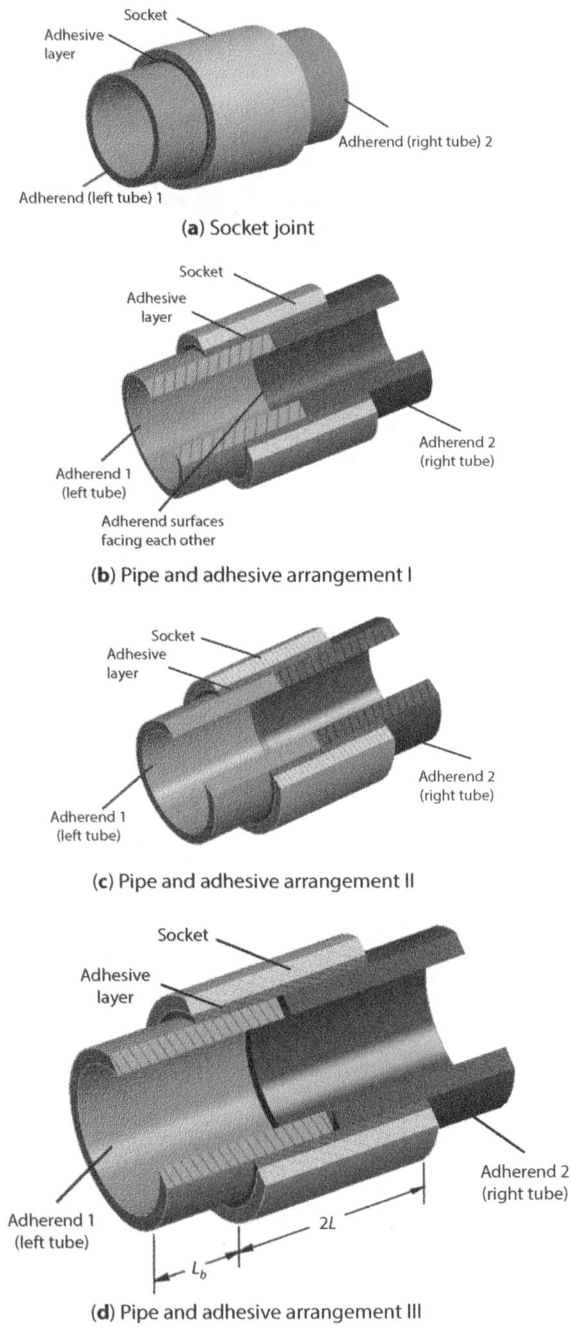

Figure 5.2 Schematic diagrams of adhesively bonded socket joints with different arrangements for the adhesive layer and adherends.

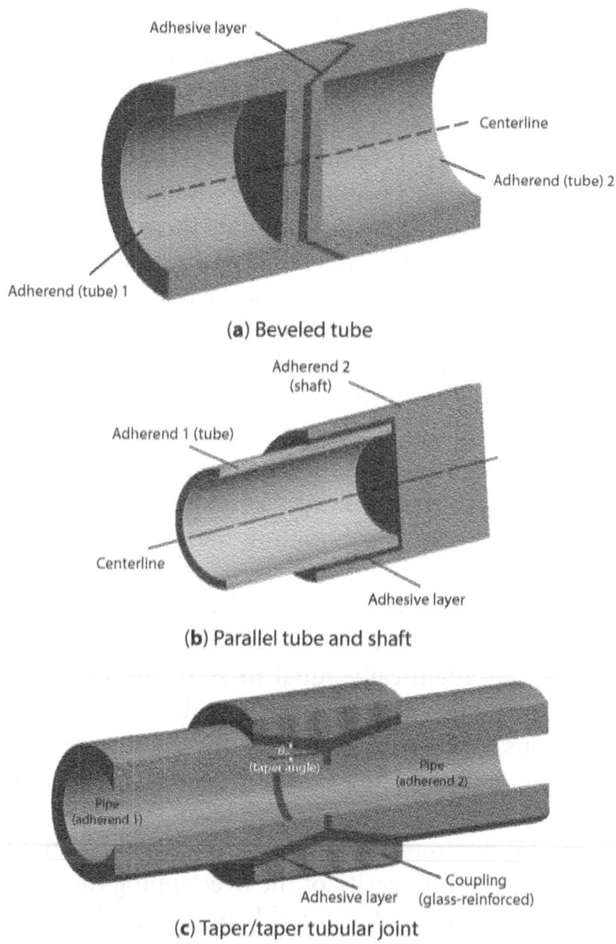

Figure 5.3 Sectional views of additional possible adhesively bonded tubular joints.

Based on these arguments, the next section briefly introduces the typical relations with which the stress components can be calculated in the adhesive layer. Additionally, the influences of these stresses on the overall performance of the joint are surveyed in other sections.

As a prerequisite, one has to know that extensive research on fiber-reinforced polymer (FRP) composite pipes has shown that FRP offers many advantages over the traditional materials such as steel in their use for construction of lightweight structural components. Moreover, corrosion resistance, dimensional tolerance, strength, abrasion resistance, and many other factors are additional advantages that make FRP suitable for construction of pipes usually joined by an adhesive layer. However, the cost of basic constituents, quality assurance, and testing, as well as their non-isotropic properties (that are a function of layup sequence and fiber orientation) are considered to be some of their main drawbacks. Some of the main barriers to the extensive use of composite pipes (tubes) have been attributed to the insufficient knowledge on stress distribution, design procedures, and proper standards. For this reason, to eliminate these barriers, a full knowledge of stress distribution in the adherends and adhesive layer is necessary.

According to many investigations, for some laminated composite joints in torsion, the shear stress is considered as the main stress component influencing the performance of the adhesive layer [5, 9, 11–16]. Many investigations have shown that the bond length and other mechanical and geometrical properties of the adherends and adhesive layer affect this stress component [13, 17–20]. These factors are considered to be the main factors influencing the adhesive layer failure. Consequently, for a safe design, a full understanding of stress distribution in the adhesive layer seems to be necessary [21].

Previous investigations have shown that for tubular joints with isotropic adherends in torsion, the radial (or normal) and shear stress components, namely σ_r and τ_{rz}, are identically equal to zero. However, for adherends with laminated composites, to satisfy the equilibrium equations and compatibility conditions in the overlap region, these stress components must be included in the analysis [20]. Additionally, for tubular joints in tension, stress concentrations in the adhesive layer can arise by three different mechanisms; (a) differential straining; (b) end effects, and (c) bending introduced by the non-collinearity of the overlapping tubes. However, for joints in torsion, the bending effects and hence the peel stresses are absent and only the end effects and differential straining need to be considered, provided there is no defect in the adherends [10, 21, 22].

In this connection, an early study by Lubkin and Reissner [23], as well as many other recent studies on the tubular lap joints [9–11, 13, 23, 24], have shown that under a tensile axial load or a bending moment, an induced shear stress component τ_{rz} and a normal stress component σ_r across the thickness of the adhesive layer are present and occur due to adherend bending. However, on the free surfaces at the ends of the adhesive layer, τ_{rz} must be zero. Consequently, a high shear stress gradient exists near the two ends of the joint, the magnitude of which depends on the geometric and mechanical properties of the adherends and adhesive layer. At this location, the interfacial shear stresses increase from zero on the free surface to some maximum value over a very short length along the bond length [9, 10, 13, 17]. As stated before, because of the stress equilibrium considerations, this high shear stress gradient is associated with a normal stress gradient across the thickness of the adhesive layer.

5.1.2 Major Factors Affecting the Peel and Shear Stresses in the Adhesive Layer and its Performance (Failure)

In general, tubular parts, as well as other types of structures, can be joined by different types of adhesives. One major concern about these joints is the magnitude of peak stresses that develop in the adhesive layer due to the applied load. It has been shown that an excessive load in the adhesive layer can result in an excessive peel or shear stress leading to the joint failure. In a very early study, McCarvill and Bell [25] developed a modified tubular butt assembly to test a modified tubular epoxy-aluminum butt joint that was subjected to torsion. Later, Alwar and Nagaraja [26] used the finite element method (FEM) to include the viscoelastic behavior of the adhesive layer in their analysis on a tubular joint under a tensile load and determined the long-term redistribution of stresses in this layer. In a different study, Adams and Peppiatt [27] applied the FEM to investigate the stress distribution in an adhesively bonded tubular joint under axial and torsional loads. Among other factors, the effects of adhesive fillet and partial tapering of the adherends on stress distribution were investigated in their work.

In the past decades, among many analytical and experimental studies performed on adhesively bonded tubular joints [7, 2, 28–32], some authors have tried to present easier or modified methods to give better estimates on the magnitudes of peak stresses that develop in the adhesive layer under different loading conditions [7, 33, 34]. In this connection, Oh [7] performed a nonlinear analysis on the adhesively bonded tubular single-lap joints to obtain the constitutive relations under a torsional load. Results showed that the peak shear stress developed at the joint end was less than the value obtained by a linear analysis. He further concluded that the torque transmissibility of the joint increased with the joint overlap length.

A close examination of the foregoing references and many others reveals that there are a few main factors that affect or control the adhesive behavior and stresses as well as its performance in joints with isotropic or composite adherends. Among these factors, one can point out the following:

1. Elastic modulus of the adhesive layer and its bond length.
2. Adhesive geometric dimensions and its mechanical properties.
3. Applied load type.

4. Presence of a defect (void, debond, or a crack) in the adhesive layer.
5. Presence of a delamination in one of the adherends (or both).

The effects of these factors will be investigated in section 3.

5.2 Governing Equations Based on Linear Elasticity

To investigate the effects of a defect, delamination, and/or the geometric and mechanical properties of the adhesive and adherends on the overall performance of the adhesive layer, first, the governing equilibrium equations must be deduced for the joint in question. Depending on the structural shape of the joint and presence of any possible defect in the adherends or adhesive layer, these equations can be different from one joint to another. For example, adopting one of the tubular joints shown in Figures 5.1–5.3, as a first step, one has to introduce the proper assumptions that are suitable for satisfaction of equilibrium equations that depend on the type of the load applied to the joint.

5.2.1 Typical Assumptions in a Tubular Lap Joint under Torsion

The allowable assumptions for derivation of governing equations in a joint depend on the load type, joint configuration, presence of the defect, geometric and mechanical properties of the joint constituents. For example, based on the linear elasticity to deduce the equilibrium equations in a defect-free tubular joint under torsional load (as shown in Figure 5.4) it can be assumed that:

1. Only the circumferential displacements in the tubes and the adhesive layer need to be considered. In other words, the axial and radial components of the displacements are considered to be zero.
2. The circumferential components of the displacements in the tubes and the adhesive layer can be assumed to be a function of axial (x) and radial (r) directions.
3. For a typical composite adherend, the stacking sequence in the adherend can be assumed to be either monoclinic [7], orthotropic or quasi-isotropic [11], transversely isotropic [35], or isotropic. Consequently, the main shear stress

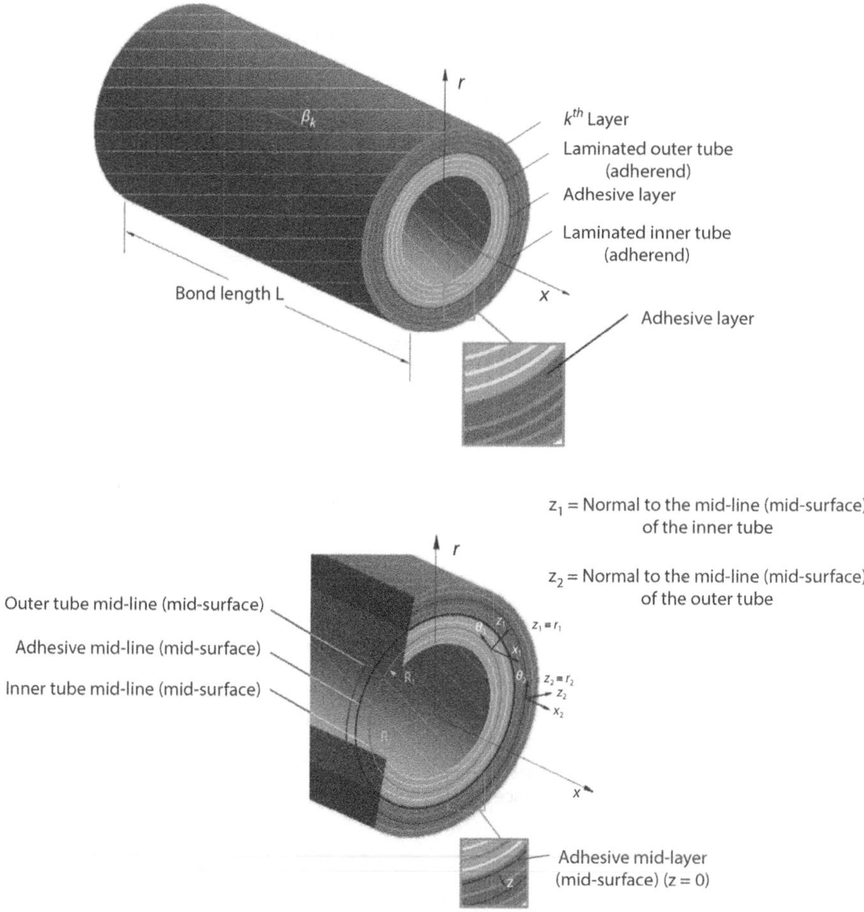

Figure 5.4 Schematic view of bonded region L in a tubular single lap joint with two composite adherends [11].

produced in the adhesive layer is the torsional shear stress ($\tau_{r\theta}$), and hence one can neglect the effects of other stress components in the model ([7, 12]).

However, in a joint with an axial force one cannot neglect the effect of peel stress σ_r in the adhesive layer due to the induced bending moment in the joint (as opposed to a tubular joint under torsion).

Now, based on a three-dimensional free-body diagram of the joint (see Figure 5.5), using the foregoing assumptions, the governing equilibrium equations in the joint can be deduced, as will be shown next. For this purpose, allow τ_1 and τ_2 to designate the interfacial shear stresses resulting from $(\tau_{r\theta})_m$

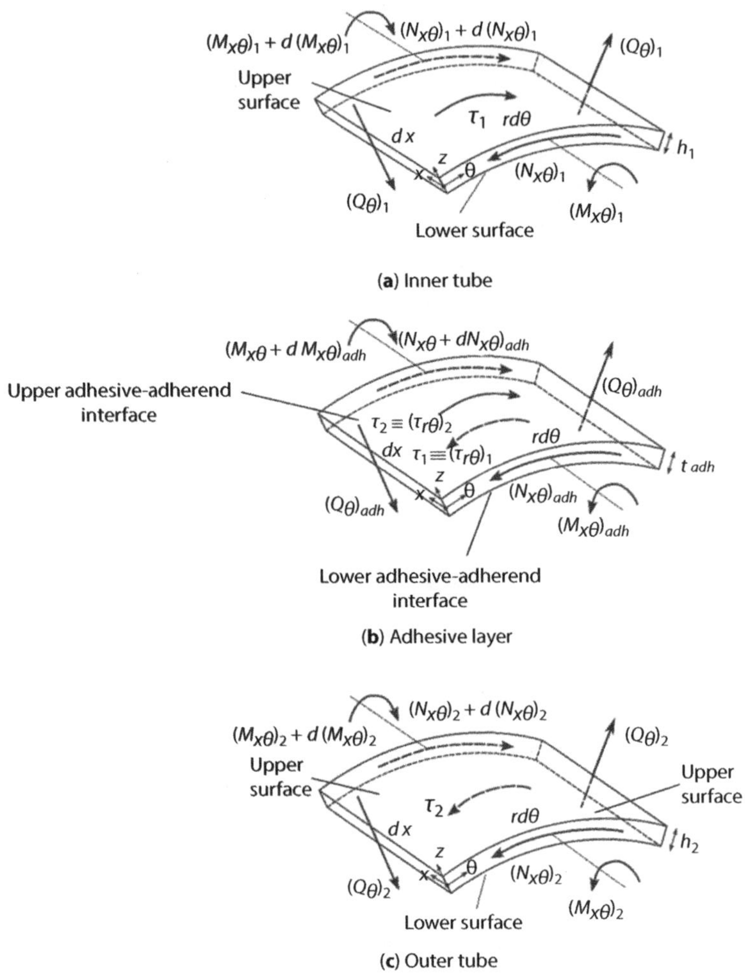

Figure 5.5 Free-body diagrams of the differential elements of the two tubes in contact with the adhesive layer [11].

($m = 1, 2$), at the lower and upper interfaces of the adhesive layer, respectively. Moreover, denoting the $(N_{x\theta})_m$, $(Q_\theta)_m$, $(M_{x\theta})_m$, and h_m ($m = 1, 2$) as the components of the resultant shear force, transverse shear force, twisting moment, and pipe thicknesses, respectively, application of the force equilibrium equations gives the governing equilibrium equations in the θ- direction as:

$$\frac{(dN_{x\theta})_1}{dx} + \frac{(Q_\theta)_1}{\bar{R}_1} = -\frac{R_2}{\bar{R}_1}\tau_1 \tag{5.1a}$$

$$\frac{(dN_{x\theta})_2}{dx}+\frac{(Q_\theta)_2}{\overline{R}_2}=\frac{R_3}{\overline{R}_2}\tau_2 \tag{5.1b}$$

It is worth noting that in each component forming the joint, the origin of local coordinate axes is located at its mid-surface (see Figure 5.4).

However, if the circumferential components of the displacement in each adherend are assumed to be a function of x and θ, then one can write:

$$V_m(x, z) = V_{om}(x) + z_m\,\varphi_{\theta m}(x)\ (m = 1, 2) \tag{5.2a}$$

$$U_m = W_m = 0\ (m = 1, 2) \tag{5.2b}$$

In Eq. (5.2a), $V_{0m}(x)$, and $\varphi_{\theta m}(x)$, are the mid-plane displacement and rotation in each pipe, respectively. Additionally, z_m is the local distance from the mid-plane of each pipe.

Introducing the normal components of the strain tensor in cylindrical coordinates as in Eq. (5.3a)

$$\varepsilon_x=\frac{\partial U}{\partial x},\quad \varepsilon_\theta=\frac{1}{r}\left(\frac{\partial V}{\partial \theta}+W\right),\quad \varepsilon_r=\frac{\partial W}{\partial r} \tag{5.3a}$$

then the shear strain components are:

$$\gamma_{r\theta}=\frac{1}{r}\left(\frac{\partial W}{\partial \theta}-V+r\frac{\partial V}{\partial r}\right),\quad \gamma_{rx}=\frac{\partial U}{\partial r}+\frac{\partial W}{\partial x},\quad \gamma_{x\theta}=\frac{\partial V}{\partial x}+\frac{1}{r}\frac{\partial U}{\partial \theta} \tag{5.3b}$$

In these equations, $r_m=\overline{R}_m+z_m, (m=1,2)$. Moreover, the constitutive relations for a composite laminate with n layers and fiber angles β_k in each layer (k is the layer number) are $\sigma=\overline{Q}\varepsilon$. In this equation, the matrices associated with the kth layer are;

$$\sigma^{(k)}=\left\{\sigma_x^{(k)}\quad \sigma_\theta^{(k)}\quad \sigma_r^{(k)}\quad \tau_{r\theta}^{(k)}\quad \tau_{rx}^{(k)}\quad \tau_{x\theta}^{(k)}\right\}^T \tag{5.3c}$$

$$\varepsilon^{(k)}=\left\{\varepsilon_x^{(k)}\quad \varepsilon_\theta^{(k)}\quad \varepsilon_r^{(k)}\quad \gamma_{r\theta}^{(k)}\quad \gamma_{rx}^{(k)}\quad \gamma_{x\theta}^{(k)}\right\}^T \tag{5.3d}$$

and,

$$\bar{\mathbf{Q}}^{(k)} = \begin{bmatrix} \bar{Q}_{11} & \bar{Q}_{12} & \bar{Q}_{13} & 0 & 0 & \bar{Q}_{16} \\ \bar{Q}_{12} & \bar{Q}_{22} & \bar{Q}_{23} & 0 & 0 & \bar{Q}_{26} \\ \bar{Q}_{13} & \bar{Q}_{23} & \bar{Q}_{33} & 0 & 0 & \bar{Q}_{36} \\ 0 & 0 & 0 & \bar{Q}_{44} & \bar{Q}_{45} & 0 \\ 0 & 0 & 0 & \bar{Q}_{45} & \bar{Q}_{55} & 0 \\ \bar{Q}_{16} & \bar{Q}_{26} & \bar{Q}_{36} & 0 & 0 & \bar{Q}_{66} \end{bmatrix}^{(k)} \tag{5.3e}$$

To obtain the elements of the reduced stiffened matrix $\bar{\mathbf{Q}}^{(k)}$ in terms of the elements of stiffness matrix $\mathbf{Q}^{(k)}$ and the engineering constants, one can refer to [36].

Now substituting Eqs. (5.2a) and (5.2b) in the strain-displacement relations we have;

$$(\varepsilon_x)_m = (\varepsilon_\theta)_m = (\varepsilon_r)_m \, (\gamma_{rx})_m = 0, \, (m = 1, 2) \tag{5.4a}$$

$$(\gamma_{r\theta})_m = \phi_{\theta m}(x) - \frac{V_{0m}(x) + z\phi_{\theta m}(x)}{\bar{R}_m + z_m}, \quad (m = 1,2) \tag{5.4b}$$

$$(\gamma_{x\theta})_m = \frac{dV_{0m}(x)}{dx} + z_m \frac{d\phi_{\theta m}(x)}{dx}, \quad (m = 1,2) \tag{5.4c}$$

On substitution of Eqs. (5.4a), (5.4.b), and (5.4c) into the constitutive relations $\sigma = \bar{Q}\varepsilon$, one can obtain the stresses in each layer of the composite pipes. However, for a single lap joint in torsion, only the force and moment components affecting the shear stress in the adhesive layer need to be calculated. These components could be different for other types of loadings. For this purpose, substituting Eqs. (5.4a), (5.4b), and (5.4c) in stress-strain relations and assuming that each pipe is composed of different layers, the resultant forces and moments may be written as:

$$(N_{x\theta})_m = \left(\sum_{k=1}^{n} \int_{z_k}^{z_{k+1}} \tau_{x\theta}^{(k)} \left(1 + \frac{z}{\bar{R}}\right) dz \right)_m, \quad (m = 1,2) \tag{5.5a}$$

$$(M_{x\theta})_m = \left(\sum_{k=1}^{n} \int_{z_k}^{z_{k+1}} \tau_{x\theta}^{(k)} \left(1 + \frac{z}{\bar{R}} \right) z dz \right)_m , \quad (m = 1,2) \tag{5.5b}$$

$$(Q_\theta)_m = \left(K_S \sum_{k=1}^{n} \int_{z_k}^{z_{k+1}} \tau_{r\theta}^{(k)} dz \right)_m , \quad (m = 1,2) \tag{5.5c}$$

In Eqs. (5.5a)–(5.5c), $(N_{x\theta})_m$, $(Q_\theta)_m$, and $(M_{x\theta})_m$ are the in-plane and transverse components of the shear loads, as well as the twisting moments, respectively. Furthermore, K_S represents the shear correction factor that can be assumed to be equal to 5/6 (see Refs. [37] and [38]), n is the total number of layers, and k is the layer number. These equations are recast in terms of derivatives of pipes' mid-plane displacements and rotations $\left(\frac{dV_{0m}(x)}{dx} \text{ and } \frac{d\varphi_{\theta m}(x)}{dx} \right)$ as:

$$(N_{x\theta})_m = \left(A_{66}^{(m)} + \frac{B_{66}^{(m)}}{\bar{R}_m} \right) \frac{dV_{0m}(x)}{dx} + \left(B_{66}^{(m)} + \frac{D_{66}^{(m)}}{\bar{R}_m} \right) \frac{d\varphi_{\theta m}(x)}{dx} \tag{5.6a}$$

$$(M_{x\theta})_m = \left(B_{66}^{(m)} + \frac{D_{66}^{(m)}}{\bar{R}_m} \right) \frac{dV_{0m}(x)}{dx} + \left(D_{66}^{(m)} + \frac{E_{66}^{(m)}}{\bar{R}_m} \right) \frac{d\varphi_{\theta m}(x)}{dx} \quad m = 1,2 \tag{5.6b}$$

$$(Q_\theta)_m = K_S F_{44}^{(m)} \left(-V_{0m}(x) + \bar{R}_m \phi_{\theta m}(x) \right) \tag{5.6c}$$

In general, A_{ij}, B_{ij}, and D_{ij} represent the elements of extensional, coupling, and stiffness matrices **A**, **B**, and **D**, respectively, values of which are given in Eqs. (5.7a)-(5.7c).

$$A_{ij} = \int_{\frac{-h}{2}}^{\frac{h}{2}} \bar{Q}_{ij}\, dz = \sum_{k=1}^{n} \bar{Q}_{ij}^{(k)} (z_{k+1} - z_k) \tag{5.7a}$$

$$B_{ij} = \int_{\frac{-h}{2}}^{\frac{h}{2}} \bar{Q}_{ij} z\, dz = \frac{1}{2} \sum_{k=1}^{n} \bar{Q}_{ij}^{(k)} \left(z^2{}_{k+1} - z^2{}_{k} \right) \tag{5.7b}$$

$$D_{ij} = \int_{\frac{-h}{2}}^{\frac{h}{2}} \bar{Q}_{ij} z^2 \, dz = \frac{1}{3} \sum_{k=1}^{n} \bar{Q}_{ij}^{(k)} \left(z^3{}_{k+1} - z^3{}_k \right) \tag{5.7c}$$

Additionally, E_{ij} and F_{ij} are expressed as:

$$E_{ij} = \int_{\frac{-h}{2}}^{\frac{h}{2}} \bar{Q}_{ij} z^3 \, dz = \frac{1}{4} \sum_{k=1}^{n} \bar{Q}_{ij}^{(k)} \left(z^4{}_{k+1} - z^4{}_k \right) \tag{5.7d}$$

$$F_{ij} = \int_{\frac{-h}{2}}^{\frac{h}{2}} \frac{\bar{Q}_{ij}}{\bar{R} + z} dz = \sum_{k=1}^{n} \bar{Q}_{ij}^{(k)} \ln \frac{\bar{R} + z_{k+1}}{\bar{R} + z_k} \tag{5.7e}$$

Although in some cases due to the nature of the problem, a constant shear stress is considered along the adhesive thickness, yet in this case to allow for variation of this stress along adhesive thickness, a second-order displacement $v_{adh}(x, r)$ is considered for the adhesive circumferential displacement such that:

$$v_{adh}(x, r) = \eta_0(r)\, v_0(x) + \eta_1(r)\, v_1(x) + \eta_2(r)\, v_2(x) \tag{5.8}$$

In Eq. (5.8) $v_0(x)$, is the mid-plane displacement of the adhesive layer while $v_1(x)$ and $v_2(x)$ are the corresponding displacements at the top and bottom surfaces, respectively. Using continuity in displacements at the tube-adhesive interfaces, it can be shown that the expressions for coefficients $\eta_0(r)$, $\eta_1(r)$ and $\eta_2(r)$ are:

$$\eta_0(r) = \frac{r^2 - r(R_2 + R_3) + R_2 R_3}{\left(R_3 - \bar{R}_{adh}\right)\left(R_2 - \bar{R}_{adh}\right)} \tag{5.9a}$$

$$\eta_1(r) = \frac{-r^2 + r\left(\bar{R}_{adh} + R_3\right) - \bar{R}_{adh} R_3}{(R_3 - R_2)\left(R_2 - \bar{R}_{adh}\right)} \tag{5.9b}$$

$$\eta_2(r) = \frac{-r^2 + r\left(\bar{R}_{adh} + R_2\right) - \bar{R}_{adh} R_2}{\left(R_3 - \bar{R}_{adh}\right)(R_2 - R_3)} \tag{5.9c}$$

The tubes radii used in Eqs. (5.9a)- (5.9c) are shown in Figures 5.4 and 5.6 where:

$$\bar{R}_{adh} = \frac{R_2 + R_3}{2} \tag{5.10a}$$

In cases in which the adhesive thickness is considered to be constant, one can approximate the magnitude of shear stress at any section by Eq. (5.11).

$$\tau = G\,t_{adh}\,(v_1 - v_2) \tag{5.11}$$

where $(v_1 - v_2)$ is the relative circumferential displacement of the composite tubes at their interfaces with the adhesive layer. G and t_{adh} are the adhesive shear modulus and thickness, respectively. However, since $(v_1 - v_2)$ is a function of both the shear moduli and polar moments of inertia of the two tubes (adherends), then it is expected that the adhesive layer experiences a nonsymmetric shear stress distribution as the load is transferred from one tube to the other.

On using the expression for the adhesive displacement, it can be shown that the expression for variation in the twisting moment in the adhesive layer is [11]:

$$T_{adh}(x) = 2\pi G \int_{R_2}^{R_3} \left(\eta_0(r) \frac{dv_0(x)}{dx} + \eta_1(r) \frac{dv_1(x)}{dx} + \eta_2(r) \frac{dv_2(x)}{dx} \right) r^2 dr \tag{5.12}$$

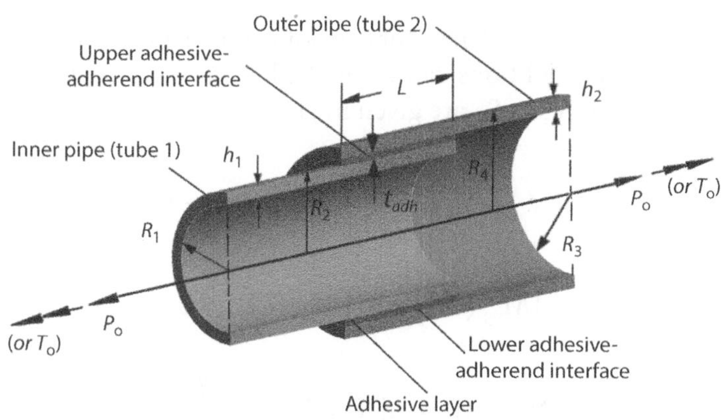

Figure 5.6 Geometric dimensions of a tubular lap joint assembly used in [11].

Now, on proper application of equilibrium equations (forces and moments), one can show that the following relations will determine the complete displacement field, and hence the shear stress distribution in the adhesive layer, provided proper boundary conditions are applied.

$$S_1 \frac{d^2V_{01}(x)}{dx^2} + S_2 \frac{d^2\varphi_{\theta 1}(x)}{dx^2} + S_3V_{01}(x) + S_4\varphi_{\theta 1}(x) + S_5V_{02}(x) + S_6\varphi_{\theta 2}(x) + S_7v_0(x) = 0 \tag{5.13a}$$

$$S_8 \frac{d^2V_{02}(x)}{dx^2} + S_9 \frac{d^2\varphi_{\theta 2}(x)}{dx^2} + S_{10}V_{01}(x) + S_{11}\varphi_{\theta 1}(x) + S_{12}V_{02}(x) + S_{13}\varphi_{\theta 2}(x) + S_{14}v_0(x) = 0 \tag{5.13b}$$

$$S_{15} \frac{d^2V_{01}(x)}{dx^2} + S_{16} \frac{d^2\varphi_{\theta 1}(x)}{dx^2} + S_{17}V_{01}(x) + S_{18}\varphi_{\theta 1}(x) + S_{19}V_{02}(x) + S_{20}\varphi_{\theta 2}(x) + S_{21}v_0(x) = 0 \tag{5.13c}$$

$$S_{22} \frac{d^2V_{02}(x)}{dx^2} + S_{23} \frac{d^2\varphi_{\theta 2}(x)}{dx^2} + S_{24}V_{01}(x) + S_{25}\varphi_{\theta 1}(x) + S_{26}V_{02}(x) + S_{27}\varphi_{\theta 2}(x) + S_{28}v_0(x) = 0 \tag{5.13d}$$

$$S_{29}\frac{dV_{01}(x)}{dx} + S_{30}\frac{d\varphi_{\theta 1}(x)}{dx} + S_{31}\frac{dV_{02}(x)}{dx} + S_{32}\frac{d\varphi_{\theta 2}(x)}{dx} + S_{33}\frac{dv_0(x)}{dx} = T_0 \tag{5.13e}$$

In this case, for a defect-free tubular joint under pure torsion, the following boundary conditions can be applied. It is worth mentioning that the expressions for S_{11} to S_{33} are given in [11].

$$[N_{x\theta}(0)]_1 = \frac{T_0}{2\pi \bar{R}_1^2}, [N_{x\theta}(L)]_1 = 0 \tag{5.14a}$$

$$[M_{x\theta}(0)]_1 = 0,\ [M_{x\theta}(L)]_1 = 0 \tag{5.14b}$$

$$[N_{x\theta}(0)]_2 = 0, [N_{x\theta}(L)]_2 = \frac{T_0}{2\pi \bar{R}_2^2} \tag{5.14c}$$

$$[M_{x\theta}(0)]_2 = 0, \ [M_{x\theta}(L)]_2 = 0 \tag{5.14d}$$

$$\left[\tau_{x\theta}\left(L, \bar{R}_{adh}\right)\right]_{adh} = 0 \tag{5.14e}$$

5.3 Factors Influencing the Adhesive Behavior and Stresses

5.3.1 The Effects of Geometric and Mechanical Properties of the Adhesive and Adherends

The effects of geometric and mechanical properties of the adhesive and adherends for a tubular lap joint under torsion were fully investigated in [11]. Here, the semi-analytical results were supported by the finite element solution. The inner pipe (tube 1) was made of steel and the outer pipe (tube 2) was assumed to be either an orthotropic carbon-epoxy composite pipe with the stacking sequence of $[\pm 15]_{10}$ (as well as other arrangements) for the layers, or a twenty-layer quasi-isotropic pipe with different layer arrangements. Mechanical properties of the adhesive layer, as well as the composite and steel tubes are given in Table 5.1.

Figure 5.7 shows the unsymmetric shear stress distributions $(\tau_{r\theta})_1$ and $(\tau_{r\theta})_2$ along the bond length with the higher peak values on the right side of the joint. The unsymmetric distributions of shear stresses for the applied torque of 150 N.m were due to the unequal torsional stiffnesses of the two pipes (adherends). According to this figure, the torsional stiffness of the $[\pm 15]_{10}$ laminated outer tube was smaller than that of the inner steel tube. As shown, the finite element findings well support the results based on semi-analytical solution [11].

Furthermore, according to Figure 5.8, as the length of the overlap region reduces, stress concentration at one end appears to affect the other, hence increasing the peak shear stresses that develop at the two ends of the overlap region. Results indicate that there is a so-called "*effective length*" for which the whole bond length remains active (shear stresses greater than zero) without a considerable rise in $(\tau_{r\theta})_{max}$ at either end of the overlap region. Therefore, to improve the joint performance it seems to be essential to select the right bond length for better transformation of the applied torque.

The adhesive mechanical properties also influence the peak shear stresses that develop at the two ends of the overlap region. According to Figure 5.9, increasing the elastic modulus of the adhesive layer results in an

Table 5.1 Geometric and mechanical properties used in [11] for comparison of results.

	Adhesive layer, EPCO 9923 epoxy	**Inner pipe (pipe 1) (steel)**	**Outer pipe (pipe 2) $[\pm 15°]_{10}$**
Inner pipe radius (mm)	14.9	11.9	15.0
Outer pipe radius (mm)	15.0	14.9	18.0
Length (mm)	20.0 (bonded length)	60.0 (total)	60.0 (total)
Elastic modulus E_1, GPa	1.3	200.0	128.0
Elastic modulus E_2, GPa	1.3	200.0	10.0
Elastic modulus E_3, GPa	1.3	200.0	10.0
Shear modulus G_{12}, GPa	0.46	76.92	4.49
Shear modulus G_{13}, GPa	0.46	76.92	4.49
Shear modulus G_{23}, GPa	0.46	76.92	3.58
Poisson's ratio υ_{12}	0.41	0.3	0.28
Poisson's ratio υ_{13}	0.41	0.3	0.28
Poisson's ratio υ_{23}	0.41	0.3	0.47

increase in adhesive peak shear stresses at the upper and lower adhesive-adherend interfaces.

The simultaneous effects of the inner tube radius R_2 and the bond length L on the adhesive peak shear stress $(\tau_1)_{max}$ are shown in Figure 5.10. Here, the adhesive thickness was assumed to be 0.1 mm, while the initial values of 11.9 mm and 12.9 mm were selected for R_1 and R_2 (see Figure 5.6 for these dimensions).

To obtain these results, the adhesive thickness was assumed to be intact while the inner and outer radii of tube 2 were calculated according to the formulae $R_3 = R_2+0.1$, and $R_4=R_3+$ 3 mm. According to Figure 5.10(a), for any bond length L the increase in the inner tube radius R_2 decreases the peak shear stress in the adhesive layer. Additionally, considering the coordinates of points A, B, C, and D shown in this figure and holding the bond length intact and equal to 70 mm, it is observed that increasing R_2 from 12.9 to 14.9 mm reduces the peak shear stress by 15.0%. However,

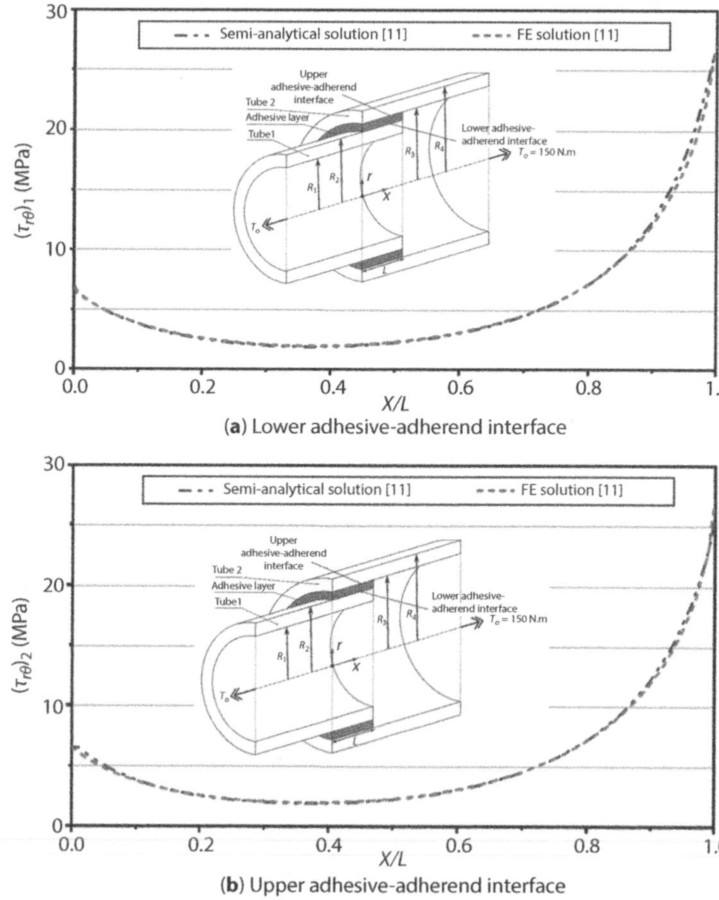

Figure 5.7 Variations in the interfacial shear stress distributions in the adhesive layer along the bond length based on the finite element and semi-analytical solutions (steel inner tube and $[\pm 15]_{10}$ carbon-epoxy laminated outer tube, $L = 20$ mm).

according to Figure 5.10(b) for a bond length $L = 20$ mm, any incremental increase in R_2 by 0.1 mm reduces the peak shear stress $(\tau_{r\theta})_{1max} \equiv (\tau_1)_{max}$ by about 1% leading to a total decrease of 15.7% at R_2 =14.9 mm.

Figure 5.11 shows the effect of adhesive thickness on the shear stress distribution in the adhesive layer. Here, the adhesive thickness was increased in increments of 0.05 mm from a value of 0.05 mm to 0.2 mm. Additionally, keeping the inside and outside radii of the inner tube constant and equal to R_1= 11.9 mm and R_2=14.9 mm, the radii of the outer tube were increased such that its wall thickness remained constant and equal to 3 mm.

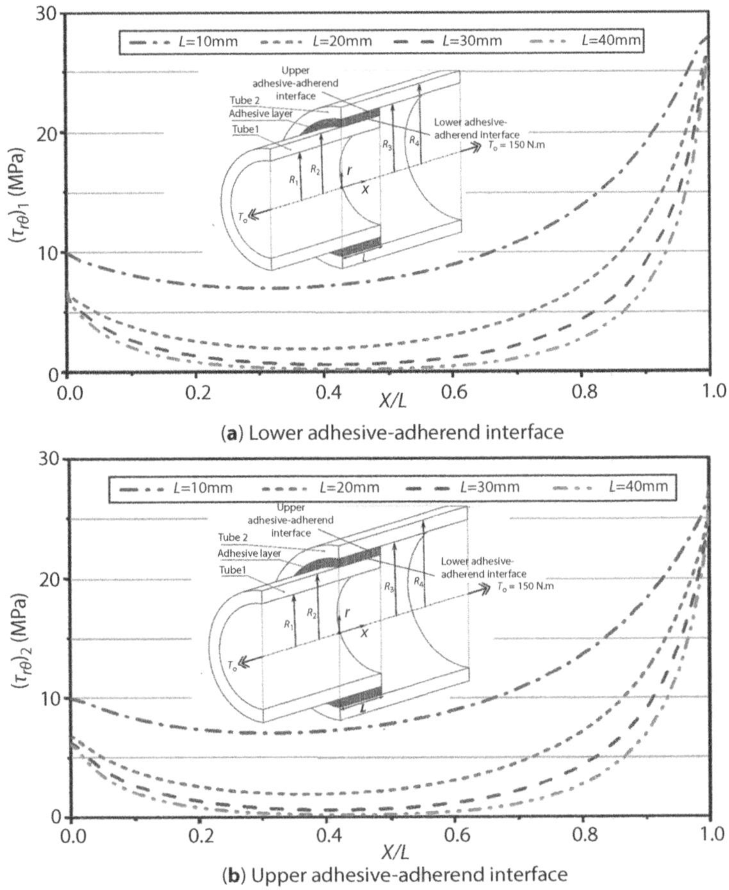

Figure 5.8 The effect of joint length on variations of the interfacial shear stresses in the adhesive layer along the bond length (steel inner tube and $[\pm 15]_{10}$ carbon-epoxy laminated outer tube).

Results in this figure show that the increase in the adhesive thickness by 0.1 mm (i.e. from 0.05 to 0.15 mm) reduces the peak shear stresses by about 37% and 40% for $(\tau_{r\theta})_1$ and $(\tau_{r\theta})_2$ respectively. However, as explained before, a similar increase in R_2 (namely 0.1 mm) reduces the peak shear stress in the adhesive layer by only 1%. This means that the peak shear stress in the adhesive layer is more susceptible to a change in the adhesive thickness in contrast to a change in the tube radii.

To seek the effects of lamination on the adhesive shear stress, Figure 5.12 was generated to explore this behavior. In Figure 5.12(a), two types of lamination setups, namely orthotropic and quasi-isotropic laminates with

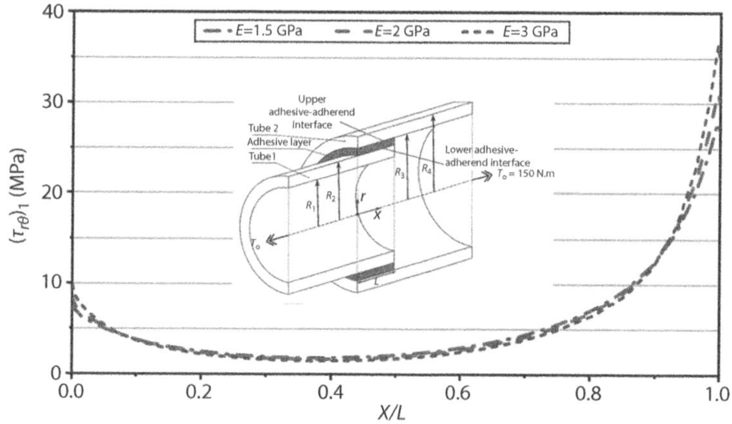

(**a**) Lower adhesive-adherend interface

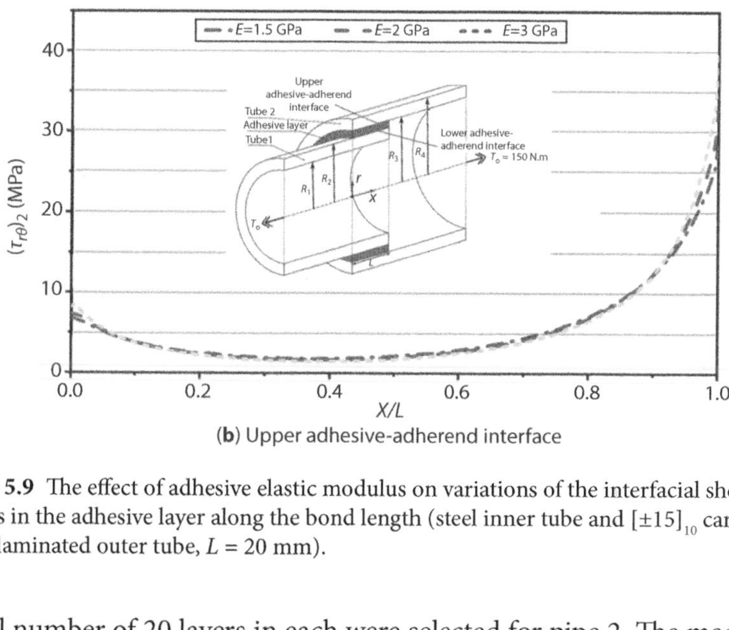

(**b**) Upper adhesive-adherend interface

Figure 5.9 The effect of adhesive elastic modulus on variations of the interfacial shear stresses in the adhesive layer along the bond length (steel inner tube and $[\pm 15]_{10}$ carbon-epoxy laminated outer tube, L = 20 mm).

a total number of 20 layers in each were selected for pipe 2. The mechanical and geometric properties of the adhesive layer, as well as the steel and carbon-epoxy tubes are given in Table 5.1.

In the first setup, tube 2 was composed of a series of orthotropic laminates with lamination codes of $[\pm 15]_{10}$, $[\pm 30]_{10}$, $[\pm\ 45_5]_s$, $[\pm\ 60]_{10}$, and $[90/0]_{10}$. However, in the second arrangement, the quasi-isotropic setups of $[0_4/90_4/\text{-}45_4/45_4]$, $[0_4/36_4/72_4/\text{-}72_4/\text{-}36_4]$, and $[0_2/36_2/72_2/\text{-}72_2/\text{-}36_2]_s$, were used for this pipe. Results indicate that for the orthotropic arrangement of $[90/0]_{10}$, the maximum peak shear stress produced in the adhesive layer is

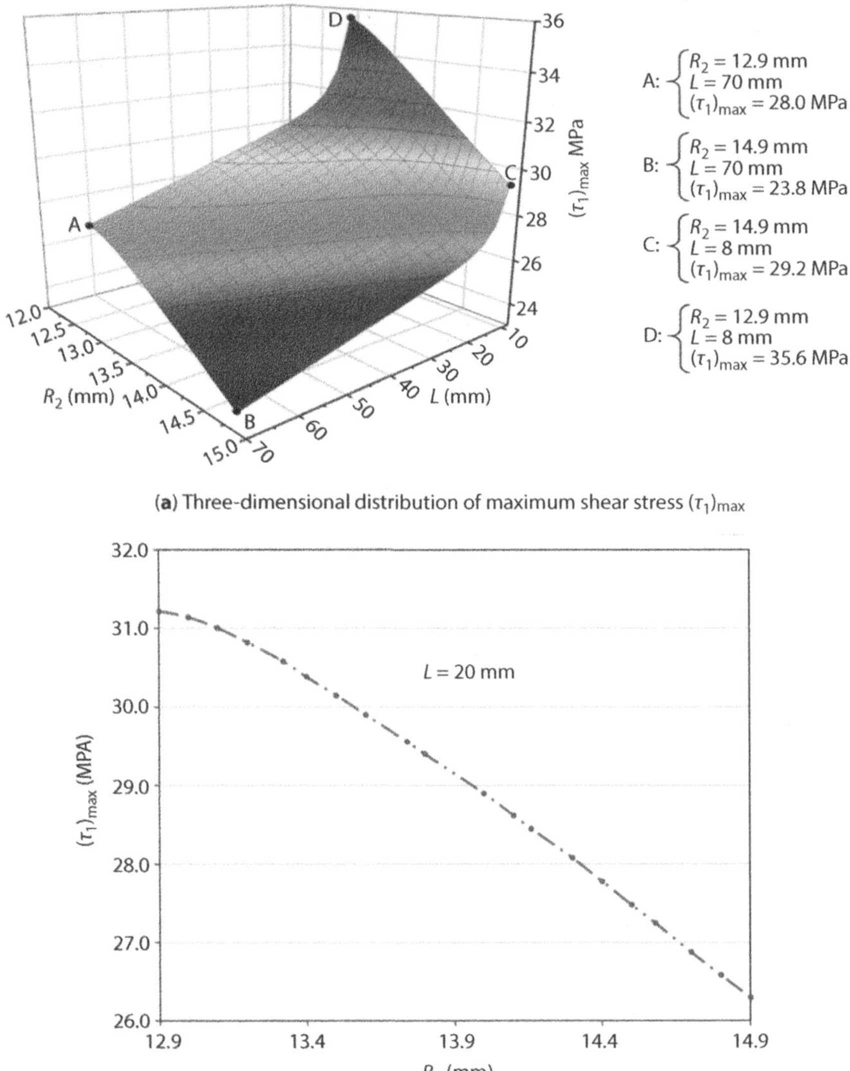

(**a**) Three-dimensional distribution of maximum shear stress $(\tau_1)_{max}$

(**b**) Variation of max shear stress in the adhesive layer with respect to R_2

Figure 5.10 The effects of tube 1 radius R_2 and bond length L on variation of the peak shear stress in the adhesive layer (T_o = 150 N.m).

45.0 MPa for $(\tau_{r\theta})_1$, while for the $[\pm 45]_{10}$ setup this value is reduced to 14 MPa. This indicates a total reduction of 221.4% in the adhesive peak shear stress $(\tau_{r\theta})_1$. However, among the quasi-isotropic arrangements of the layers, the $[0_5/90_5/\text{-}45_5/45_5]$ setup produces higher peak shear stress $(\tau_{r\theta})_{1max}$ compared to $[0_2/36_2/72_2/\text{-}72_2/\text{-}36_2]_s$ arrangement. Selecting the latter setup,

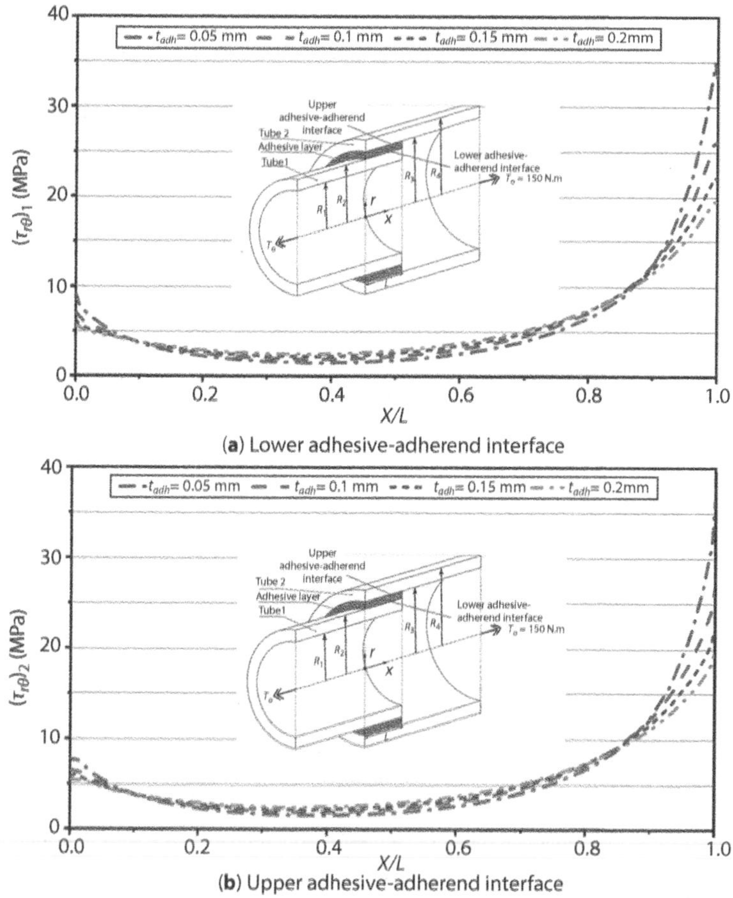

Figure 5.11 The effect of adhesive thickness on variations of the interfacial shear stresses in the adhesive layer along the bond length (steel inner tube and $[\pm 15]_{10}$ carbon-epoxy laminated outer tube, $L = 20$ mm).

the peak shear stress at the right end of the overlap region can be reduced by 26%. However, according to Figure 5.12(b), this effect is reversed for $(\tau_{r\theta})_{2max}$. Comparison of the results in Figure 5.12 (a) indicates that the best lamination setup for pipe 2 for the transfer of torsional load is the orthotropic pipe with a stacking sequence of $[\pm 45]_{10}$ for the layers. This will produce the minimum peak shear stress in the adhesive layer.

Based on this argument, compared with the quasi-isotropic setup of $[0_2/36_2/72_2/-72_2/-36_2]_s$, the selection of an orthotropic tube with the layer arrangement of $[\pm 45]_{10}$ can reduce the maximum shear stress $(\tau_{r\theta})_{1max}$ in the adhesive layer by an additional 30%. On the other hand, Figure 5.12(c)

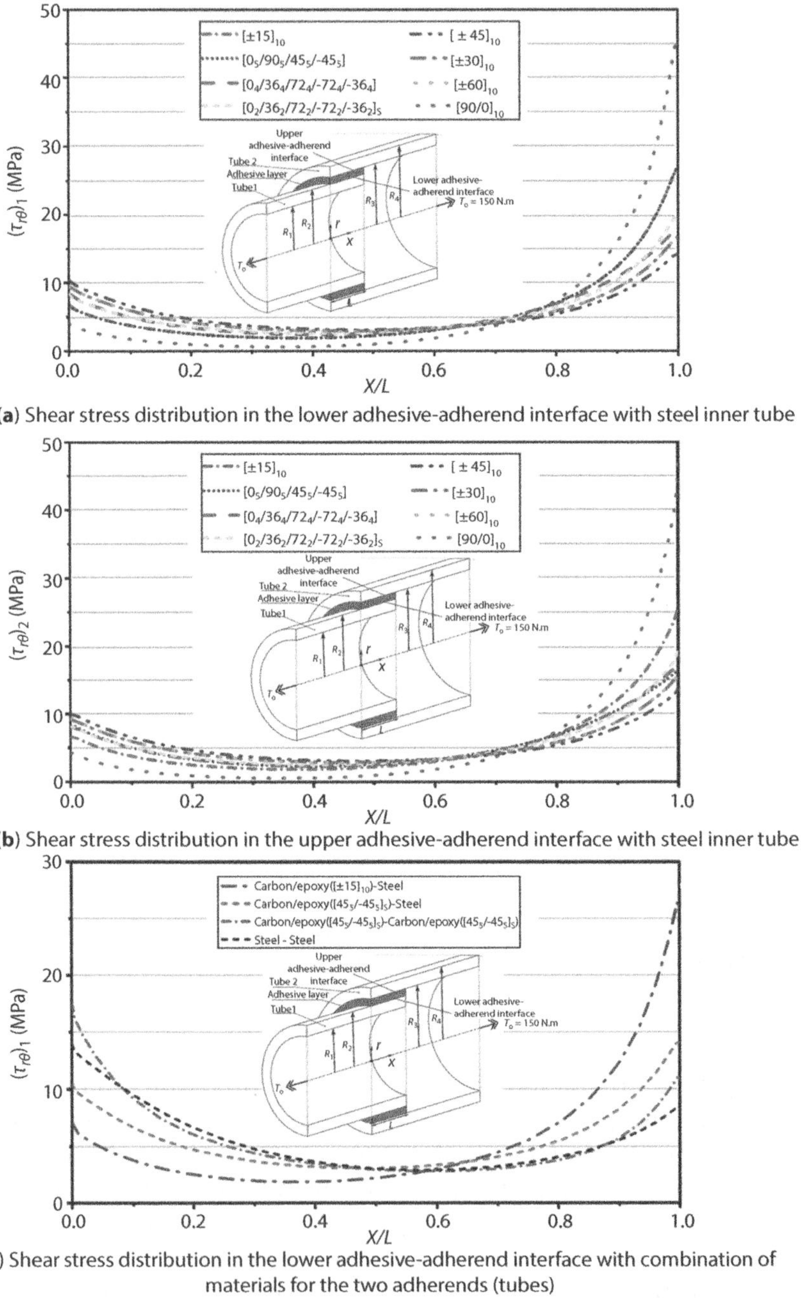

(**a**) Shear stress distribution in the lower adhesive-adherend interface with steel inner tube

(**b**) Shear stress distribution in the upper adhesive-adherend interface with steel inner tube

(**c**) Shear stress distribution in the lower adhesive-adherend interface with combination of materials for the two adherends (tubes)

Figure 5.12 Variation in the adhesive shear stress along the bond length in the upper and lower adhesive-adherend interfaces for various lamination setups (except for the two $[\pm 45_5]_s$ carbon-epoxy adherends in (c), steel is used for pipe 1 material).

indicates that among the two materials selected for the two adherends, although selection of $[\pm 45]_{10}$ carbon-epoxy composite for both tubes can considerably reduce the structural weight of the joint, yet it increases the peak shear stress in the adhesive layer by 22% in comparison with the steel adherends (steel-steel joint). Additionally, as shown, selection of a carbon-epoxy-steel joint instead of the steel-steel joint only shifts the peak shear stress from one end of the overlap region to the other without causing a considerable change in its value. Examination of the results in this figure will allow one to decide on the possibility of replacing the steel tubes with the laminated composite pipes to reduce the structural weight of the tubes without a considerable rise in the adhesive layer shear stress.

The effects of delamination on stress distribution in the adhesive layer of a tubular lap joint (see Figure 5.1) as well as a tubular socket joint (see Figure 5.2) were fully investigated in [10] using different finite element models. In the single lap joint model, adherend 1 (or the inner pipe) was made of composite material, while adherend 2 (or the outer pipe) was considered to be either aluminum or a laminated composite. In the socket joint model, the two adherends were identical and both adherends and the socket were made of composite materials. Furthermore, any material nonlinearity in the adhesive layer was neglected.

Experimental results on adhesively-bonded double lap joints [32] (see Figure 5.1(c)) have shown that the torque transmission capability of the joint structure can be 2.7 times as large as that of the single lap joint.

5.3.2 The Effects of Load Type on the Adhesive Stresses and Behavior

The effect of a torsional load on the adhesive layer stresses was discussed in the previous section. To study the effect of a bending moment on the adhesive layer, Yang *et al.* [39] used a generalized model to generate three types of permanent composite pipe joints. The three main components, namely the pipes, the coupling, and the adhesive layer used to generate these joints were selected as: (1) for the adhesively bonded socket joint, the two pipes were joined using a socket and an adhesive layer; (2) in the butt-and-strap joint model, the two pipes were joined by a fiber-reinforced coupling using resin matrix as the adhesive layer; and (3) in the heat-activated coupling joint the epoxy resin was used as the adhesive layer to join the two pipes and the epoxy prepreg as the coupling. The properties of these constituents are given in Table 5.2. The first-order laminated anisotropic plate theory was used to express the kinematics and constitutive relations in the

Table 5.2 Geometric and mechanical properties of the joint models used in [39].

	Pipe (E-Glass/ Derakane 470, 50% glass content by weight)	Epoxy adhesive	Coupling (E-Glass/ Derakane 470, 50% glass content by weight)
E_{xx} (GPa)	25.2	0.96	25.2
E_{yy} (GPa)	7.5	0.96	7.5
G_{xy} (GPa)	2.4	0.96	2.4
υ_{xy}	0.32	0.412	0.32
Total bonded length $(2L)^*$ = 50.8 mm, $(L_b)^*$ = 25.4 mm, inner pipe diameter = 101.6 mm, applied bending moment = 113 N-m, adhesive thickness = 0.254 mm, shear correction factor K_s = 5/6.			

* See Figure 5.2(d)

composite pipe and coupling. Using separation of variables, the original system of partial differential equations was converted to the ordinary differential equations, solutions of which were obtained using Maple V. The obtained solutions on variations of peel and shear stresses in the adhesive layer were supported by the finite element analysis using ABAQUS software program.

Good correlation was found between the results obtained from finite element findings and those of analytical solution. Furthermore, there were no discernible differences in the adhesive peel or shear stress distributions resulting from the two solution methods.

The effects of axial load on the shear and peel stress distributions are studied by other investigators in [11], [31], and [40]. In [11], the developments of adhesive shear and peel stresses were investigated in a tubular defect-free single lap joint, as well as a joint hosting a debond/void in the adhesive layer. The applied load produced an axial stress of 100 MPa in the pipes. Aluminum was used for the inner pipe, while the outer pipe was made of glass-epoxy composite laminate with Araldite epoxy as the adhesive layer. The geometric and mechanical properties of the selected adherends and adhesive layer are given in Table 5.3. For the defect-free bond, a lamination setup of $[\pm 45_2]_s$ was selected for the orthotropic composite outer pipe.

The effects of axial load on the shear stress distribution at the upper and lower interfaces of the defect-free adhesive layer are shown in

Table 5.3 Geometric and mechanical properties of the adherends and adhesive layer used in [11].

	Inner pipe	Outer pipe	Adhesive layer
Inner diameter (mm)	10	12.2	12
Outer diameter (mm)	12	13.2	12.2
Overall length (mm)	90	90	30*
Elastic modulus E_1 (GPa)	75	26.6	2.5
Elastic modulus E_2 (GPa)	75	11.6	2.5
Elastic modulus E_3 (GPa)	75	11.6	2.5
Shear modulus G_{12} (GPa)	28.85	4.8	1.0
Shear modulus G_{13} (GPa)	28.85	2.6	1.0
Shear modulus G_{23} (GPa)	28.85	2.6	1.0
Poisson's ratio υ_{12}	0.3	0.1	0.35
Poisson's ratio υ_{13}	0.3	0.15	0.35
Poisson's ratio υ_{23}	0.3	0.20	0.35

* The overall length of the adhesive layer in the single lap joint is equal to the overlap (bond) length L that is equal to 30 mm.

Figures 5.13(a) and 5.13(b). For comparison, other results from [40] are superimposed on both figures. Although the results obtained in [40] well predict the zero shear stresses at the two ends of the overlap region (due to the applied method used in formulation of the problem), yet the predicted values of the peak shear stresses are lower than those predicted by the semi-analytical and finite element solutions introduced in [11]. Additionally, although the shear stresses near the two ends must show a tendency for the steep edge effects, yet the structural mechanics (rather than continuum) approach used in [11] does not address this type of singularity at these locations. Nonetheless, the values of peak shear stresses predicted in [11] are closer to the finite element results and give an overestimate of this stress component.

The effects of lamination setup (stacking sequence) on the adhesive shear and peel stresses are shown in Figure 5.14. As shown, contrary to the joint in torsion, the especially orthotropic lamination setup of $[0_4]$ appears

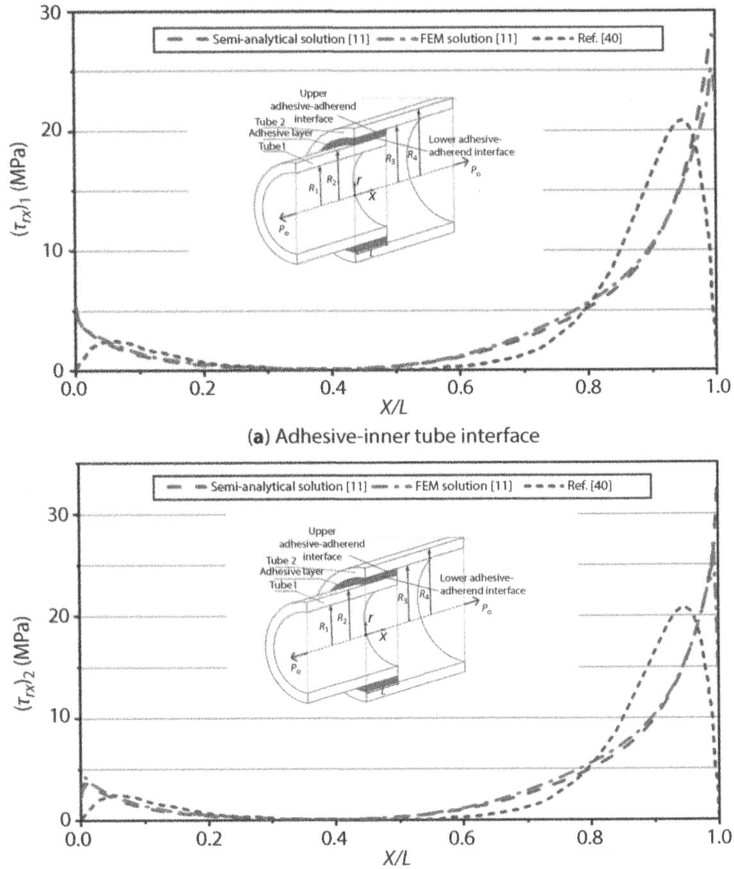

(a) Adhesive-inner tube interface

(b) Adhesive-outer tube interface

Figure 5.13 The effect of axial load on variations of the interfacial shear stresses in the adhesive layer along the bond length (aluminum inner tube and $[\pm 45_2]_s$ glass-epoxy outer tube).

to produce the least shear and peel stresses in the adhesive layer compared to other layer arrangements.

Using differential quadrature method (DQM), Mohieddin Ghomshei and Shahi [16] studied the effects of torsional and hygrothermal loads on stress distribution in a tubular single-lap adhesively bonded composite joint. In this analysis, the effects of several parameters such as adhesive thickness, overlap (bond) length, temperature change, and relative humidity were studied on the joint behavior. Their results were comparable to those from finite element findings and the previously published data.

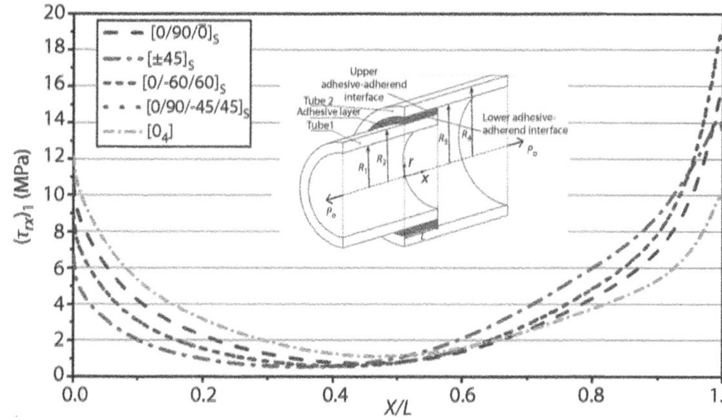

(**a**) Shear stress distribution

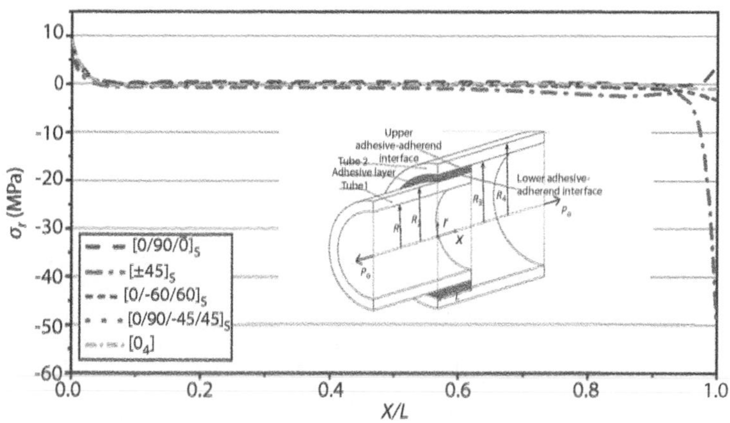

(**b**) Peel stress distribution

Figure 5.14 The effect of stacking sequence of the layers on variations of the shear and peel stresses along the bond length in a joint under a tensile load (aluminum inner tube and glass-epoxy outer tube).

5.3.3 The Effects of Damages due to Voids, Debonds, or Delaminations

A defect in a joint may be due to one of the following sources:

1. A void or a debond in the adhesive layer. These types of defects may arise from a partial or total separation between the adhesive layer and the adherends (see Figure 5.15). In the case of a void, a partial or total circumferential gap exists between the two tubes. The defect may also be due to a crack

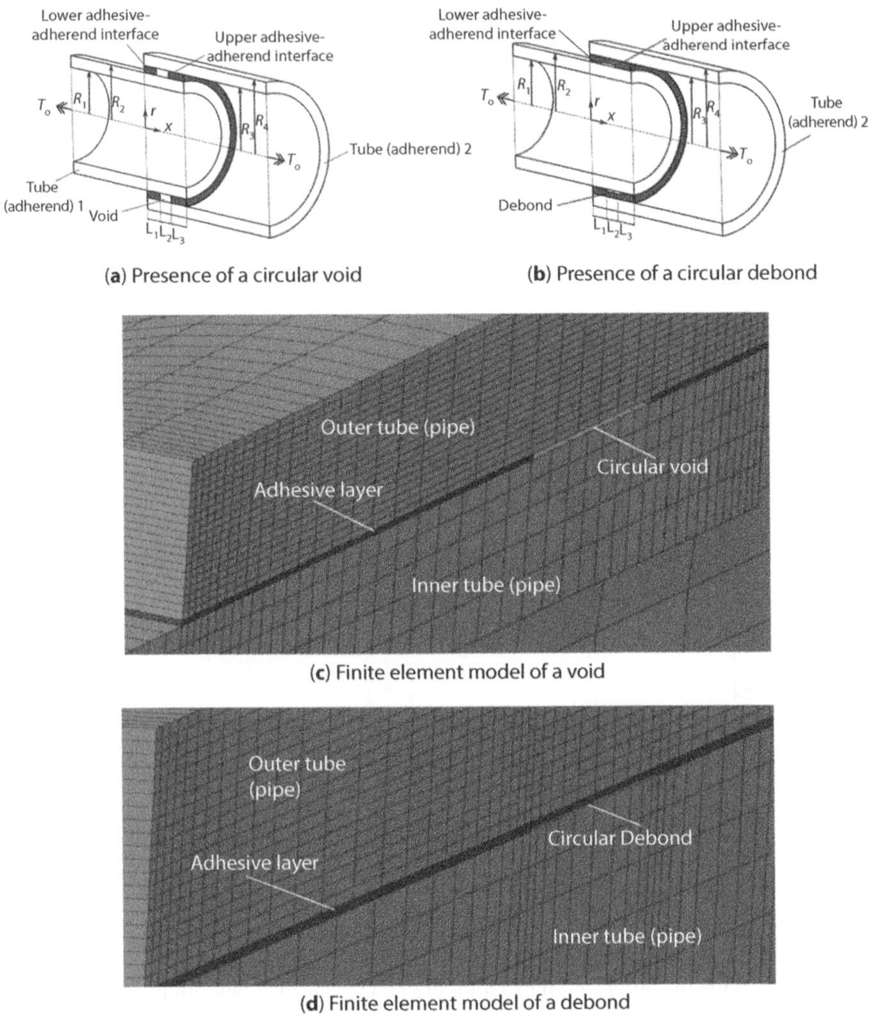

Figure 5.15 Schematic diagrams of a circular void and a circular debond and their finite element models in a tubular single-lap joint.

in the adhesive layer or a separation between the adhesive and adherends at one of their interfaces.

2. Delamination in the composite adherend(s) or in the socket. This type of defect can exist in at least one tube (or in the socket), completely outside the overlap region or partially or completely inside the bond length [10] (see Figures 5.16 and 5.17).

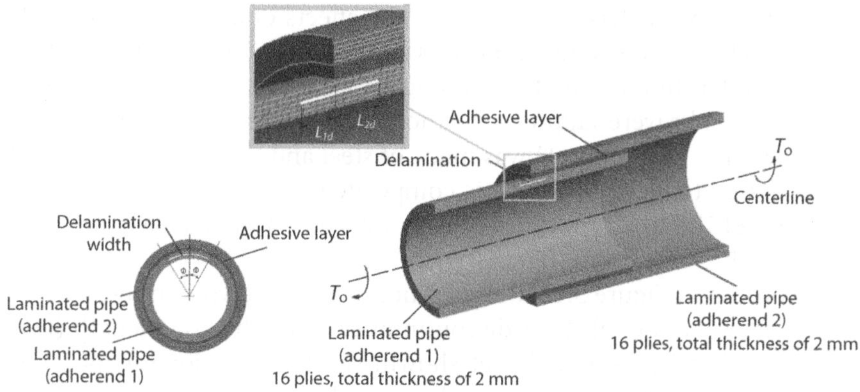

Figure 5.16 Cross-sectional view of a tubular composite socket joint with a delaminated adherend used in [10].

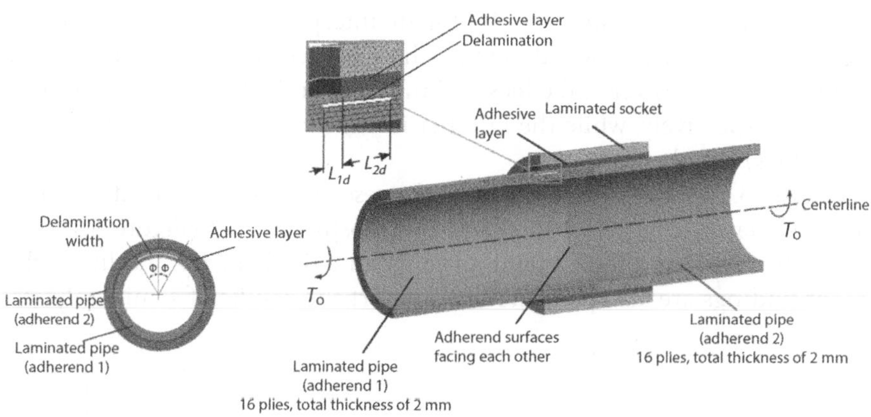

Figure 5.17 Cross-sectional view and load type in a tubular composite lap joint with a delaminated adherend used in [10].

In these cases, the presence of a defect can alter the shear stress distribution in the adhesive layer. For joints in torsion, it has been shown [11, 17] that the presence of a void or a debond in the adhesive layer can increase the peak shear stress by 26% and 21% (or more), respectively, as the defect gets close to the critical end of the bond length (the end experiencing maximum shear stress). Furthermore, for voids or debonds far enough from the adhesive ends, their effects on the peak shear stress in the adhesive layer vanish.

Additionally, in Refs. [11] and [13] the effects of a circular void and a circular debond were studied on stress distributions in the adhesive layer of a single lap tubular joint sustaining an axial load P_o. Here, the semi-analytical results were supported by those of finite element solution. In Ref. [11], the inner pipe (tube 1) was made of steel and the outer pipe (tube 2) was assumed to be an orthotropic composite tube with the layer arrangement of $[\pm 15]_{10}$. The properties of each tube and adhesive layer are given in Table 5.1.

According to Figure 5.18, the presence of a central void in the adhesive layer creates additional stress discontinuities along the bond length (here, in terms of $(\tau_{rx})_1$). Although not shown, similar behavior was observed for $(\tau_{rx})_2$. However, comparison of these results with those given in Figure 5.13(a) for a defect-free joint shows 17.9% increase in the peak shear stress considering the fact that location of the peak shear stress still occurs at the right end of overlap region (instead of the void's vertical edges). One has to realize that in this figure (and others yet to come in the next paragraphs), the three numbers i.e. 4-22-4 are interpreted as follows: the numbers on the right and left designate the distances (in mm) between the void's (or debond) vertical edges and the right and left ends of the overlap region, respectively, while the number in the middle designates the void's (debond's) length.

Figure 5.19 shows the resulting peel stress distributions in a defect-free adhesive layer (Figure 5.19(a)) and that hosting a void (Figure 5.19(b)). In both cases, the tubular joint sustains an axial load P_o. The finite element findings are also presented to support the results of semi-analytical

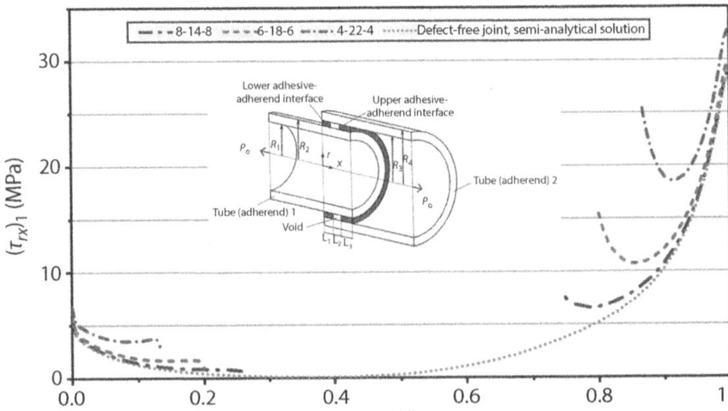

Figure 5.18 The effect of a central void size on variation of shear stress in the adhesive layer along the bond length [11].

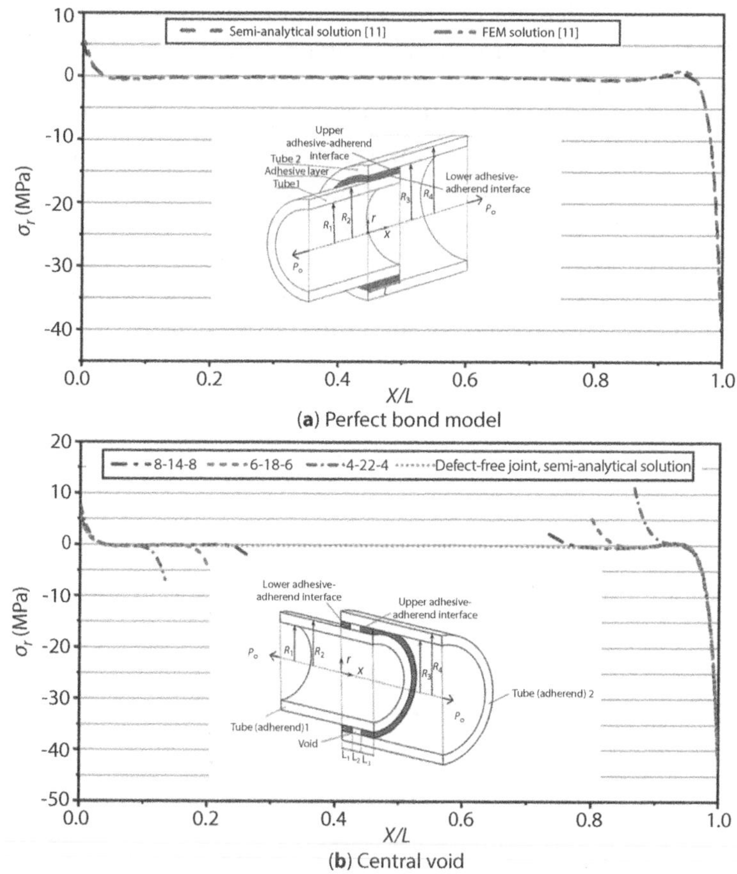

Figure 5.19 Distribution of peel stress along the bond length in (a) defect-free joint, and (b) defective joint with a central void of different sizes [11].

solutions. In contrast to a torsional load, application of an axial load produces a bending moment in the joint resulting in a peel stress the magnitude of which can exceed the limiting strength of the adhesive layer (if not controlled) and hence causing a joint failure. However, according to Figure 5.19(b), presence of a central void as large as 75% of the total debond length increases the peak peel stress, in terms of a positive traction force, by 36%. Examination of the results in this figure shows that the positive value of this stress at the right edge of the void is larger than its value at the left edge of the bonded region, indicating the adverse effect of a void via introducing multi-discontinuities in the shear stress (with relatively high stress values) along the bond length.

The effects of a debond on the shear and peel stresses in the adhesive layer due to an axial load are shown in Figures 5.20(a) and 5.20(b), respectively. A close comparison of Figures 5.18 and 5.20(a) shows the difference in adverse effect of a debond and a void on the magnitude and the location of the peak shear stress in the adhesive layer. In contrast to a central void with a size being 75% of the overlap length, under the same loading condition, a debond of the same size creates a peak shear stress that occurs at the right vertical edge of the debond rather than at the right end of the overlap region. However, according to Figures 5.19(b) and 20(b), similar effects are observed on the peel stress resulting from the two defects.

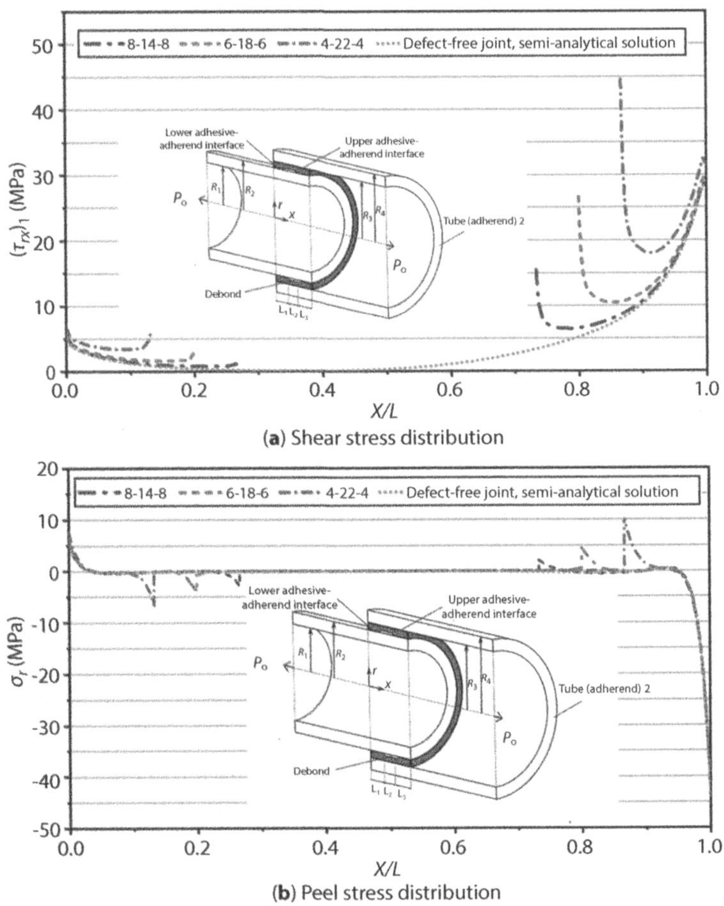

Figure 5.20 The effect of a central debond with different sizes on the peel stress distribution along the bond length [11].

Table 5.4 shows the percentage change (decrease or increase) in the adhesive peak shear stresses $(\tau_{rx})_1$ and $(\tau_{rx})_2$, as well as any change in the absolute value of maximum peel stress σ_r with respect to any deviation in the elastic modulus of the adhesive layer from E = 2.5 MPa. These stress components are developed as a result of the axial load P_o applied to the joint. Similarly, Table 5.5 shows the effect of any variations in the bond length L on the magnitudes of these stress components.

The effects of delamination on the stress distribution in the adhesive layer of a lap joint, as well as a socket joint, were fully investigated in [10] using a thorough finite element analysis. Here, for the single lap joint (see Figure 5.16), adherend 1 (or the inner pipe) was made of composite material, while adherend 2 (or the outer pipe) was considered to be either aluminum or a laminated composite tube. Additionally, in the socket joint arrangement (see Figure 5.17) considered in [10], the two adherends were identical and both adherends and the socket were made of composite materials. Furthermore, any material nonlinearity in the adhesive layer was neglected. The study included a thorough finite element-based analysis to

Table 5.4 The effect of adhesive elastic modulus E on the peak shear and peel stresses in a defect-free lap joint due to the application of an axial load P_o.

Stress component	**E = 3.5 GPa**	**E = 1.0 GPa**
Percentage change in $(\tau_{rx})_1$	10.52 (increase)	25.72 (decrease)
Percentage change $(\tau_{rx})_2$	13.83 (increase)	30.25 (decrease)
Percentage change σ_r	18.49 (increase)	36.69 (decrease)

Table 5.5 The effect of overlap length on the peak shear and peel stresses at the two ends of a defect-free lap joint in presence of an axial load P_o.

Stress component	**L = 30 mm**	**L = 10 mm**
Interfacial shear stress $(\tau_{rx})_1$	5.46 MPa (left end)	8.9 MPa (left end)
	27.83 GPa (right end)	28.67 GPa (right end)
Interfacial shear stress $(\tau_{rx})_2$	2.88 MPa (left end)	6.07 MPa (left end)
	32.39 GPa (right end)	33.81 GPa (right end)
Peel stress σ_r	5 MPa (left end)	12 MPa (left end)
	-43 MPa (right end)	-42 MPa (right end)

account for the effects of delamination's spatial location, width, length, and the applied load on the shear and peel stresses developed in the adhesive layer. The selected properties for the glass-epoxy and graphite-epoxy pipes used in this study are given in Table 5.6.

Additional investigations on the effect of adhesive nonlinear behavior on stress distribution in the adhesive layer in presence of a delamination in one of the adherends can be found in [41].

Based on the selected data in Table 5.6 and an applied torque of 100 N.m, a thorough finite element-based investigation in this reference showed that for a single tubular lap joint in torsion, the spatial location of the delamination plays an important role in the stress behavior of a defective adhesively bonded joint. Results indicated that for a delamination completely outside the overlap region (see Figure 5.17) its effect on the peel and shear stresses in the adhesive layer can be neglected, provided it is far from the left end of the overlap region.

However, for delaminations being totally or partially located in the bond length, their presence affects both foregoing stresses. Consequently, this location was introduced as the worst-case scenario. To account for the effect of delamination length on stress distribution in the adhesive layer, the length L_{1d} shown in Figure 5.17 was kept intact and equal to 2 mm, while L_{2d} (shown in the same figure) was varied in increments of 2 mm, from 2 mm to 8 mm.

Additional results showed that the presence of a delamination with a length L_{2d} = 2 mm, and $\Phi = 45°$, leads to a drop in the magnitude of peel stress from 0.416 MPa (positive traction) to - 2.17 MPa (negative traction). However, this value remained intact provided the delamination

Table 5.6 Geometric and mechanical properties of the constituents used in [10].

Graphite-epoxy pipe	Glass-epoxy pipe
E_1 = 155.0 GPa	E_1 = 50.0 GPa
$E_2 = E_3$ = 12.1 GPa	$E_2 = E_3$ = 15.2 GPa
υ_{23} = 0.458	υ_{23} = 0.428
$\upsilon_{12} = \upsilon_{13}$ = 0.248	$\upsilon_{12} = \upsilon_{13}$ = 0.254
G_{23} = 3.20 GPa	G_{23} = 3.28 GPa
$G_{12} = G_{13}$ = 4.4 GPa	$G_{12} = G_{13}$ = 0.47 GPa

length was increased only up to L_{2d} = 4 mm. Nonetheless, the peel stress dropped to -4.85 MPa, as L_{2d} reached a value of 6 mm. Further investigations showed that the effect of the delamination's circumferential width on the peel and shear stresses is negligible for angles ranging from Φ = 15° to 45°. Moreover, the maximum decrease in the peak values of peel and shear stresses in the adhesive layer occurred for a delamination spacially located near the adherend surface and adjacent to the adhesive layer. Considering stresses in the adhesive layer, although their magnitudes were significantly different for a stiffer composite joint (graphite-epoxy as opposed to the glass–epoxy), yet the trends for their distributions were similar. Figure 5.21 shows the distribution of shear stress in the adhesive layer along the overlap length and its peak values as a function of Φ, when delamination spans over the entire length of the glass-epoxy adherend.

Reference [10] presents a similar study on the influence of adherend's delamination on the stress distribution in a single lap joint as well as a socket joint under torsion. Compared to a lap joint, in a socket joint arrangement not only stress concentrations were developed at the two ends of the bonded region, but also at a section where the two adherend end surfaces were facing each other (see Figure 5.16), causing an additional stress discontinuity at this location. Furthermore, in the socket joint arrangement, the orientation of fibers did not have considerable effect on the peel stress except when the joint was constructed of $[45]_{16}$ layup. Moreover, the maximum decrease in the peak shear stress occurred for the $[+45/-45]_8$ layup.

Further analysis on delamination effect on the adhesive stress behavior in a laminated adhesively bonded tubular single lap joint made of fiber-reinforced polymer composite can be found in [42].

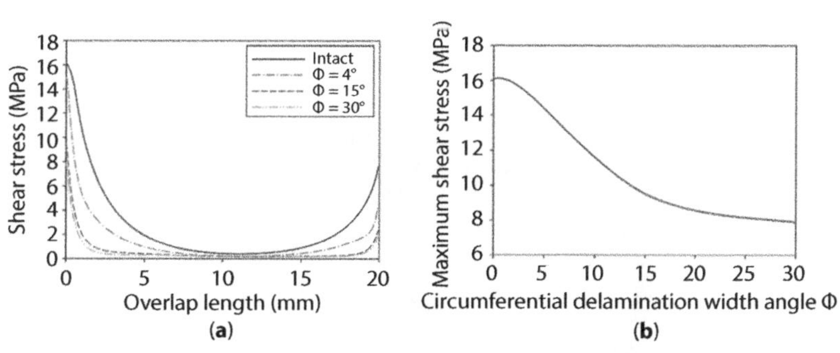

Figure 5.21 Distribution of (a) shear stress along the overlap length, and (b) maximum shear stress as a function of delamination width angle Φ, when delamination spans over the entire length of the glass–epoxy adherend (applied torsional moment = 100 N.m) [41]. For circumferential width associated with angle Φ refer to Figure 5.16 or 5.17). (Reproduced with permission from Taylor and Francis).

5.3.4 Additional Factors Influencing the Adhesive Behavior and Its Performance

As explained before, one of the major factors influencing the adhesive stresses is the properties of constituents used in construction of the joint. In this connection, Nimje and Panigrahi [14] investigated the effects of a functionally graded adhesive layer on the failure of a socket joint with laminated FRP composite tubes. In a related study, Aimmanee and Hongpimolmas [15] used elasticity theory to analyze the behavior of an adhesively bonded tubular-coupler joint, with an optimum variable-stiffness composite coupler, subjected to pure torsion. In their study, the fiber angles were assumed to vary along the coupling length. However, to obtain a solution, the coupling length was divided into small segments in which the fiber angle was assumed to be constant.

In a related work, the effect of axial load on the structural integrity of a composite tubular joint was experimentally and numerically studied in [24]. The mechanical tests were focused on the relatively small pipes (100 mm in diameter) while the finite element-based investigation was applied to the larger pipes (up to 400 mm). Additionally, an adhesive layer was used to join the two pipes in a shaved taper/taper arrangement, as shown in Figure 5.22. The analysis was performed on the perfect joints as well as those suffering from a debond.

In these joints, the optimized values of the overlap lengths between the pipes and coupling (or the socket) that varied with the pipe diameter were determined by the manufacturers ([24]). Two types of joints were used in this study. In type 1, a glass fiber-reinforced epoxy (GRE) coupling was used to join the two GRE pipes using an adhesive layer at a taper angle θ_o.

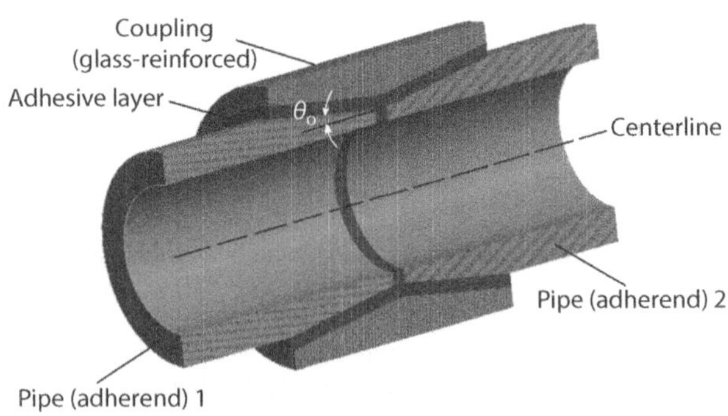

Figure 5.22 Cross-sectional view of a typical shaved taper/taper joint arrangement used in [24] (dimensions not to scale).

This model was used to investigate the joint behavior as well as the influence of the defect size on the integrity of the adhesive bond under tension. However, in type 2, using mild steel strip elements, an idealized double shear steel joint was introduced to study the effect of pipe diameter on the joint strength and determine the critical adhesive stresses based on the same type of loading. The overlap length was allowed to be well above its optimum value, in a sense that the failure load would not increase with the increase in this length. The elastic properties of the adhesive were; $E =$ 3.5 GPa, $G = 1.5$ GPa, $\upsilon = 0.37$. Values of 90 MPa and 125 MPa were selected for the in-plane shear strength and the in-plane transverse strength of the GRE lamina, respectively.

Furthermore, type 1 specimens that were produced in the laboratory included shaving of their ends to the required dimensions for a taper/taper joint with $\theta_o = 2.5°$. The experimental and numerical results revealed that the use of double lap shear joint for simulation of the tubular joint behavior can be relevant but requires more investigation. Additionally, in the GRE pipe joints, there appeared to be a size-dependent failure mode transition from pipe delamination to the failure of the adhesive layer. Furthermore, in the GRE pipes, the taper/taper connections gained considerable strength at their interface with the adhesive layer under a tensile load, such that their failure stresses were 30% lower than those of the equivalent steel joints.

One other possibility for improvement of the joint performance is the simultaneous use of mechanical fasteners and an adhesive layer (hybrid joint) [43, 44] in the joint. In this connection, mechanical fasteners (rivets), a layer of adhesive, as well as a combination of the two (riveted/bonded) were used to investigate the overall performance of the joint. To study this behavior, the structures in [43] were subjected to pure torsion to investigate (numerically and experimentally) the mechanical behavior of bonded lap joints. In this work, the mechanical behavior of the joint as well as its maximum torque capacity and strength were investigated. To search for the possible benefits of a hybrid joint, test models were prepared using mechanical fasteners, a layer of adhesive, and a combination of the two called a hybrid (riveted/bonded) joint. Each classification was analyzed three times and the average value was reported as the torque-rotational angle diagram. Additional finite element results were generated to support the test results. Figure 5.23 shows the geometric view and dimensions of the aluminum to aluminum bonded joint.

In aluminum to aluminum specimens, Al 7076-T6 was used for the inner and outer tubes while super special (S. S.) adhesive manufactured by Ghafari Chemical Company, Tehran, Iran, was selected as the bonding agent. However, in the hybrid joint, the filament winding technique was

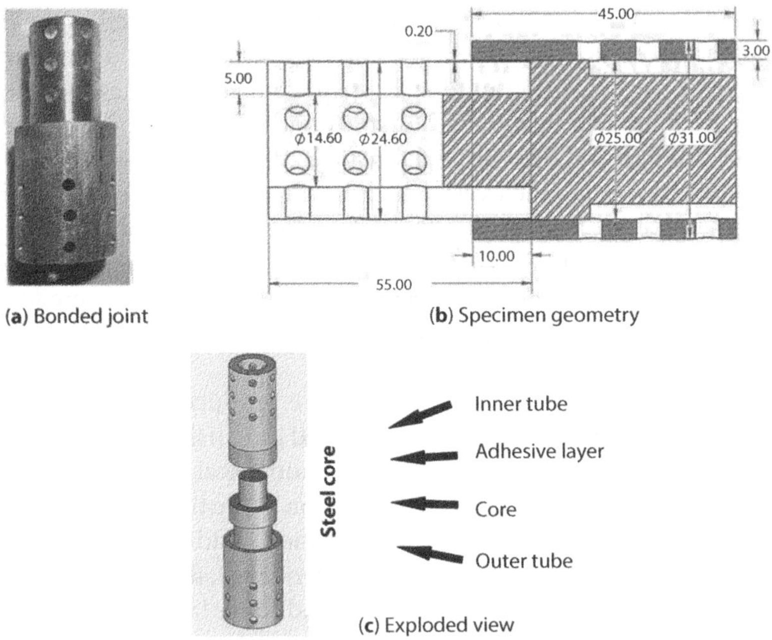

Figure 5.23 Dimensions (in mm) and geometric view of the aluminum to aluminum bonded joint [43].

adopted to fabricate the composite tube using glass fiber roving 2400 tex, and Huntsman Araldite LY 5052/Aradur 5052 as the adhesive in combination with aluminum 5050 for the rivets (see Figure 5.24 for the assembled structure and finite element model of this joint). Based on [1], the overlap length was assigned to be 20 mm.

The mechanical properties of cohesive zone material are given in Table 5.7. Additionally, mechanical properties of the adhesive layer and aluminum adherends are given in Table 5.8.

Figure 5.25 shows the shear stress distribution in the adhesive layer resulting from the finite element simulation for two different bond lengths of 10 mm and 20 mm. According to the test results, the hybrid riveted/bonded joint showed 250% and 40% higher torque capacity in comparison with the adhesively bonded and mechanically fastened (riveted) joints, respectively [44]. Consequently, it was concluded that due to a lower joint strength between the adhesive layer and E-glass-epoxy composite tube, selecting a hybrid riveted/bonded joint in place of an adhesively bonded joint might be helpful. Furthermore, defining the peak point in each torque-rotational angle diagram as the capacity of the joint, it was

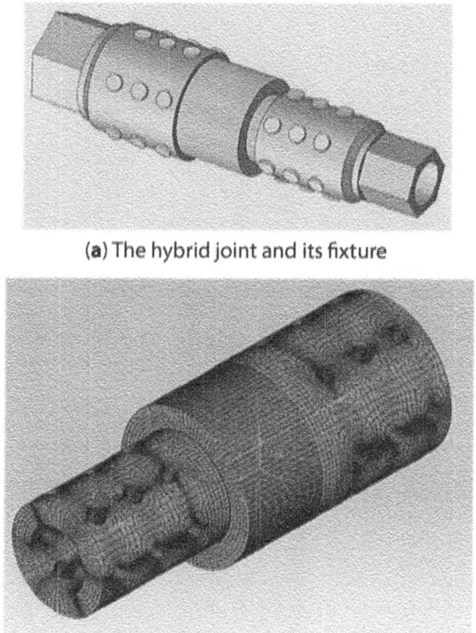

(a) The hybrid joint and its fixture

(b) Finite element model

Figure 5.24 Schematic diagram and finite element model of the hybrid joint [43].

Table 5.7 Cohesive zone mechanical properties [43].

Property	Value
Young's modulus	2100 (MPa)
Shear modulus	699 (MPa)
Tensile failure strength	28.9 (MPa)
Shear failure strength	17.8 (MPa)
Toughness	0.9 N/mm^2

Table 5.8 Mechanical properties of the adhesive layer and aluminum adherends [43].

Material	Elastic modulus (MPa)	Poisson's ratio
S. S.[1] adhesive	2100	0.4
Al 7075-T6	72000	0.33
Al 5052	70000	0.33

[1] Super special.

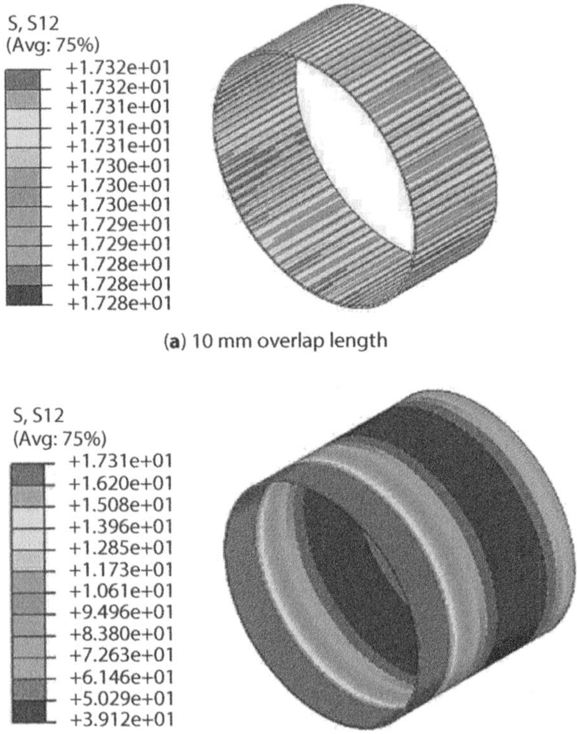

(**a**) 10 mm overlap length

(**b**) 20 mm overlap length

Figure 5.25 Shear stress distribution in the adhesive layer based on finite element solution for two different bond lengths of 10 mm and 20 mm [43].

concluded that the riveted/bonded joint has a peak point energy 4 times higher than that of an adhesively bonded joint in torsion.

Another type of adhesively bonded tubular hybrid joint might include a combination of brittle (with high stiffness and high strength) and ductile (high elongation and high fracture toughness) materials as the adhesive layer, as used in [45]. In this reference, the importance of adhesive end geometry (see Figure 5.26), as well as the geometry and properties of adhesive layer (either ductile Hysol 9361, brittle Hysol 9360 or their combination as a hybrid adhesive) for joining carbon fiber reinforced polymer (CFRP) to titanium was verified. Results showed that the application of hybrid adhesive significantly increases the joint strength due to its ability to redistribute the induced stresses while manifesting a complex damage progression. Additional results indicated that although the brittle-ductile-brittle (BDB) hybrid adhesive design yields similar strength results as the ductile-brittle-ductile case, yet the adhesive layer never completely failed at the peak force.

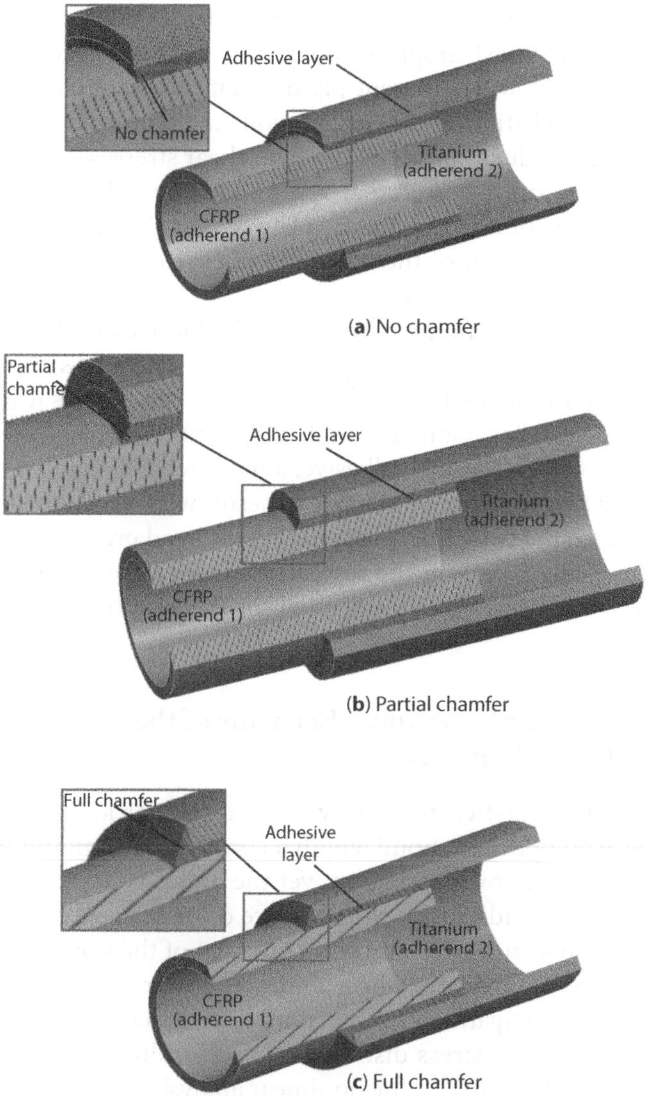

Figure 5.26 Schematic diagrams of a tubular single lap joint with three different end arrangements for the adhesive layer used in [45].

In [46] it was shown that application of hybrid composite adherends and thickness modification of the plies can reduce the stresses in a double lap tubular joint under an axial force. In this study, three-dimensional finite element analysis was used to include the nonlinear behavior of the adhesive layer and obtain a solution for stress distribution in this layer.

Materials used in this analysis included Adekit A140, Carbon-epoxy, Boron-epoxy, T300/934 graphite/epoxy, and Aramid-epoxy composites. Different laminates with different ply orientations were used to seek the effects of adherend stiffness, ply orientation, and the use of hybrid adherends on possible reductions in the peel and shear stresses developed in the adhesive layer. Results showed that application of a hybrid composite (by changing the type of fibers in the two layers near the adhesive layer) and a thickness modification of the adherend plies lead to the best results in reducing the adhesive stresses.

Using heat-activated prepreg welding and adhesive bonding, the development of a hybrid joining technique for tubular joints was explained and studied in [47]. In this study, four different types of adhesives with different mechanical properties, namely a milled glass fiber-filled polyester putty, a two-part epoxy- polyamide adhesive, a modified two-part methacrylate adhesive, and a two-part methacrylate adhesive were used to seal the existing gap between the two pipes. Glass fiber reinforced prepreg was also used to wrap the pipes. Based on the results obtained, a flexible adhesive with a moderate strength was recommended for the proposed hybrid joining approach.

5.3.5 The Effect of Nonlinear Behavior of the Adhesive on Its Performance

Although in a defect-free adhesively bonded tubular joint in tension (or torsion) with an optimum bond length a considerable amount of the load is transferred to the overlap region, yet the adhesive layer usually starts to fail at one of its ends due to the presence of high stress concentrations. Therefore, it is instructive to examine the effect of the nonlinear behavior of the adhesive layer on the shear and/or peel stress distributions. For this purpose, investigations on the nonlinear behavior of the adhesive layer and its effect on stress distribution in the adhesive layer were performed in [7] and [8]. In [7], the nonlinear analysis of a tubular single-lap adhesive joint under general loading (inner pressure p_i, outer pressure p_o, twisting moment T_o, axial load P_o, and thermal load ΔT) was performed by including the nonlinear behavior of the adhesive layer bonding the steel and composite adherends. Stress distributions in the adhesive layer were explored for different cases and compared with their values using a linear analysis. According to the results, the nonlinear analysis produced a more accurate prediction of the joint strength by relieving the stress concentration at the edge of the bondline, due to the stiffness-softening effect of the adhesive.

The effects of both material and geometric nonlinearities in governing the bond efficiency in joints with composite/metallic adherends experiencing high strain fields due to torsion were studied in [8]. In this study, the *Ramberg-Osgood* elastoplastic model was selected to represent the adhesive nonlinear behavior. Results indicated that in a joint whose overlap length has surpassed a certain limit, the nonlinear behavior in the adherend materials at high strain magnitudes plays an important role in the bond efficiency. However, beyond a certain overlap length, due to possible fabrication-related anomalies such as the presence of the void in the adhesive layer, regardless of the overlap length, the increase in the joint capacity was limited. Moreover, it was concluded that if the nonlinear design approach is to be adopted for the joint, it is recommended that the overlap length be limited to a specific length, above which there is no remarkable change in the joint capacity unless an improved fabrication process can be applied. Further analyses on the effect of nonlinear behavior of the adhesive layer on the joint performance can be found in [1, 5, 6, 24] and many others.

Furthermore, to investigate the effect of nonlinearity on stress distributions in joints with composite adherends two adhesively bonded dissimilar orthotropic cylindrical shells exposed to tensional, shear, and torsional loads, as well as bending moments were investigated in [48]. The adhesive material was modeled based on two different theories of elasticity. In the first model, a linear elastic behavior was assumed for its performance, while in the second, a more realistic approach was considered to account for its inelastic behavior. The elastostatic and nonlinear analyses were adopted to solve the problem. For this purpose, the laminated E-glass fiber reinforced plastic (GRP) tubes were assumed to be identical and composed of four E-glass/epoxy layers with $[0°/90°]_S$ arrangement. Additionally, the adherend geometric properties were $R_1 = 60.0$ mm, $R_2 = 63.1$ mm, $h_1 = h_2 = 3.0$ mm, $L = 20.0$ mm, $(L_1)_{adherend} = (L_2)_{adherend} = 40.0$ mm, and $t_{adh} = 0.1$ mm (see Figure 5.1(c) for geometric dimensions). Each adherend was composed of unidirectional E-glass/epoxy plies with $(E_{11})_m = 41.4$ GPa, $(E_{22})_m = 10.3$ GPa, $(\upsilon_{12})_m = 1.25$, $(G_{12})_m = (G_{31})_m = 5.1$ GPa, and $(G_{23})_m = 41$ GPa where $m = 1, 2$. For the linear elastostatic analysis, the AY 103 two-component epoxy from Ciba-Geigy was selected as the adhesive with $E = 2.8$ GPa and $\upsilon = 0.4$. However, to illustrate the effect of nonlinear adhesive behavior, FM73 (from American Cyanamid) rubber-modified epoxy with a ductile structural behavior was selected for this layer. Using uniaxial tension and compression tests, the mechanical properties of this material were determined as, $E = 2210.0$ MPa, $\upsilon = 0.43$, proportional limit $S_{prop} = 20.0$ MPa, ratio of compressive to tensile strength $\lambda = 1.4$, ultimate tensile strength $S_{ult} = 46.2$ MPa, and ultimate tensile strain $e_{ult} = 0.057$.

Results showed that the nonlinear behavior assumption for the adhesive layer affects the adhesive stress distribution, even at low levels of external loading. Furthermore, selection of a nonlinear behavior for the adhesive layer considerably reduced the adhesive peak stresses compared to the elastostatic solution. Consequently, it was concluded that in general the adhesive nonlinearity cannot be ignored if reliable ultimate load predictions as well as more accurate adhesive stress distributions are required.

5.3.6 Factors Influencing the Failure Behavior of the Adhesive Layer

In the study performed by Das and Pradhan [2], a finite element-based failure analysis was performed on the adhesively bonded composite pipe joints. They used Tsai-Wu theory to obtain an appropriate length for the joint and also identified regions prone to initiation of failure or damage. In the study performed by Xu and Li [49], the finite difference method was used to obtain the three-dimensional solution of stresses developed in an adhesively bonded composite joint subjected to torsion. In their analysis, the stress variation along the adhesive thickness was considered. Results showed that for tubular joints in torsion, the peel stress in the adhesive layer can be neglected provided the pipes are isotropic and/or orthotropic. In another study performed in [50], Spaggiari and Dragoni managed to regulate the torsional stresses in a tubular lap joint using functionally graded adhesives, properties of which were considered to vary along the axial direction. Furthermore, they solved the constitutive relations to obtain constant stress along the joint and to reduce the stress concentration at the joint ends. Further analyses on the adhesive failure in joints with other types of loadings and adherend properties can be found in [2, 3, 14, 18, 21], [32], and [51–57].

The fatigue resistance and axial strength of thick-walled adhesively bonded E-glass composite-to-aluminum tubular lap joints were measured in [52] for compressive and tensile loadings. The analysis was performed on tapered and untapered aluminum adherends according to two methodologies. In the first approach, the adhesive layer was assumed to be uncracked while behaving as an elastic-perfectly plastic material. In the second approach, the linear elastic fracture mechanics was used to investigate the joint behavior. Test results showed that joint failures generated during monotonic loading were quite similar to those generated by cyclic loading. Additionally, for all loadings, the initiation of joint failure occurred at the inner bond end (where the aluminum adherend terminates) and advanced in the adhesive adjacent to its interface for tensile and compressive monotonic and cyclic loadings. However, results in [32] have shown that the fatigue life limit of a double lap joint in torsion can be 16 times as large as that of a single lap joint.

5.4 Design Aspects Regarding the Selection of Adhesive Layer

Other researchers have tried to introduce proper expressions for the geometric and load parameters required in the design process of adhesively bonded joints. Ouyang and Li [1] used the following assumptions to derive the constraint equation for debonding process of a socket pipe joint with the isotropic pipe and coupler under a torsional load.

1. Linear elasticity in conjunction with the classical torsion theory and small deformation can be adopted.
2. The adhesive thickness is much smaller than the pipe thickness, where the pipe thickness is very small compared to the pipe radius.
3. The external torsional load is assumed to be resisted only by the two adherends (pipes or tubes). This means that the torsional load carried by the thin and soft adhesive layer *can be ignored.*
4. The local bending effects in the joint due to a torsional load can be neglected and hence the bonded joint experiences pure mode III loading condition.
5. The cohesive zone or debonding path is assumed to develop only along the adhesive-adherend interface.
6. The two adherends (the main pipe and the coupler) are made of homogeneous and isotropic materials.

Defining ϕ as the relative interface rotation between the inner pipe (pipe 1) and the socket (coupler) considered as the second tube, then $\phi = \phi_1 - \phi_2$. Using linear elasticity theory, it was shown that:

$$\frac{G_2 J_2 G_1 J_1}{G_2 J_2 + G_1 J_1}\phi'' = 2\pi R_{adh}^2 \tau\delta \tag{5.15}$$

In this equation, δ or the circumferential slip between the adherend and adhesive at their interface was equal to $\delta = R_{adh}(\phi_1 - \phi_2) = R_{adh}\,\phi$. Additionally, $\tau \equiv \tau(x)$ was considered to be the interfacial shear stress along the circumferential direction, G_1, and G_2 were the shear moduli of the two adherends, and J_1, and J_2, corresponded to the polar moments of inertia of the two pipes (adherends). Moreover, h_1 and h_2 were the pipe wall thicknesses and R_{adh} (see Figure 5.27(a)) was defined as the mean radius of the adhesive layer as:

$$R_{adh} = \frac{\left(R_{1ave} + \dfrac{h_1}{2}\right) + \left(R_{2ave} - \dfrac{h_2}{2}\right)}{2} \qquad (R_{2ave} \succ R_{1ave}) \qquad (5.16)$$

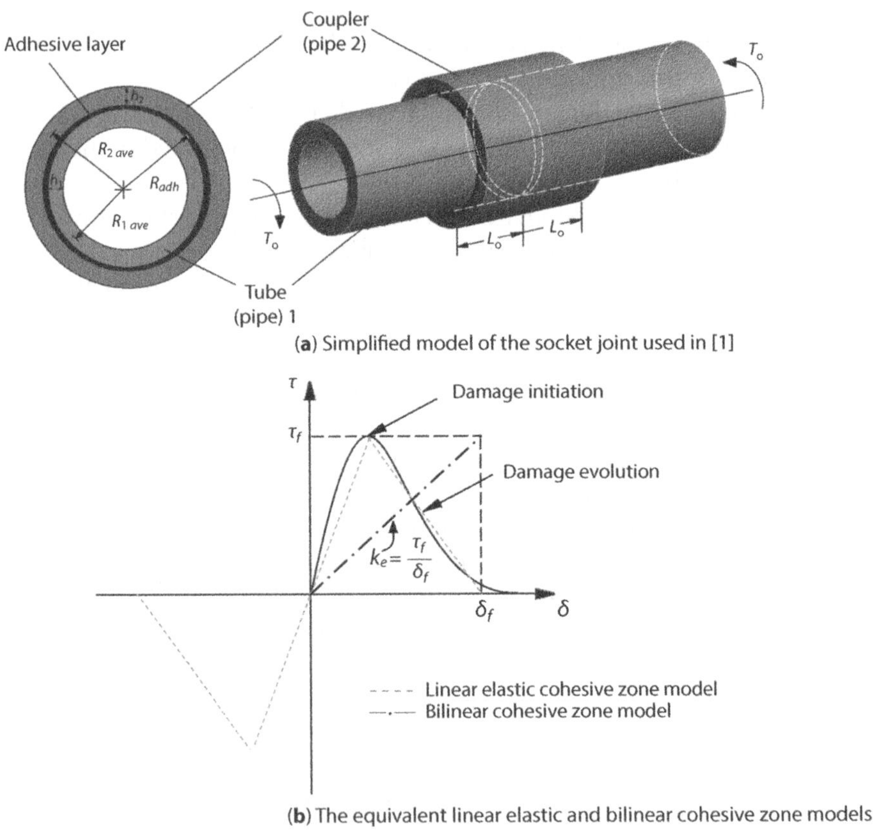

(a) Simplified model of the socket joint used in [1]

(b) The equivalent linear elastic and bilinear cohesive zone models

Figure 5.27 A simplified model of the socket joint and the equivalent linear elastic and bilinear cohesive zone models used in [1] for the derivation of design parameters.

It was shown that based on an equivalent linear elastic or bilinear cohesive zone model shown in Figure 5.27(b), one can determine the governing equation for the relative interface rotation ϕ in terms of other geometric and mechanical parameters of the constituents as follows.

Based on the equivalent linear elastic cohesive zone model, the interface shear stress can be correlated with the interface slip δ as:

$$\tau(x) = k_e \delta = k_e R_{adh} \phi \; for \; (0 \le \delta \le \delta_f) \tag{5.17}$$

where as shown in Figure 5.27, δ_f was the value of circumferential interface slip associated with $\tau = \tau_f$. Additionally, k_e was defined as $\frac{\tau_f^2}{2G_f}$ where G_f, or the interface fracture energy under mode III shear loading, was equal to $\int_0^{\delta_f} \tau \, d\delta$.

On using Eqs. (5.15) and (5.17) and applying the boundary conditions for the joined pipes in torsion, it was shown that the expression for $\phi(x)$ is:

$$\phi(x) = T_o \left(\frac{\frac{1}{\alpha_o G_1 J_1} + \frac{e^{(\alpha_o L_o)}}{\alpha G_2 J_2}}{e^{(\alpha_o L_o)} - e^{(-\alpha_o L_o)}} \right) e^{(-\alpha_o x)} + T_o \left(\frac{\frac{1}{\alpha_o G_1 J_1} + \frac{e^{(-\alpha_o L_o)}}{\alpha_o G_2 J_2}}{e^{(\alpha_o L_o)} - e^{(-\alpha_o L_o)}} \right) e^{(\alpha_o x)} \tag{5.18}$$

Where;

$$\alpha_o = \sqrt{2\pi R_{adh}^3 k_e \frac{G_1 J_1 + G_2 J_2}{G_1 J_1 G_2 J_2}} \tag{5.19}$$

In Eq. (5.19), T_o is the applied torque and L_o is the half-length of the bonded region. Furthermore, it was shown that the maximum torque capacity of the pipe socket joint under pure torsion could be estimated by;

$$T_{\max} = \phi_f \left(\frac{e^{(\alpha_o L_o)} - e^{(-\alpha_o L_o)}}{\frac{2}{\alpha_o G_2 J_2} + \frac{1}{\alpha_o G_1 J_1} (e^{(\alpha_o L_o)} - e^{(-\alpha_o L_o)})} \right), \quad (for \; G_2 J_2 \ge G_1 J_1) \tag{5.20}$$

And

$$T_{max} = \phi_f \left(\frac{e^{(\alpha_o L_o)} - e^{(-\alpha_o L_o)}}{\frac{2}{\alpha_o G_1 J_1} + \frac{1}{\alpha_o G_2 J_2}\left(e^{(\alpha_o L_o)} - e^{(-\alpha_o L_o)}\right)} \right), \quad (for\, G_2J_2 \le G_1J_1) \tag{5.21}$$

However, defining the theoretical maximum torsion as the torsional capacity of the joint for the bond length approaching infinity (long bond lengths), then according to [1], its value can be calculated as:

$$(T_T)_{max} = \sqrt{4\pi R_{adh} G_f \frac{G_1J_1}{G_2J_2}(G_1J_1 + G_2J_2)} \quad (for\; G_2J_2 \succ G_1J_1) \tag{5.22}$$

$$(T_T)_{max} = \sqrt{4\pi R_{adh} G_f \frac{G_2J_2}{G_1J_1}(G_1J_1 + G_2J_2)} \quad (for\, G_2J_2 \prec G_1J_1) \tag{5.23}$$

Based on Eqs. (5.22) and (5.23), for long enough bond lengths, the maximum transferable torsional load is only a function of R_{adh}, torsional stiffnesses of the two pipes (G_1J_1 and G_2J_2), and the interface fracture energy G_f.

Based on the argument presented in [1], the magnitude of l_e (defined as the bond length required to attain 99% of the theoretical maximum transferable torsional load) can be calculated as:

$$l_e = \frac{1}{\alpha_o}\left(\frac{0.99\xi + \sqrt{(0.99\xi)^2 + 1 - (0.99)^2}}{1 - 0.99} \right) \tag{5.24}$$

Where:

$$\xi = \begin{cases} \dfrac{G_1J_1}{G_2J_2} & if\; G_2J_2 \ge G_1J_1 \\ \dfrac{G_2J_2}{G_1J} & if\; G_2J_2 \le G_1J_1 \end{cases} \tag{5.25}$$

Additional relations based on the bilinear cohesive zone model are derived in [1].

Nonetheless, to predict the failure onset in a tubular socket joint, one has to realize that an adhesively bonded tubular socket joint can experience two vital modes of mechanical failure as; (1) cohesive failure within the adhesive bond layer, and (2) failure onset at the tube-adhesive (or socket-adhesive) interface (also called interfacial failure) due to the excessive shear and/or peel stresses arising from the applied load on the joint structure. Based on this knowledge, in the study performed on the effect of a functionally graded adhesive on the failure of a socket joint with laminated FRP composite tubes [14], Tsai–Wu coupled stress failure criterion [58] was used to identify the initiation of interfacial failure by computing the failure indices "*e*" which took into account the interaction of all six stress components. Since in the tubular socket joint the interfacial failure was mainly due to the interlaminar stresses, so only the interlaminar shear stresses $\tau_{r\theta}$ and τ_{rz}, as well as the through-thickness normal stress σ_r were required to predict the initiation of such failure. Based on this argument, the Tsai-Wu criterion was simplified as follows:

$$\left(\frac{\sigma_r}{S_T}\right)^2+\left(\frac{\tau_{r\theta}}{S_{r\theta}}\right)^2+\left(\frac{\tau_{rz}}{S_{rz}}\right)^2=e^2 \qquad \begin{array}{l} e\geq 1 \text{ failure occurs} \\ e\prec 1 \text{ no failure} \end{array} \tag{5.26}$$

In this equation, S_T, $S_{r\theta}$, and S_{rz} were the interlaminar, normal and shear strengths, respectively, where $S_{r\theta}$, = S_{rz}. Similarly, the failure index in the adhesive layer was formulated using the cohesive failure-based criterion. For this purpose, the parabolic yield criterion (see [59] and [60]) used to predict the location of cohesive failure initiation was expressed as:

$$(\sigma_1-\sigma_2)^2+(\sigma_2-\sigma_3)^2+(\sigma_3-\sigma_1)^2+2\,(\sigma_1+\sigma_2+\sigma_3)\,(|Y_c|-Y_T)=2eY_T\,|Y_c| \tag{5.27}$$

In Eq. (5.27), σ_1, σ_2, and σ_3 were the principal stresses at the location of interest while Y_T and Y_c are the yield strengths of the adhesive layer in tension and compression, respectively. For an adhesively bonded tubular socket joint subjected to an axial load P_0, the failure indices calculated from Eqs. (5.26) and (5.27) could be used to determine the possible location of interfacial failure initiation. For more information regarding this issue refer to [14].

5.5 Summary

The study on the effects of shear and peel stress distributions on the behavior of structural adhesives in tubular composite joints has shown that the main factors affecting the shear and peel stress components in the adhesive layer are the type of external load applied to the joint (i.e. axial or torsional loads, as well as bending moment or their combinations), the adherend fabrication (i.e. isotropic, quasi-isotropic or orthotropic), type of the joint (i.e. lap joint, socket joint, or any other type introduced before), and mechanical and geometric properties and dimensions of the adhesive and adherends (tubes). Results have shown that in tubular lap joints in torsion the effect of bending moment in the joint can be neglected provided there is no defect in the joint, and hence shear stress is the main stress component that controls the adhesive failure. However, for tubular joints in bending or tension, both shear and peel stresses govern the stress state in the joint. Additionally, depending on the magnitude of the applied load(s) (torsional, axial, bending, or their combinations) stress components that develop in the adhesive layer (and/or the adherend) may cause adhesion failure if they are not accurately calculated or properly controlled. These failures can be caused by the progression of stresses beyond the adhesion strength values, causing cohesive failure or a separation between the adhesive and adherend at their interface. Furthermore, if an initial crack exists in the adhesive layer, it can introduce additional stresses, leading to an early failure if its presence is not considered. However, experimental results on stress distribution in lap joints under torsion have shown that the torque transmission capacity and fatigue life limit of a single lap joint can be considerably improved by replacing the single lap joint with a double lap joint.

Additional investigations on the composite-metal (as the adherends) joints have shown that it is quite possible to replace the steel-steel tubular joints in torsion with laminated carbon-epoxy-steel joints with almost similar peak shear stresses in the adhesive layer while considerably reducing the structural weight of the joint.

Moreover, related studies on GRE pipe joints under a tensile load indicate that there is a size-dependent failure mode transition from pipe delamination to adhesion failure. Additional studies on these joints with tapered end shapes for the adherends (see Figure 5.3(c)) show that the taper/taper connection can considerably modify the joint strength under tension such that the stresses at failure are 30% lower than those of an equivalent steel joint.

Considering experimental and numerical results on the mechanical behavior of tubular hybrid riveted/bonded single lap joints with E-glass epoxy composite and aluminum adherends, it was concluded that the hybrid riveted/bonded joints show 250% and 40% higher torque capacities in comparison with the plain adhesively bonded and mechanically fastened (riveted) joints, respectively. Therefore, it can be concluded that for higher torque transmissibility, selecting a hybrid riveted/bonded joint, in place of adhesively bonded or mechanically fastened joint, can improve the joint load bearing capacity. In addition, application of a hybrid adhesive (i.e. a combination of three parts brittle-ductile-brittle adhesive) has shown to significantly increase the joint strength due to its ability to redistribute the induced stresses with a complex damage progression. Moreover, a ductile material encapsulated by brittle materials as a hybrid brittle-ductile-brittle adhesive layer has yielded similar joint strength results as the ductile-brittle-ductile case.

In addition, results on the nonlinear behavior assumption for the adhesive layer have revealed that selecting such behavior considerably reduces the adhesive peak stresses compared with the elastostatic assumption. Therefore, in general, the adhesive nonlinearity cannot be ignored if reliable ultimate load predictions as well as more accurate adhesive stress distributions are required. Furthermore, in a joint whose overlap length has surpassed a certain limit, the nonlinear behavior for the adherend materials at high strain magnitudes plays an important role in the bond efficiency. However, beyond a certain overlap length, due to possible fabrication-related anomalies (such as presence of voids in the adhesive layer), the increase in the joint capacity is limited regardless of the overlap (bond) length.

Acknowledgement

The author would like to thankfully acknowledge Dr. Kashmiri L. Mittal for his useful comments and advice during preparation of this chapter.

Nomenclature

A	Extensional stiffness matrix
A_{ij}	Elements of extensional stiffness matrix

B	Coupling matrix
B_{ij}	Elements of coupling matrix
D	Bending stiffness matrix
D_{ij}	Elements of bending stiffness matrix
E	Elastic modulus of the isotropic adhesive layer
E_1, E_2, E_3	Elastic moduli of the composite adherends (or adhesive layer) in the principal material axes (for an isotropic adhesive layer $E_1, E_2, E_3 = E$)
E_{ij}	Variables defined in Eq. (5.7d)
F_{ij}	Variables defined in Eq. (5.7e)
E_{xx}, E_{yy}	Elastic moduli along the natural body axes.
G	Shear modulus of the isotropic adhesive layer
G_1, G_2	Shear moduli of the isotropic adherends 1 and 2, respectively.
G_{12}, G_{13}, G_{23}	Shear moduli of the composite adherends (or adhesive).
G_f	interface fracture energy under mode III shear loading
G_{xy}	Shear modulus in natural material axes.
$h_m (m = 1, 2)$	Pipe (adherend) thicknesses.
J_1, J_2	Polar moments of inertia of the two adherends (tubes).
K_S	Shear correction factor.
k	Layer number.
L	Adhesive-adherend bond length.
L_o	Half-length of the bonded region used in Eqs. (5.18), (5.20), and (5.21).
$(L_m)_{adherend}$ $(m = 1, 2)$	Portion of the each adherend length outside of the bonded region (see Figure 5.1(c).

L_{1d}, L_{2d}	Portions of the delamination length outside and inside of the bond line L, respectively (see Figure 5.16 or 5.17)
l_e	Bond length parameter defined in Eqs. (5.18), (5.20, 5.21)
$M_{x\theta}$	Twisting moment per unit length
$N_{x\theta}$	Shear force per unit length
n	Total number of layers in each composite tube.
P_o	Applied axial load
Q_θ	Transverse shear force per unit length
R_1, R_2, R_3, R_4	Inner and outer radii of tubes 1 and 2, respectively [11] (see Figure 5.6).
R_{adh}	Mean radius of adhesive layer defined in Eq. (5.16)
$\bar{R}_{adh}$	Radial location in the adhesive layer defined in Eq. (5.10a).
R_{1ave}, R_{2ave}	Average radii of the pipe and the coupler (see [1])
$\bar{R}_m(m=1,2)$	Radial locations in terms of r_m and z_m
$r_m = (m = 1, 2)$	Radial distance.
S_m $(m = 1, 33)$	Parameters defined in Eqs. (5.13a)-(5.13e) (see [11]).
$S_{r\theta}$, S_{rz}	Interlaminar shear strengths
S_T	Interlaminar normal strength
T_o	Applied torque.
$T_{adh}(x)$	Twisting moment in the adhesive layer.
t_{adh}	Adhesive thickness used in Eq. (5.11).
T_{max}	Maximum torque capacity of the pipe and socket under pure torsion (see Eqs. (5.20) and (5.21)

$(T_T)_{max}$	Theoretical maximum torsion or torsional capacity of the joint for bond length approaching infinity (see Eq. (5.23))
U_m $(m = 1, 2)$	Axial displacement component in the adherends.
V_m (x, z) $(m = 1, 2)$	Circumferential displacement component in the adherends.
V_{om} (x) $m = 1, 2$	Mid-plane displacement in the pipes (adherends).
v_1, v_2	Circumferential displacement components in the lower and upper adhesive surfaces, respectively.
v_{adh} (x, r)	Circumferential displacement in the adhesive layer.
W_m $(m = 1, 2)$	Radial displacement component in each adherend (tube).
x	Axial direction along each pipe (adherend).
Y_c, Y_T	Yield strengths of the adhesive layer in compression and tension, respectively
α_o	Parameter defined in Eq. (5.19)
β_k	Fiber angle in each layer
δ	Circumferential slip between the adherend and adhesive at their interface
δ_f	Circumferential interface slip associated with $\tau = \tau_f$
e	Strain matrix
$\boldsymbol{\varepsilon}^{(k)}$	Strain matrix in the kth layer
$\varepsilon_x^{(k)}, \varepsilon_\theta^{(k)}, \varepsilon_r^{(k)}$	Normal strain components in the kth layer
$(\varepsilon_x)_m$, $(\varepsilon_\theta)_m$, $(\varepsilon_r)_m$ $m = 1, 2$	Normal strains components in each adherend
Φ	Angle defining one-half of the circumferential width of the delaminated region.
ϕ	Relative interface rotation

ϕ_1, ϕ_2	Interface rotations of the inner tube and the socket
$\gamma_{r\theta}^{(k)}, \gamma_{rx}^{(k)}, \gamma_{x\theta}^{(k)}$	Components of shear strain in the kth layer
$(\gamma_{r\theta})_m, (\gamma_{rx})_m, (\gamma_{x\theta})_m$ $m = 1, 2$	Shear strain components in each adherend
η_o, η_1, η_2	Parameters defined in Eqs. 5.9(a)-5.9(c)
$\varphi_{\theta m}(x)$ $m = 1, 2$	Mid-plane rotation in each pipe (adherend).
λ	Ratio of compressive to tensile strength
σ	Stress matrix
σ_m $(m = 1, 2, 3)$	Principal stress components
$\sigma_x^{(k)}, \sigma_\theta^{(k)}, \sigma_r^{(k)}$	Normal stress components in the kth layer
$\sigma_x, \sigma_\theta, \sigma_r$	Normal stress components
$(\tau_{r\theta})_m$ $m = 1, 2$	Components of shear stress at the lower and upper adhesive-adherend interfaces ($\tau_1 = (\tau_{r\theta})_1$, $\tau_2 = (\tau_{r\theta})_2$)
$\tau_{r\theta}^{(k)}, \tau_{rx}^{(k)}, \tau_{x\theta}^{(k)}$	Components of the shear stress in the kth layer
$\upsilon_{12}, \upsilon_{13}, \upsilon_{23}$	Poisson's ratios in principal material axes
υ_{xy}	Poisson's ratio in natural body axes
ξ	Torsional stiffness ratio defined in Eq. (5.25)

References

1. Z. Ouyang and G. Li, Cohesive zone model based analytical solutions for adhesively bonded pipe joints under torsional loading, *Int. J. Solids. Struct.*, 46, 1205–1217 (2009).
2. R. R. Das and B. Pradhan, Adhesion failure analyses of bonded tubular single lap joints in laminated fibre reinforced plastic composites, *Int. J. Adhesion Adhesives*, 30, 425–438 (2010).
3. I. Y. Sulu and S. Temiz, Failure and stress analysis of internal pressurized composite pipes joined with sleeves, *J. Adhesion Sci. Technol.*, 32, 816-832 (2018).

4. R. R. Das and B. Pradhan, Finite element based design and adhesion failure analysis of bonded tubular socket joints made with laminated FRP composites, *J. Adhesion. Sci. Technol.*, 25, 41–67 (2011).
5. J. H. Oh, Strength prediction of tubular composite adhesive joints under torsion, *Composites Sci. Technol.*, 67, 1340–1347 (2007).
6. M. M. Abdel Wahab and G. De Roeck, A fracture mechanics approach for designing an adhesive joint for glass fibre-reinforced pipes, *J. Adhesion Sci. Technol.*, 11, 719-733 (1997).
7. J. H. Oh, Nonlinear analysis of adhesively bonded tubular single-lap joints for composites in torsion, *Composites Sci. Technol.*, 67, 1320–1329 (2007).
8. R. Hosseinzadeh and F. Taheri, Nonlinear investigation of overlap length effect on torsional capacity of tubular adhesively bonded joints, *Composite Struct.*, 91, 186–195 (2009).
9. G.P. Zou and F. Taheri, Stress analysis of adhesively bonded sandwich pipe joints subjected to torsional loading, *Int. J. Solids. Struct.*, 43, 5953- 5968 (2006).
10. R. A. Esmaeel and F. Taheri, Influence of adherend's delamination on the response of single lap and socket tubular adhesively bonded joints subjected to torsion, *Composite Struct.*, 93, 1765–1774 (2011).
11. M. Ghamarian, Stress analysis of adhesively bonded laminated composite pipes with a tubular void or debond, M. Sc. thesis, Shahid Chamran University of Ahvaz, Ahvaz, Iran (2021).
12. J. H. Choi and D. G. Lee, An experimental study of the static torque capacity of the adhesively-bonded tubular single lap joint, *J. Adhesion*, 55, 245–260 (1996).
13. M. Shishesaz and S. Tehrani, Interfacial shear stress distribution in the adhesively bonded tubular joints under tension with a circumferential void or debond, *J. Adhesion Sci. Technol.*, 34, 1172-1205 (2020).
14. S.V. Nimje and S.K. Panigrahi, Effects of functionally graded adhesive on failures of socket joint of laminated FRP composite tubes, *Int. J. Damage Mech.*, 26, 1170-1189 (2017).
15. S. Aimmanee and P. Hongpimolmas, Stress analysis of adhesive-bonded tubular-coupler joints with optimum variable-stiffness composite adherend under torsion, *Composite Struct.*, 164, 76–89 (2017).
16. M. M. Mohieddin Ghomshei and R. Shahi, Stress analysis of single-lap bonded joints in composite tubes under torsion and hygrothermal effects using DQM, *J. Sci. Technol. Composites*, 4, 375–385 (2018).
17. M. Shishesaz and S. Tehrani, The effects of circumferential voids or debonds on stress distribution in tubular adhesive joints under torsion, *J. Adhesion*, 96, 1396-1430 (2019).
18. N. Baishya, R. R. Das, and S. K. Panigrahi, Failure analysis of adhesively bonded tubular joints of laminated FRP composites subjected to combined internal pressure and torsional loading, *J. Adhesion Sci. Technol.*, 31, 2139-2163 (2017).

19. S. R. Graves and D. F. Adams, Analysis of a bonded joint in a composite tube under torsion, *J. Composite Mater.*, 15, 211-224 (1981).
20. C. T. Chon, Analysis of tubular lap joint in torsion, *J. Composite Mater.*, 16, 268-284 (1982)
21. M. Shishesaz and M. Hosseini, Effects of joint geometry and material on stress distribution, strength and failure of bonded composite joints: an overview, *J. Adhesion*, 96, 1053-1121 (2020).
22. R. D. Adams, J. Comyn, and W. C. Wake, *Structural Adhesive Joints in Engineering*, 2nd Edition, Chapman & Hall, London, UK (1997).
23. J. L. Lubkin and E. Reissner, Stress distribution and design data for adhesive lap joints between circular tubes, *J. Appl. Mech., Trans. ASME*, 78, 1213-1221 (1956).
24. S. A. Hashim, M. J. Cowling, and S. Lafferty, The integrity of bonded joints in large composite pipes, *Int. J. Adhesion Adhesives*, 18, 421-429 (1998).
25. W. T. McCarvill and J. P. Bell, Torsional test method for adhesive joints. *J. Adhesion*, 6, 185–193 (1974).
26. R. S. Alwar and Y. R. Nagaraja, Viscoelastic analysis of an adhesive tubular joint, *J. Adhesion*, 8, 79–92 (1976).
27. R. D. Adams and N. A. Peppiatt, Stress analysis of adhesive bonded tubular lap joints, *J. Adhesion*, 19, 1–18 (1977).
28. P. J. Hipol, Analysis and optimization of a tubular lap joint subjected to torsion, *J. Composite Mater.*, 18, 298-311 (1984).
29. Y. R. Nagaraja and R. S. Alwar, Nonlinear stress analysis of an adhesive tubular lap joint, *J. Adhesion*, 10, 97–106 (1979).
30. J. D. Whitcom and K. Woo, Analysis of debond growth in tubular joints subjected to tension and flexural loads, *Computers Structures*, 46, 323-329 (1993).
31. Y. P. Shi and S. Cheng, Analysis of adhesive-bonded cylindrical lap joints subjected to axial load, *J. Eng. Mech.*, 119, 584-602 (1993).
32. J. H. Choi and D. G. Lee, The torque transmission capabilities of the adhesively-bonded tubular single lap joint and the double lap joint, *J. Adhesion*, 44, 197–212 (1994).
33. C. Yang, Design and analysis of composite pipe joints under tensile loading, *J. Composite Mater.*, 34, 332–349 (2000).
34. O. Nemes, F. Lachaud, and A. Mojtabi, Contribution to the study of cylindrical adhesive joining, *Int. J. Adhesion Adhesives*, 26, 474-480 (2006).
35. S. Kumar and M. A. Khan, An elastic solution for adhesive stresses in multi-material cylindrical joints, *Int. J. Adhesions Adhesives*, 64, 142-152 (2016).
36. C. T. Herakovich, *Mechanics of Fibrous Composites*, John Wiley and Sons, West Sussex, UK (1998).
37. J. N. Reddy, *Mechanics of Laminated Composite Plates and Shells: Theory and Analysis*, Second Edition. CRC Press, New York (2004).

38. J. M. Whitney and N. J. Pagano, Shear deformation in heterogeneous anisotropic plates, *J. Appl. Mech.* 37, 1031–1036 (1970).
39. C. Yang, H. Huang, and Z. Guan, Stress model of composite pipe joints under bending, *J. Composite Mater.*, 36, 1331-1348 (2002).
40. S. Aimmanee, P. Hongpimolmas, and K. Ruangjirakit, Simplified analytical model for adhesive-bonded tubular joints with isotropic and composite adherends subjected to tension, *Int. J. Adhesion Adhesives,* 86, 59–72 (2018).
41. R. A. Esmaeel and F. Taheri, Stress analysis of tubular adhesive joints with delaminated adherend, *J. Adhesion Sci. Technol.*, 23, 1827–1844 (2009).
42. R. R. Das and B. Pradhan, Delamination damage analysis of laminated bonded tubular single lap joint made of fiber-reinforced polymer composite, *Int. J. Damage Mech.*, 23, 772-790 (2014).
43. M. Yousefi, S. Rahnama, and M. Farhadi Nia, Theoretical and experimental investigation on mechanical behavior of aluminum to aluminum tubular bonded lap joint under pure torsion and a finite element comparison with hybrid rivet/bonded joint, *J. Appl. Computational Mech.*, 8, 485-492 (2022).
44. M. Yousefi, S. Rahnama, and M. Farhadi Nia, Experimental investigation on the mechanical behavior of hybrid bonded (rivet/adhesive) tubular lap joint (composite to metal) under pure torsion, *J. Adhesion Sci. Technol.*, 34, 2510-2521 (2020).
45. I. Kaiser and K. T. Tan, Damage and strength analysis of carbon fiber reinforced polymer and titanium tubular-lap joint using hybrid adhesive design, *Int. J. Adhesion Adhesives,* 103, 102710 (2020).
46. M. Mokhtar, K. Madani1, M. Belhouari, S. Touzain, X. Feaugas, and M. Ratwani, Effects of composite adherend properties on stresses in double lap bonded joints, *Mater. Design*, 44, 633–639 (2013).
47. G. Li, D. Davis, C. Stewafu, J. Peck, and S-S. Pang, Joining composite pipes using hybrid prepreg welding and adhesive bonding, *Polym. Composites,* 24, 697-705 (2003).
48. O. T. Thomsen, Elasto-static and elasto-plastic stress analysis of adhesive bonded tubular lap joints, *Composite Struct.*, 21, 249-259 (1992).
49. W. Xu and G. Li, Finite difference three-dimensional solution of stresses in adhesively bonded composite tubular joint subjected to torsion, *Int. J. Adhesion Adhesives*, 30, 191–199 (2010).
50. A. Spaggiari and E. Dragoni, Regularization of torsional stresses in tubular lap bonded joints by means of functionally graded adhesives, *Int. J. Adhesion Adhesives*, 53, 23–28 (2014).
51. R. R. Das and N. Baishya, Failure analysis of bonded composite pipe joints subjected to internal pressure and axial loading, *Procedia Eng.*, 144, 1047–1054 (2016).
52. E. D. Reedy Jr. and T. R. Guess, Composite-to-metal tubular lap joints: strength and fatigue resistance, *Int. J. Fracture,* 63, 351-367 (1993).

53. D. G. Lee, K. S. Kim, and Y.-T. Im, An experimental study of fatigue strength for adhesively bonded tubular single lap joints, *J. Adhesion,* 35, 39-53 (1991).
54. J. W. Kwon and D. G. Lee, The effects of surface roughness and bond thickness on the fatigue life of adhesively bonded tubular single lap joints, *J. Adhesion Sci. Technol.*, 14, 1085–1102 (2000).
55. O. E. Canyurt, Fatigue strength estimation of adhesively bonded tubular joint using genetic algorithm approach, *Int. J. Mech. Sci.*, 46, 359–370 (2004).
56. U. Kumar and R. R. Das, Adhesion failure analyses of laminated FRP composite made bonded tubular T-joints with axially compressed brace, *Composite. Struct.*, 258, 113386, (2021).
57. H. C. Biscaia, M. S. Carvalho, A. P. Martins, and R. Micaelo, Interfacial failure of circular or tubular hybrid bonded joints: A theoretical description, *Eng. Failure. Anal.*, 132, 105936 (2022).
58. S. W. Tsai and E. M. Wu, A general theory of strength for anisotropic materials, *J. Composite Mater.*, 5, 58–80 (1971).
59. R. Raghava, R. M. Caddel, and G. S. Y. Yeh, The macroscopic yield behavior of polymers. *J. Mater. Science,* 8, 225–232 (1973).
60. R. D. Adams, Strength predictions for lap joints, especially with composite adherends: A review, *J. Adhesion,* 30, 219–242 (1989).

6

Inelastic Response of Structural Aerospace Adhesives

Yi Chen and Lloyd Smith*

Washington State University, Pullman, WA, USA

Abstract
Most bonding in the aerospace industry relies on structural adhesives with high strength to achieve good bond quality in load bearing structures. As polymers, the mechanical behavior of adhesives, especially inelastic deformations corresponding to the durability of the bonded system, are complicated and in need of study. Considering both experimental and theoretical work, this chapter reviews the inelastic response of structural aerospace adhesives from time-independent plastic behavior to time-dependent inelastic behavior focusing on creep and cyclic responses. The influences of temperature and moisture on inelastic behaviors are also discussed.

Keywords: Structural adhesive, plasticity, viscoelasticity, viscoplasticity, creep, ratcheting, temperature, moisture

List of Symbols

$\varepsilon, \dot{\varepsilon}$	Tensile strain and tensile strain rate
t	Time
D_0	Instantaneous tensile compliance
ΔD	Transient tensile compliance
σ	Tensile stress

**Corresponding author*: lvsmith@wsu.edu

K.L. Mittal and S.K. Panigrahi (eds.) Structural Adhesives: Properties, Characterization and Applications, (255–290) © 2023 Scrivener Publishing LLC

D_i	i^{th} Prony compliance
λ_i	i^{th} retardation time
E	Stiffness of an elastic spring
η	Viscosity
$a_{1i}, a_{2i}, a_{3i}, b_{1i}, b_{2i}$	Constants in the modified Burger model
g_0, g_1, g_2, a_σ	Nonlinear viscoelastic parameters
ψ^t	Reduced time
e_{ij}	Deviatoric strain tensor
ε_{kk}, θ	Volumetric strain
S_{ij}	Deviatoric stress tensor
J_0, B_0	Instantaneous shear and bulk compliances
$\Delta J, \Delta B$	Transient shear and bulk compliances
f	Yield function
n, Y	Constants
a_T	Temperature shift factor
C_1, C_2	Material constants in the Williams-Landel-Ferry equation
T, T_0	Temperature and reference temperature
ΔT	Change in temperature
μ	Shear modulus
K	Bulk modulus
e_{eff}	Effective strain
δ	Parameter relating the change in free volume due to mechanical dilation
c	Solvent concentration
Δc	Change in solvent concentration
α	Volumetric thermal expansion of the free volume

β	Volumetric expansion due to a change in moisture content
f_d	Reference value of free volume under dilatation
f_s	Reference value of free volume under shear
B_d, B_s	Parameters in the free volume approach
D	Diffusion coefficient
F	Flux of solvent
x, y, z	Spatial coordinates

6.1 Introduction

A structural adhesive is applied to join surfaces together when the load required to cause separation is substantial [1]. Structural adhesives are used in many applications to bond a wide range of similar and dissimilar materials providing flexibility in joint design and repair, especially in the aerospace industry where a high strength to weight ratio is of primary importance. Adhesive bonding distributes load over a large area to reduce stress concentrations, while bolt or rivet joining methods introduce defects and weakness near the bolt or rivet holes. As composite materials and sandwich structures are increasingly used for airframe and primary structures, adhesive bonding has become popular due to its good bonding appearance and strength between dissimilar materials [2]. Moreover, adhesively bonded joints provide excellent damping and vibration properties, and improved resistance to fatigue and corrosion. With these advantages, structural adhesives are now replacing the conventional joining methods to bond or repair critical parts for long aircraft life and reduced maintenance cost.

No single adhesive works for all purposes in the aerospace industry. Various adhesives play different roles for varying environments, load-bearing requirements, and substrate types. Most aerospace adhesives can be classified by their chemistries into the following categories [3]: acrylic, amino-formaldehyde, ceramic, epoxy, cyanate ester, phenolic, polyimide, polyurethane, and silicone. Structural bonding mostly uses epoxies, phenolics, and polyimides. Common types of adhesively bonded joints are shown in Figure 6.1.

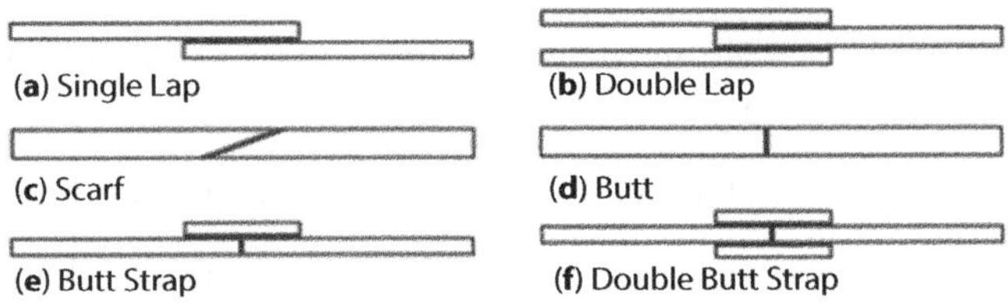

Figure 6.1 Common adhesively bonded joints [5].

As most adhesives are synthetic polymers, they generally exhibit complex mechanical behavior including time-dependent and plastic features which are associated with reduced adhesion. Adding to the complexity are complicated joint structures and the difficulty of measuring strain in thin adhesive layers that may behave differently from bulk adhesives [4]. Experiments with bulk adhesives and bonded joints have been developed to characterize the mechanical behaviors of adhesives. As some physical experiments are expensive and difficult to manipulate, analytical and numerical models are often needed to assist the design and optimization of bonded joints. Therefore, understanding and accurately predicting the inelastic behavior of adhesives in bonded joints is needed in adhesive design.

This chapter reviews the inelastic behavior of structural aerospace adhesives and is divided into three sections. The first section considers classical plasticity and compares the performance of bonded joints experimentally and introduces commonly used yield criteria. In the second section, time/rate-dependent behaviors corresponding to viscoelasticity and viscoplasticity are discussed. Responses to creep and cyclic loading are introduced using examples of aerospace adhesives. Viscoelastic and viscoplastic constitutive models are discussed and model selection is suggested. The last section discusses the effects of temperature and moisture on the inelastic behavior of adhesives. Examples throughout the chapter involve aerospace structural adhesives. However, epoxy (the most popular aerospace adhesive) is also used in cases where insufficient aerospace studies are found. The phenomena and models may not be suitable for all aerospace applications.

6.2 Time-Independent Plasticity

6.2.1 Yield Stress

Adhesives experience yielding and introduce unrecoverable deformation as loading is increased to failure. Although adhesives are time-dependent, plasticity with analytical solutions can describe the stress-strain response of joints subjected to short-term loading and for preliminary study and design.

Plasticity requires a definition of the yield surface from the stress-strain response obtained from monotonic tests. Some will use a 0.2%/0.3% strain offset and the intersection of the initial and final slopes of the stress-strain curve (as shown in Figure 6.2) to determine the yield stress from the monotonic stress-strain curves. However, the yielding phenomenon of adhesives is often gradual without a significant change between the elastic and plastic zones because of the temporal effect introducing gradual and recoverable deformation to mix with plastic yielding under quasi-static load. Therefore, the measurement of strain recovery (where yielding starts when residual strain is observed after sufficient recovery) is preferred [6].

The tensile and compressive yield stresses are obtained by loading the adhesive of thin bulk [6] (Figure 6.3c) and cylindrical bulk [7] (Figure 6.3a & 3b) in tension and compression, respectively. There also have been some studies using butt and tubular joints to measure the yield stress in tension and compression, respectively. However, adhesives in bonded joints usually undergo complicated stress conditions, such as stress concentrations at the edges of bonds and radial stress components triggered by constraints from stiff adherends [8]. Thus, tubular and butt joints are not recommended for obtaining tensile and compressive properties.

The tests using bulk adhesives are sensitive to defects such as voids during processing. Moreover, the ratio of bulk thickness to the adhesive thickness in a bonded joint can be larger than 100 which results in different mechanical behaviors. To analyze and predict adhesive behaviors in practical applications, experiments on adhesively bonded joints under shear and combined stresses are required.

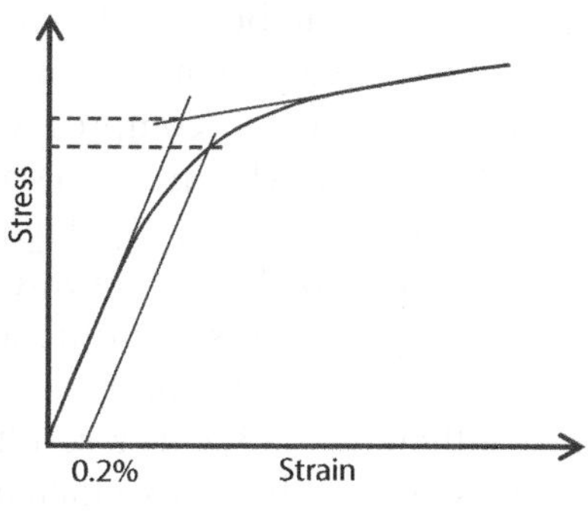

Figure 6.2 Determination of yield stress using the 0.2% offset method and the intersection of initial and final slopes.

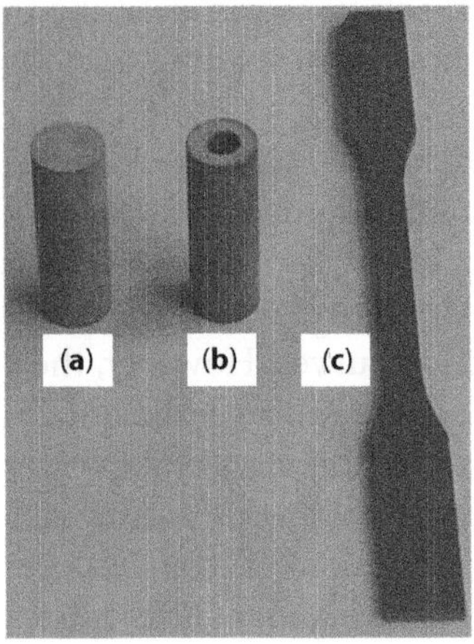

Figure 6.3 Bulk specimens for tests in (a) & (b) compression, and (c) tension [7].

Single and double lap shear joints [9] are used to characterize the shear properties of adhesives but both show stress concentration near the edges of joints, which may produce yielding in the adhesive near the edge (edge effect) and are reflected in the load-displacement response resulting in a smaller estimation of the yield stress. By studying the effects of adherend thickness and overlap length on the stress distribution along the bondline, the thick-adherend shear test (TAST) [10] is proposed as a modification of lap shear joints with thicker adherends to achieve reduced stress concentrations. The tubular butt joint under torsional loading [11] is also used to analyze the shear behavior of adhesives and shows a more even stress distribution than lap and thick adherend shear joints. Scarf joints with small angles are also used to provide a uniform shear stress condition for shear studies.

Sometimes the yield surface has to be considered with hydrostatic and von Mises equivalent stresses since polymers may behave differently in tension and compression. Thus tension/compression-shear mixed load condition presented by Arcan [12] (Figure 6.4a) and scarf [13] joints with varying angles and Iosipescu joints [14] (Figure 6.4b) are used to obtain yield stress under biaxial load with varying ratios of normal to shear stress. Figure 6.5 shows yield stress as a function of the hydrostatic stress for a bulk epoxy adhesive, Arcan joints, and Althof test (modified TAST with thinner adherends [15]). The result shows an apparent effect of hydrostatic stress on yielding. To efficiently minimize stress concentrations and edge effects in Arcan joints, beaks are machined in the adherends adjacent to the adhesive [16] as shown in Figure 6.6.

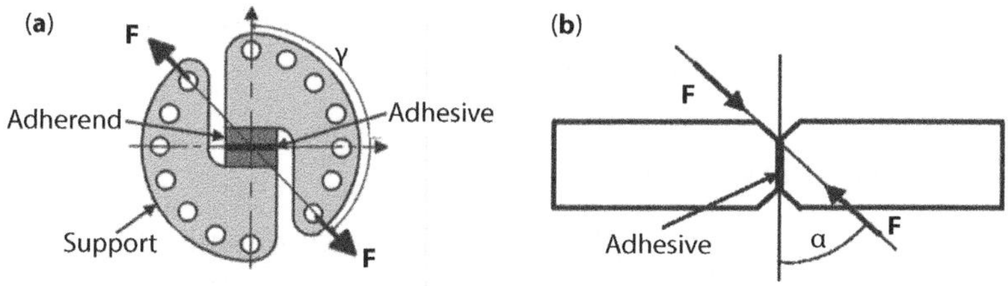

Figure 6.4 Mixed-load test specimens: (a) an Arcan test fixture [12] and (b) an Iosipescu specimen under tensile-shear loading.

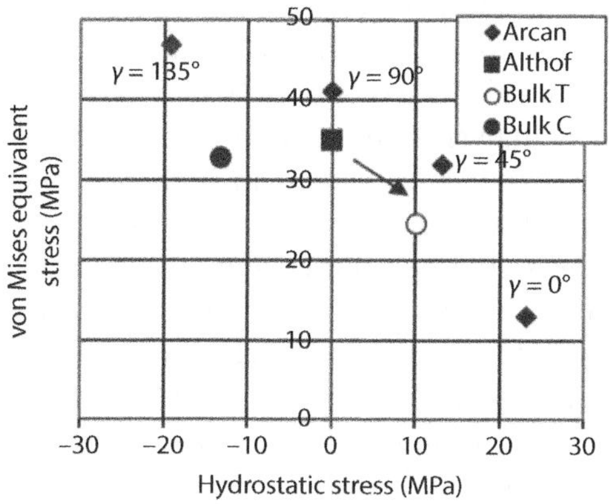

Figure 6.5 Yield stress in the hydrostatic stress-von Mises equivalent stress plot for an epoxy adhesive [7]. 'Bulk T' and 'Bulk C' represent bulk test under tensile and compression, respectively. γ is defined in Figure 6.4a and represents the type of tensile-shear loading for the Arcan test. Due to the complex stress state, the point for the Althof test (solid square) is plotted using only the shear stress while the arrow indicates a likely accurate position for the Althof test.

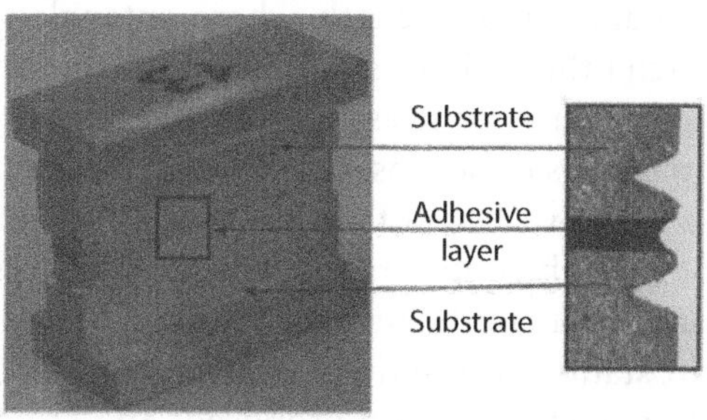

Figure 6.6 Arcan joint with beaks in substrates adjacent to the adhesive [16].

This geometry is also suitable for thick adherend joints [17] to achieve a more uniform stress distribution.

6.2.2 Elasto-Plastic Models

As mentioned above, elasto-plastic models are commonly used for stress analysis of adhesively bonded joints. In elasto-plasticity, the stress-strain relation is linear elastic at low stress. Plastic strain occurs when the stress goes beyond the first yield stress. The total strain can be described as sum of the elastic and plastic components.

The yield criterion determines the initial yield surface at which elastic deformation will be accompanied by plastic deformation. Deviatoric yield criteria including von Mises and Tresca yield criteria are frequently used for adhesives [12, 18, 19]. The von Mises yield criterion has been frequently employed for stress analysis of bonded joints [20–23]. Employing the von Mises criterion Sayman [21] presented an analytical solution to determine the shear stress distribution along the bondline of a single lap joint with a ductile adhesive layer. The one-dimensional analytical solution had good agreement with the numerical solution and showed that the peak shear stress increased with the decreasing thickness of the adherend. Plasticity also occurs from peel stress (or tensile stress) of bonded joints which is related to the adhesion strength. Thus, an efficient method of dividing the adhesive layer into elastic and plastic zones and considering the adherend with a bending-shear coupling stiffness was employed by Yang *et al.* [20] and Mardani *et al.* [23] to obtain the peel and normal stress distributions in lap shear joints. The plastic zone was located at the edge of the overlap and described by the von Mises criterion in plane strain. The predicted stresses were in good agreement with finite element analysis.

Deviatoric yield criteria are not adequate for adhesives where the yield behavior is sensitive to hydrostatic stress presenting a difference between the tensile and compressive yield stresses. Polymer yielding is often associated with secondary bond strength where entangled polymer chains slide relative to each other. Hydrostatic compression inhibits this sliding motion leading to higher yield stress [24]. In addition, most adhesives are toughened with particles or scrims. Cavitation can occur in the adhesive under tension (a negative hydrostatic pressure) but not under shear or compression [25, 26], which lowers the tensile yield stress. In many joint geometries, failure occurs in the region with high peel stress that leads to a significant hydrostatic stress and volumetric strain [27]. Consequently, hydrostatic yield criteria have been proposed. The Drucker-Prager yield criterion, taken from the von Mises criterion by adding a hydrostatic component, is widely used for adhesives sensitive to hydrostatic stress [28, 29].

Cognard *et al.* [30] plotted yield stress in the von Mises - hydrostatic stress diagram for different tensile-shear tests and employed an exponential Drucker-Prager yield criterion to consider the nonlinear relation between the hydrostatic stress and the von Mises stress on the yield surface. The exponential Drucker-Prager yield criterion was also verified to reproduce the yield behavior of adhesives tested at high stress levels [30] and high temperatures [31], and under high hydrostatic stress induced by the adherends' constraint [14, 26].

6.3 Time-Dependent Inelasticity

To pursue a long service life the adhesive behaviors under long time loading should be understood and accurately predicted. The dynamic response of adhesively bonded joints has received attention since numerous adhesives are tested and working under impact or cyclic loading where they exhibit time/rate-dependent deformations which can be described by viscoelasticity or viscoplasticity. Viscoelasticity is a phenomenon where materials behave as both an elastic solid and a viscous fluid. Adhesives manifest viscoelasticity in various ways, such as creep behavior where deformation continues under a constant load and a time-dependent deformation recovery taking place after load removal. If the deformation following stress removal is not fully recoverable after a sufficiently long time and the unrecoverable deformation exhibits rate- or time-dependent characteristics, then this unrecoverable deformation is defined as viscoplastic. Viscoelasticity occurs as a load is applied and continues through the load-unload process while viscoplasticity occurs when a threshold load, or yield stress, is exceeded.

6.3.1 Creep Loading

Adhesively bonded joints often need to withstand continuous load. Under long-term loading adhesive properties will decrease with time, and the stress-strain response will be different from short duration [32]. Figure 6.7 shows a typical creep-recovery response to a constant load. In the primary stage creep increases at a decreasing rate while it proceeds at a steady rate in the secondary stage. As the time of loading increases, the creep rate increases in the tertiary stage until failure. The first two stages have been experimentally and theoretically studied in detail.

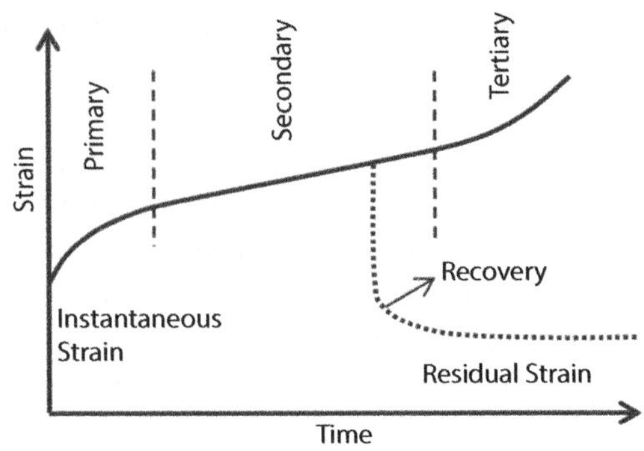

Figure 6.7 Typical creep-recovery curve for a viscoelastic material.

The study of adhesive creep response under varying stress levels and load durations is needed to characterize the time-dependent features and is used as basis to develop corresponding constitutive models. The creep behavior of adhesives at low stress levels can be described by linear visco-elasticity, that is confirmed experimentally by two properties,

1. Proportionality: strain response is proportional to the stress level at a given time, such that compliance is dependent on time
2. Superposition: the response to each incremental load is uncoupled and contributes together to the total response

Increasing stress levels can give rise to significant nonlinear viscoelasticity where the compliance of adhesives is not only dependent on time but also on stress. Creep tests with different levels of constant stress are used to determine the threshold between linear and nonlinear viscoelasticity [33]. Yu *et al.* [34] presented a series of experimental data on creep-recovery in tension and compression for an epoxy adhesive, the behavior of which is typical of many classes of structural adhesives. The tensile creep response is shown in Figure 6.8a where the three characteristic stages of creep can be seen. Moreover, the data also display the nature of nonlinear viscoelasticity. Comparing the tensile and compressive creep data in Figure 6.8b, creep developed more rapidly with time in tension than in compression, which indicates creep can also depend on the hydrostatic stress. This hydrostatic sensitivity is also found in other published articles [32].

Adhesive strain and stress can be redistributed during creep. Su and Mackie [35] investigated the creep behaviors of two epoxy-based structural

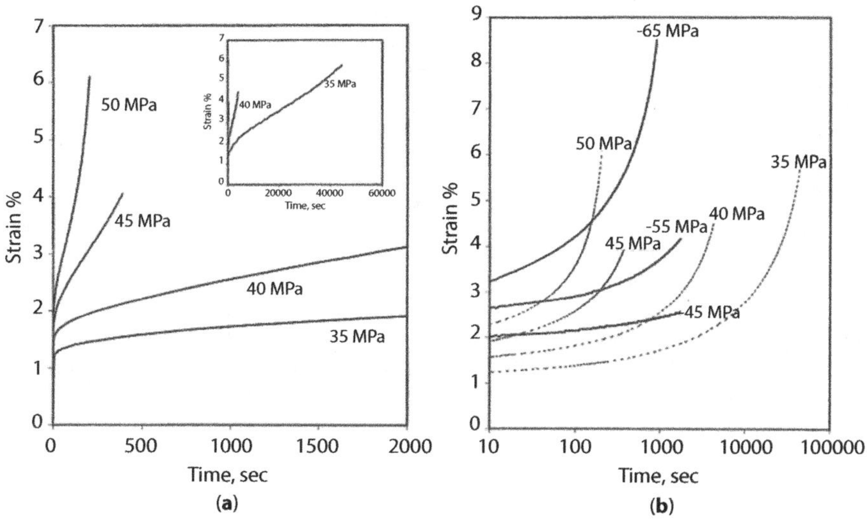

Figure 6.8 (a)Tensile creep data and (b) comparison of compressive (bold) and tensile (dashed) creep data for an epoxy adhesive [34].

adhesives in the TAST. A viscoelastic model showed that creep helped reduce the stress concentrations along the bondline and led to a more even stress distribution with reduced peak normal and shear stresses. However, the decrease in peak stress was accompanied by an increase in shear strain. To ensure a good performance in the long-term service, a balance between stress and strain development should be achieved and strong creep behavior should be limited.

At high stress levels, some adhesives exhibit not only viscoelastic but also viscoplastic behavior which further increases the nonlinearity of creep. Viscoplastic deformation can be examined by residual strain of the adhesive under different load durations. Chen and Smith [36] conducted a series of creep tests on two adhesives with scarf joints and measured the residual strain after recovery. The residual strain developed nonlinearly with increasing load duration that was described by viscoplasticity, as shown in Figure 6.9.

Cyclic creep-recovery test can also be used to study viscoplastic behavior. Bidaud *et al.* [37] studied the viscous effects of a polyurethane adhesive in both normal and shear using cyclic creep-recovery tests with several load levels, followed by recovery, as shown in Figure 6.10. At low loading levels, creep was fully recoverable while at higher loads, time-dependent residual strain appeared, indicating viscoplastic flow.

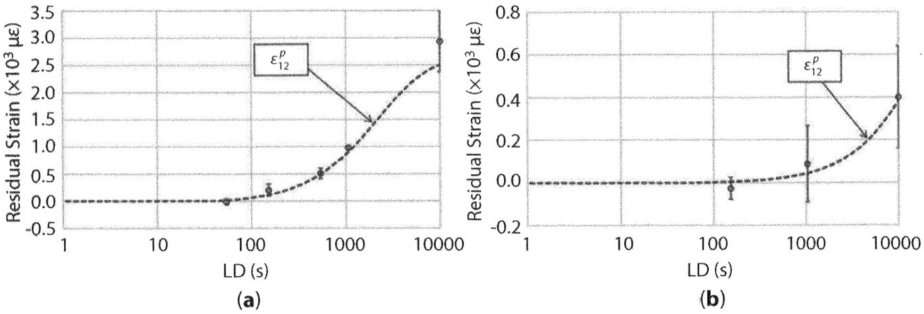

Figure 6.9 Viscoplastic development under creep loading for (a) a toughened and (b) standard epoxy-based adhesive in scarf joints [36]. Circles represent the measured residual strain for varying load durations (LD). The dashed line $\left(\varepsilon_{12}^{p}\right)$ is the simulated viscoplastic shear strain fitting to the experiment.

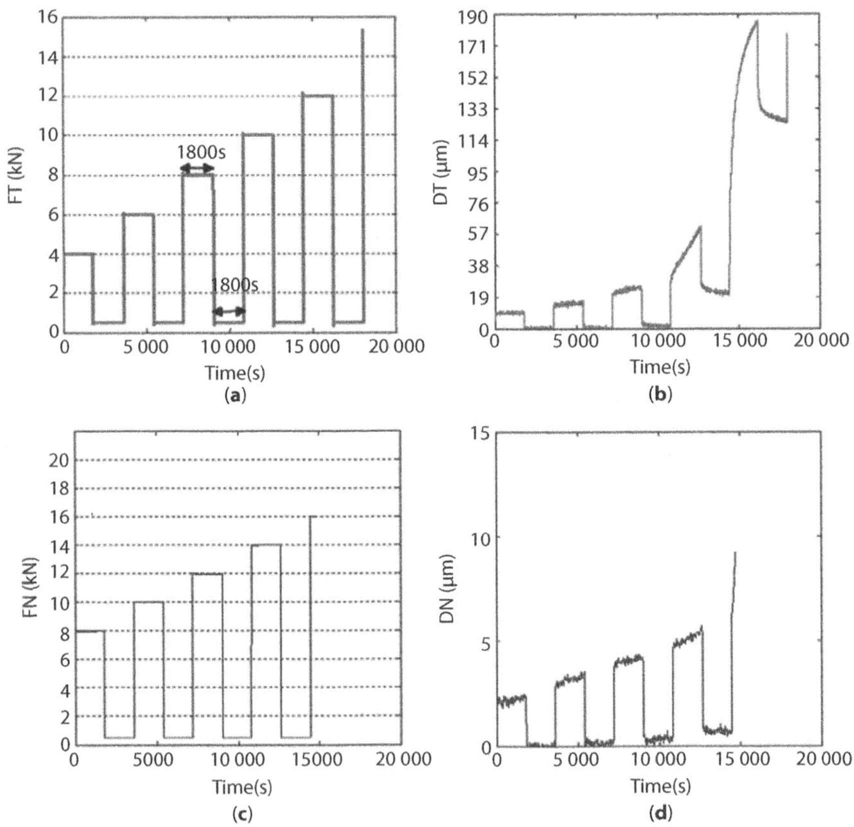

Figure 6.10 Cyclic tensile shear creep/recovery tests: FT and FN represent the applied loads in tangential and normal directions, respectively. DT and DN are the displacements in tangential and normal directions, respectively [37].

6.3.2 Cyclic Loading

A large number of adhesives are used in both load-carrying and secondary structures, such as wing assemblies, which are tested or working under cyclic and even reversed loading.

Due to time-dependent characteristics, ratcheting, a phenomenon of inelastic strain accumulation when the material is subjected to a cyclic stress, is observed in adhesively bonded joints and is receiving attention since this accumulation of inelastic deformation can cause damage and reduce life. While ratcheting and fatigue both involve cyclic loading, the former is associated with strain accumulation and the latter is associated with crack initiation and propagation.

Different adhesives in bulk [33], tubular butt joints [38] and scarf joints [39], have been tested in uniaxial cyclic tension to study the effects of stress amplitude, mean stress and frequency on ratcheting. For adhesives the ratcheting strain increases with increasing stress amplitude and mean stress, which results in a decreased life. Frequency influences ratcheting more significantly in the tertiary stage than in the transient and steady stages. Figure 6.11 shows an example of the influences of stress amplitude, mean stress and frequency on the ratcheting strain and life for a tubular butt joint.

Adhesives under cyclic loading have not only nonlinear viscoelasticity, but also viscoplasticity. Chen and Smith [40] studied the evolution of plastic deformation under cyclic loading at various frequencies from the residual strain of scarf joints. The plastic deformation showed a significant temporal effect at short time scales which diminished at long time scales, and was accurately represented by a viscoplastic model.

The multiaxial cyclic behavior of polymers is usually investigated using Arcan joints and Iosipescu samples. Thevenet *et al.* [41] analyzed the cyclic behavior of Arcan joints subjected to tensile/compression-shear loading with varying load amplitudes and mean loads. Longer fatigue life was achieved under compression-shear loading where large accumulative displacement at failure was attributed to the friction load under compression.

For metals, ratcheting occurs when the cyclic load has non-zero mean stress and plastic deformation occurs. For adhesives, ratcheting can occur under fully reversed loading. Krause and Smith [39] observed an enhanced ratcheting phenomenon in a toughened adhesive tested in scarf joints subjected to reversed cyclic loading, where higher ratcheting strain was observed under reversed cyclic loading than tensile cyclic loading. Moreover, under cyclic loading with reversed yielding, some adhesive plastic deformation was accompanied by the Bauschinger effect

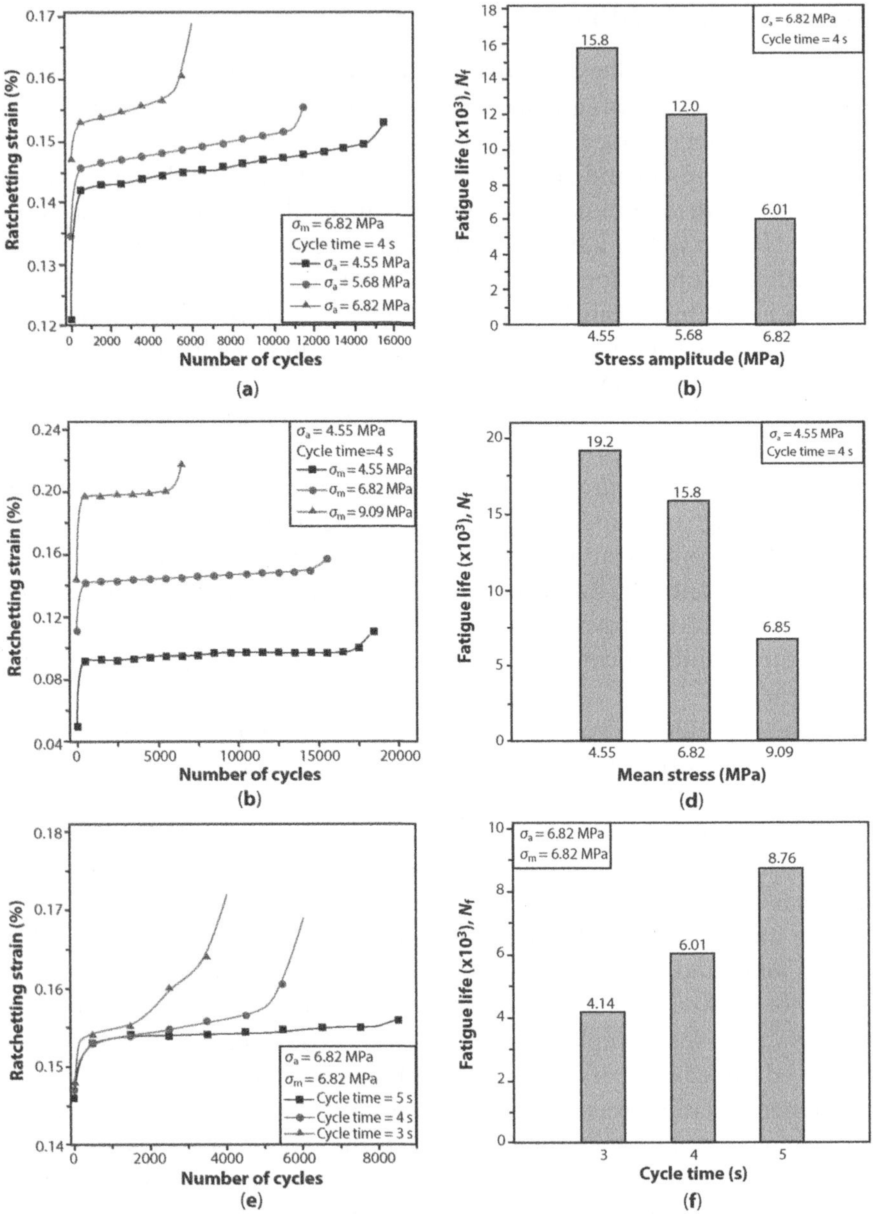

Figure 6.11 The effect of stress amplitude (σ_a), mean stress (σ_m) and frequency (cycle time) on (a, c and e) uniaxial ratcheting and (b, d and f) fatigue life of a tubular butt joint [38].

(a phenomenon of reduced yield stress in compression after the material yields in the initial tensile load). Chalkley and Chiu [42] performed a strain rate controlled cyclic test on a thick adherend joint and observed a substantial Bauschinger effect during unloading and subsequent reversed loading from the measured stress-strain response as shown in Figure 6.12.

Viscoelasticity describes energy dissipating in the internal structure or returning to the surroundings by the complex modulus that is decomposed into a loss and a storage modulus. These properties can be experimentally characterized by dynamic mechanical analysis (DMA) where cyclic loads with a range of frequencies are applied. A phase lag between the input load and the output strain response shows that energy dissipation determines the loss and storage moduli. The phase lag of a viscoelastic material is between 0° (elastic) and 90° (purely viscous). DMA is often used to characterize the viscoelastic behavior of adhesives in tension [44–46], torsion [47, 48], and bending [49] modes.

Except for creep and cyclic loads, the time-dependent behavior of adhesives under quasi-static and dynamic stress/strain states can be studied by stress relaxation tests [50, 51] at varying constant strain and monotonic tests at varying stress/strain rates [50, 52, 53]. Shear stress-strain curves [42] (Figure 6.13) of a rubber toughened adhesive tested using thick adherend joints subjected to different strain rates showed a strong rate dependent yield stress which suggested that the adhesive was viscoplastic in shear.

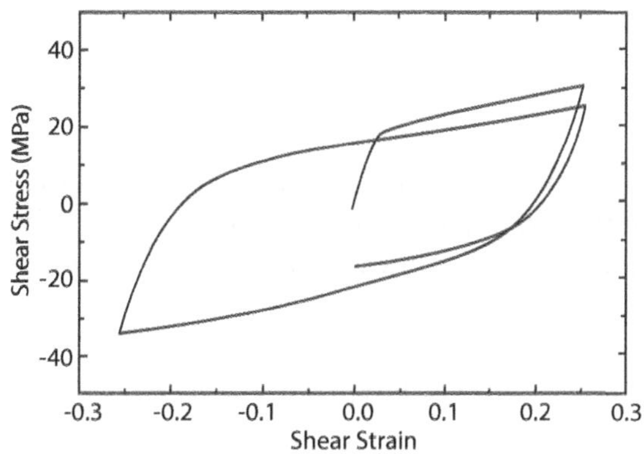

Figure 6.12 Shear stress-strain curve for the thin film adhesive FM73 under cyclic loading [43].

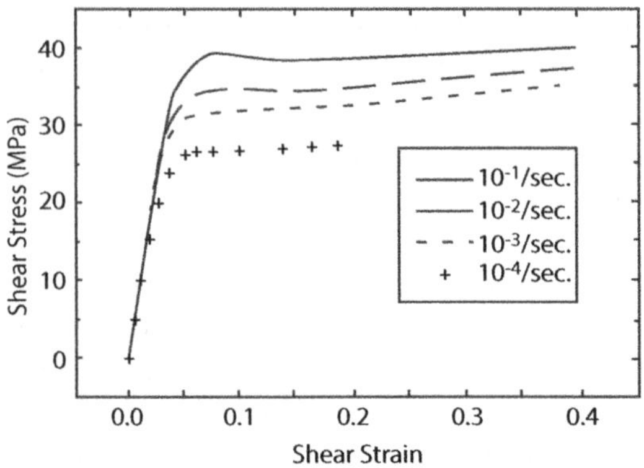

Figure 6.13 Shear stress-strain curves conducted at different strain rates [42].

6.3.3 Time-Dependent Models

No single model is able to describe all time-dependent behaviors. Temporal models consider viscoelasticity, viscoplasticity and their combination to describe the time-dependent phenomena that are observed.

6.3.3.1 *Modeling of Creep*

There have been several studies of creep behaviors of adhesives with linear viscoelastic models [54–56]. Linear viscoelasticity is characterized by proportionality and superposition described previously, and the latter leads to the Boltzmann superposition integral:

$$\varepsilon^t = D_0\sigma^t + \int_0^t \Delta D(t-\tau)\frac{d\sigma^\tau}{d\tau}d\tau \tag{6.1}$$

where D_0 is the instantaneous compliance and ΔD(t) is the transient compliance that is found from creep tests. The power law and Prony series represent creep behaviors in the linear viscoelastic range for many adhesives [57–59], expressed as

$$\text{Power law: } \Delta D(t) = \Delta D t^n \tag{6.2}$$

$$\text{Prony series: } \Delta D(t) = \sum_{i=1}^{N} D_i(1-\exp(-\lambda_i t)) \tag{6.3}$$

where D_i and λ_i are the ith Prony compliance and retardation time, respectively.

Besides the integral form of constitutive relations, there is another type of constitutive model for viscoelastic materials, the differential representation. A common approach is the rheological model presenting viscoelasticity with elastic springs and viscous dashpots for solid and fluid response, respectively, as shown in Figure 6.14. Spring deformation follows Hooke's law with stiffness E while the dashpot is assumed to behave as Newtonian fluid with viscosity η, as

$$\text{Spring: } \varepsilon = \frac{\sigma}{E} \tag{6.4}$$

$$\text{Dashpot: } \dot{\varepsilon} = \frac{\sigma}{\eta} \tag{6.5}$$

Maxwell and Kelvin-Voigt models [60] are the backbones of this approach where many variants are constructed from these two models in parallel or series. The Prony series is derived from a series of Kelvin elements [61–64].

To present plastic deformation during creep at high stress, sliding friction or yield elements have been added to rheological viscoelastic models. For example, Brinson proposed a modified Bingham model [65] (Figure 6.15a) allowing for a linear elastic, a viscoelastic and a successive plastic region to describe the tensile and shear creep and strain rate-dependent behaviors of structural adhesives in bulk and bonded structures [66, 67]. A similar model developed by Chase and Goldsmith [68] (Figure 6.15b) was used to model creep of a thermoplastic adhesive in lap shear joints [69, 70].

At high stress levels, the response becomes increasingly nonlinear and linear viscoelastic analysis is no longer adequate and consequently is modified to account for stress dependence. Constant material parameters, such

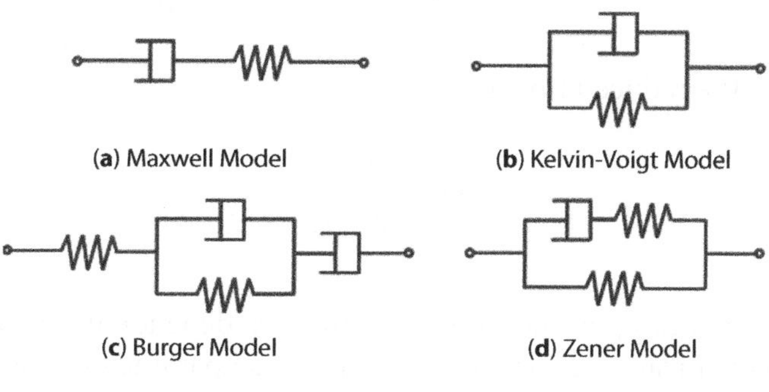

Figure 6.14 Commonly used rheological models for linear viscoelasticity.

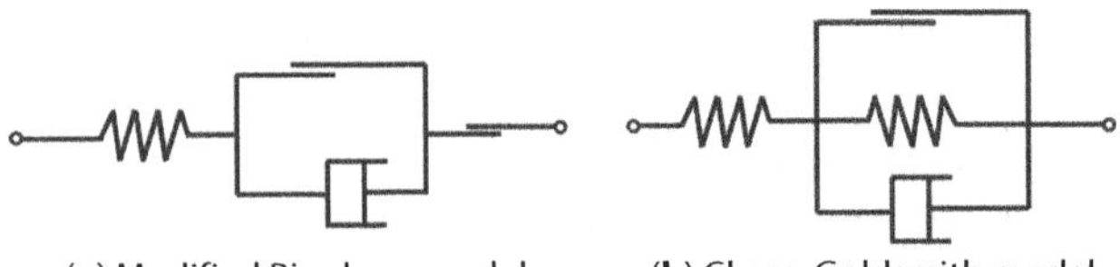

Figure 6.15 Viscoelastic-plastic models represented with spring, dashpot and slider elements.

as modulus and viscosity, are replaced with nonlinear functions of stress [71–73]. For example, Majda and Skrodzewicz [74] modified the Burger model [75] (Figure 6.14c) with

$$E_i = a_{1i}\,\sigma^2 + a_{2i}\sigma + a_{3i} \tag{6.6.1}$$

$$\eta_i = e^{b_{1i}+b_{2i}\sigma} \tag{6.6.2}$$

providing a dependence of material properties on the current stress with constants a_{ij} and b_{ij}. The model was able to describe the uniaxial creep deformation of an epoxy adhesive. In general, the more elements a rheological model has, the better agreement with experiment will be obtained. However, the more complex the model is, the more parameters should be determined by experiment and the more difficult it is to obtain its multiaxial form for complex loading conditions.

Among all the nonlinear viscoelastic theories for polymers and adhesives, the most popular is the Schapery single integral model developed from the concept of irreversible thermodynamics [76]. This one-dimensional constitutive relation may be expressed as

$$\varepsilon(t) = g_0 D_0 \sigma + g_1 \int_0^t \Delta D(\psi^t - \psi^\tau) \frac{d(g_2\sigma)}{d\tau} d\tau \tag{6.7}$$

where ψ^t is the reduced time given by

$$\psi^t = \int_0^t \frac{d\tau}{a_\sigma} \tag{6.8}$$

and a_σ, g_0, g_1 and g_2 are nonlinear parameters dependent on the current stress σ. For linear viscoelastic materials, these parameters are unity and Eq. 6.7 reduces to Eq. 6.1.

The compliance can be expressed in the form of a power law [57, 58] or Prony series [59], and characterized by creep-recovery experiments [77]. Henriksen [78] extended the Schapery model for a multiaxial stress state resulting in a constitutive matrix that consisted of an instantaneous compliance, Poisson's ratio and hereditary strains. The approach showed a good agreement with the experimental creep data in uniaxial tension for the FM73 adhesive system. However, the model did not consider the different deformation mechanisms due to the deviatoric and hydrostatic stresses. To present the effect of hydrostatic stress on compressive response, Lai and Bakker [79] modified Henriksen's model by decomposing strain into a deviatoric and a hydrostatic component, using the Schapery single integral. The model for isotropic materials can be expressed as

$$\varepsilon_{ij}(\mathrm{t}) = \mathrm{e}_{ij}(\mathrm{t}) + \frac{1}{3}\varepsilon_{kk}(t)\delta_{ij} \tag{6.9.1}$$

$$\mathrm{e}_{ij}(\mathrm{t}) = \frac{1}{2}g_0 J_0 \mathrm{S}_{ij} + \frac{1}{2}g_1\int_0^t \Delta J(\psi^t - \psi^\tau)\frac{d(g_2 S_{ij})}{d\tau}d\tau \tag{6.9.2}$$

$$\varepsilon_{kk}(t) = \frac{1}{3}g_0 B_0 \sigma_{kk}\delta_{ij} + \frac{1}{3}g_1\int_0^t \Delta B(\psi^t - \psi^\tau)\frac{d(g_2\sigma_{kk})}{d\tau}d\tau\delta_{ij} \tag{6.9.3}$$

where δ_{ij} is Kronecker delta. $e_{ij}(\mathrm{t})$, $\varepsilon_{kk}(\mathrm{t})$ and S_{ij} are the deviatoric strain, volumetric strain and deviatoric stress at the current time t, respectively. And $J_0/\Delta J(\psi^t)$ and $B_0/\Delta B(\psi^t)$ are the instantaneous/transient shear compliance and instantaneous/transient bulk compliance, respectively. Though the Schapery model is excellent in modeling creep behavior for adhesives [80] and polymeric materials [81], transferring the model into a finite element package is challenging since the entire calculation history must be stored for each increment which is expensive for time and computational cost.

In some cases of nonlinear creep, empirical equations in power law [82, 83], exponential form [84, 85], and hyperbolic sine function [86] are preferred and widely implemented in commercial finite element software. There have been several studies using empirical equations to represent nonlinear creep behavior for structural adhesives [87, 88].

Viscoplastic models are often employed to describe creep behavior with rate/time-dependent plastic deformation. Since the viscoplastic model has a yield stress beyond which the model is activated, it is generally combined with the viscoelastic model to describe the time-dependence during the whole creep procedure. Accordingly, the total strain could be presented by

$$\varepsilon^{tot} = \varepsilon^{ve} + \varepsilon^{vp} \tag{6.10}$$

where superscripts *tot, ve* and *vp* denote total, viscoelastic, and viscoplastic, respectively.

The overstress theory describes the viscoplastic component as a function of the stress above the yield point (overstress) and is frequently used in developing viscoplastic models [89–91]. Among these the Perzyna model [90] has been frequently used due to its ability to accommodate different yield criteria, which can be written as

$$\dot{\varepsilon} = \eta \langle \phi(f) \rangle^n \frac{\partial f}{\partial \sigma} = \begin{cases} \eta \left(\dfrac{f}{\Upsilon} \right)^n \dfrac{\partial f}{\partial \sigma} & \text{if } f > 0 \\ 0 & \text{if } f \leq 0 \end{cases} \tag{6.11}$$

where η and n are a viscosity and rate-sensitive parameter, respectively, determining the rate of viscoplastic deformation. Υ is commonly chosen as the initial yield stress. And $\varphi(f)$ is the overstress function expressed with the yield function *f*. Ilioni *et al.* [92] combined the Perzyna model with a nonlinear viscoelastic model where the viscoelasticity was divided into several viscous mechanisms and used to model the behaviors of Arcan joints subjected to multi-level creep loadings in tension and shear. Chen and Smith proposed a nonlinear viscoelastic-viscoplastic model [36, 93] consisting of the Schapery and the Perzyna models. The model was able to reproduce the creep behavior of bulk adhesives and bonded joints. Observing time-dependent residual strain of an adhesive from a cyclic shear creep-recovery test on Arcan joints, Bidaud *et al.* [94] combined a rheological viscoelastic model with the Perzyna model considering a Drucker-Prager yield criterion to simulate the creep-recovery responses in shear, tension, and tensile-shear. The model showed satisfactory results in both creep and recovery stages.

6.3.3.2 Modeling of Ratcheting

The ratcheting behavior of polymers is complex and can involve time-dependent, plastic, viscoelastic, and viscoplastic deformations. A nonlinear viscoelastic formulation is often needed to describe the cyclic behavior of adhesives [95, 96]. Rocha *et al.* [96] developed a rheological linear viscoelastic-viscoplastic model. The model was able to simulate the rate-dependent plasticity of an epoxy resin, but overestimated the ratcheting strain under a tensile cyclic load. It relied on viscoplasticity to represent the nonlinear stress-strain response which was in fact a combination of viscoplasticity and nonlinear viscoelasticity. Xia and coworkers [97, 98] found that the Schapery model [79] was not able to describe hysteresis for an epoxy and the underestimation of the ratcheting strain became significant at high stress levels. They proposed a mechanical analogue in which a yield criterion distinguished between loading and unloading cases to change the corresponding modulus under cyclic loading. The model showed good agreement with monotonic and cyclic experiments. Xia *et al.* [99] further modified the loading/unloading rule to incorporate the modified Schapery model [79] and the modified free volume model from Popelar and Liechti [100]. Compared with the Schapery and free volume models, the combined model better agreed with the measured ratcheting deformation and hysteresis under cyclic loading.

Due to the chain structure and nucleation of cavities, the deformation of most adhesives is sensitive to hydrostatic stress. This effect can be magnified in response to cyclic loading involving both tension and compression. Some work has shown that the mechanical behavior of adhesives is better simulated when considering hydrostatic stress, especially for ductile adhesives. Krairi and Doghri [95] developed a constitutive model coupling viscoelasticity, viscoplasticity and ductile damage, where the viscoplastic component was described by the Perzyna model with the von Mises criterion considering isotropic and kinematic hardening. In a strain-controlled cyclic test, the model overestimated the peak stress and underestimated the valley stress. The discrepancies were attributed to neglecting the effect of hydrostatic stress. Da Costa Mattos and coworkers performed a cyclic load-unload test with increasing force per cycle on polymeric materials [52, 101, 102] and observed that strain hardening was dependent on both the accumulated strain and loading rate. They modified the Lemaitre-Chaboche elasto-viscoplastic model [103] in conjunction with a hydrostatic yield criterion. The model was able to reproduce the uniaxial stress-controlled cyclic tensile tests and the corresponding hysteresis. However, the model was one-dimensional. Frank and Brockman [104] developed a nonlinear viscoelastic-viscoplastic model where the viscoplastic component was described by a modified Bodner-Partom viscoplastic model [105, 106] accounting for the effect of pressure on yielding. It was able to capture most of the time-dependent response for glassy polymers subjected to monotonic and cyclic loads. While the above models all considered the influence of hydrostatic stress on yield, Lai and Bakker [79] stated that the hydrostatic stress could also affect the viscoelastic behavior of polymers that exhibited different tensile and compression responses. Though there is limited work on the hydrostatic sensitivity to viscoelasticity, its potential effects should be noted.

6.4 Environmental Factors

Aerospace concerns for structural adhesives are not only load-bearing requirements that bonds withstand static and dynamic loads, but also environment requirements that adhesives are subjected to over a long-term service. Environmental factors include temperature, moisture, pressure, ultraviolet radiation and chemical exposure in distinct locations. This section focuses on the effects of temperature and moisture since they are the most relevant to general aerospace adhesives. Ambient temperature for adhesives in aerospace applications ranges from -100 to 120°C [107]. In some cases humidity can reach 100% with temperatures of 60-80°C. Some adhesives work with multifunctional fluids, such as fuels, lubricants, de-icing fluid and bilge water. These severe environments may accelerate the degradation of bond strength [108].

6.4.1 Temperature

At low temperature most adhesives have reduced toughness as molecular motion is reduced, while at high temperature ductility increases [109]. The glass-transition temperature (T_g), above which a large change in modulus occurs, is an important threshold for adhesives. It is therefore important to understand the adhesives' inelastic behaviors over a wide range of temperature.

Generally, the adhesive yield stress, modulus and strength are reduced at high temperature [110]. Increasing temperature enhances the strain rate sensitivity [31]. Under cyclic load, some studies show that the effect of loading frequency is temperature-dependent for adhesives with high time-dependence [111–113]. Reis *et al.* [113] performed fatigue tests at 2, 10 and 40 Hz on lap shear joints with an epoxy adhesive and observed the lowest and highest fatigue lives at 2 and 10 Hz, respectively, while the fatigue life at 40 Hz was between 2 and 10 Hz. This phenomenon was attributed to a large temperature increase at 40 Hz by internal friction.

Nonlinearity and creep effects are larger at higher temperature [69, 114]. This characteristic leads to the time-temperature superposition principle

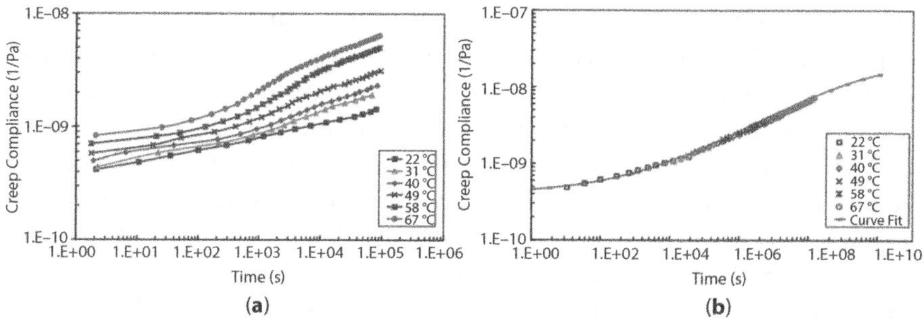

Figure 6.16 The creep compliance curves of model systems: (a) individual creep curves and (b) the master curve at 22°C [54].

[115] where the effect of increasing temperature on creep or relaxation is equivalent to expanding the loading time. This principle is widely used to predict long-term creep response from short-term creep experiments at different temperatures. Creep responses at varying temperatures are horizontally shifted along a logarithmic time scale by a shift factor a_T to form a master curve describing the long-term performance. Figure 6.16 shows a shift procedure of a master curve for a structural epoxy adhesive at the reference temperature of 22°C [54]. The shift factor a_T is often described by the Williams-Landel-Ferry equation [116] (above the glass transition) and the Arrhenius law [117] (below the glass transition).

6.4.2 Moisture

Abdel Wahab *et al.* [118] conducted tensile tests on dry and wet epoxy adhesives to study the influence of moisture on the bulk mechanical behavior. The modulus, yield stress and ultimate strength of the adhesive decreased in the wet condition, while the failure strain increased. They also studied the effect of water uptake on the strength of single lap and tensile butt joints immersed in water for varying durations, where the butt joint was observed to degrade faster. Both joints showed nonuniform water absorption that caused swelling strain and reduced strength.

When adhesively bonded joints are exposed to moisture, the strength reduction depends on the adhesive and the interfacial bond, since moisture diffusion can occur either through the adhesive layer leading to plasticization, swelling, increase of cracks, and hydrolysis [119], or at the adhesive-adherend interface resulting in bond degradation [120].

Temperature and moisture factors may affect the adhesive mechanical behavior. Feng *et al.* [54] conducted a series of creep tests at varying temperatures for structural epoxy adhesives under dry and wet conditions. Experimental results showed that the presence of moisture in adhesives accelerated creep development. Moisture can result in a similar creep response in the dry condition with a higher temperature.

Analogous to temperature, time-moisture superposition is used to obtain a master curve of long-term creep behavior from short-term performance with moisture since moisture accelerates the time-dependent deformation. The shift factor a_H can also be described by the Williams-Landel-Ferry equation [116] with moisture concentration dependence [121].

6.4.3 Modeling

To describe the effects of temperature and moisture on the viscoelastic behavior of an adhesive, it is convenient to implement these factors into models by making material parameters a function of them.

To introduce nonlinear viscoelasticity into rheological models, moduli of elastic springs and viscosities of dashpots take stress dependence into account. Using this approach the effects of temperature [87] and moisture [122] on the creep of adhesives were incorporated into rheological models.

Lai and Bakker [79] modified the Schapery model to incorporate the effects of temperature and physical aging through the definition of reduced time. The reduced time in Eq. 6.8 is modified with

$$\psi^t = \int_0^t \frac{d\tau}{a_\sigma a_T} \tag{6.12}$$

where the temperature shift factor a_T is given by the Williams-Landel-Ferry equation, as

$$\log(a_T) = \frac{C_1(T - T_0)}{C_2 + (T - T_0)} \tag{6.13}$$

where C_1 and C_2 are constant and T_0 is the reference temperature. In addition to the stress and temperature effects on reduced time, moisture can be involved by adding the time-scaling function to Eq. 6.12 [123].

A special case of the Schapery model is the so-called free volume approach of Knauss and Emri [124], [125]. It is based on the concept that the free volume affects the internal time scale of materials by allowing movements in the polymer chain. However, this approach is based on dilatational energy and was not able to describe highly nonlinear responses under shear. Popelar and Liechti [126] extended this approach to take distortional effects into account by introducing deviatoric strain into the free volume. The deviatoric stress S_{ij} and hydrostatic stress σ_{kk} are given by

$$S_{ij}(t) = 2\int_{-\infty}^{t} \mu(t' - \tau')\frac{\partial e_{ij}}{\partial \tau}d\tau \tag{6.14}$$

$$\sigma_{kk}(t) = 3\int_{-\infty}^{t} K(\psi^{t} - \psi^{\tau})\frac{\partial}{\partial \tau}(\delta\theta + \alpha\Delta T + \beta\Delta c +)d\tau \tag{6.15}$$

where μ(t) and K(t) are relaxation moduli in shear and volume deformation, respectively, $\theta=\varepsilon_{kk}$, δ relates the change in free volume due to mechanical dilation, ΔT is the change in temperature, α is the volumetric thermal expansion of the free volume, and β is the volumetric expansion due to a change in solvent concentration, Δc. Similarly to the Schapery model (Eq. 6.8), the reduced time ψ^t is a function of a shift factor a and is dependent on the temperature, moisture concentration, mechanical dilatation and effective strain ε_{eff}, as

$$\psi^{t} = \int_{0}^{t} \frac{d\tau}{a[T(\tau), c(\tau), \theta(\tau), \varepsilon_{eff}]} \tag{6.16}$$

$$\log(a) = -\frac{B_d}{2.303 f_d}\left(\frac{\theta + \alpha\Delta T + \beta\Delta c}{f_d + \theta + \alpha\Delta T + \beta\Delta c}\right) - \frac{B_s}{2.303 f_s}\left(\frac{\varepsilon_{eff}}{f_s + \varepsilon_{eff}}\right) \tag{6.17}$$

where f_d and f_s are the reference values of free volumes under dilatation and shear, respectively. B_d and B_s are material parameters under dilatation and shear, respectively. The model was able to describe an epoxy subjected to a wide range of multiaxial stress states. The terms of $\alpha\Delta T$ and $\beta\Delta c$ enable the model to describe nonlinear viscoelastic behaviors for different temperatures and moisture conditions [127].

In an adhesive joint, moisture content is frequently nonuniform, which causes a gradient in the mechanical properties of the adhesive. Thus, it may be necessary to describe the moisture content of an adhesive to quantify moisture sensitive parameters. Fick's law is commonly used to model diffusion and assumes no interaction between the absorbed water molecules and the polymer chains [128]. Fick's first law states that the flux (F) is proportional to the concentration gradient ($\partial c/\partial x$) as

$$F = -D\frac{\partial c}{\partial x} \tag{6.18}$$

where D is the diffusion coefficient.

Fick's second law describes the change in concentration as the sum of the fluxes across the six faces of an infinitesimal volume as

$$\frac{\partial c}{\partial t} = \frac{\partial}{\partial x}\left(D\frac{\partial c}{\partial x}\right) + \frac{\partial}{\partial y}\left(D\frac{\partial c}{\partial y}\right) + \frac{\partial}{\partial z}\left(D\frac{\partial c}{\partial z}\right) \tag{6.19}$$

Roy and Reddy [129] presented a two-dimensional numerical analysis of stress and strain for bonded joints subjected to the combined effect of moisture, temperature, and stress. Fick's second law was used to model moisture content and solved in conjunction with the Schapery integral model using shift factors to present the effects of temperature and stress. Fickian diffusion models are often used to describe moisture degradation of polymers [130] and adhesives [127, 131, 132]. As non-Fickian diffusion is accompanied by extensive swelling and chemical reactions and is promoted by higher temperature/humidity and thinner adhesive layers [133], the dual-Fickian [134, 135] and Langmuir Law [136, 137] are widely used.

6.5 Summary

This chapter reviewed approaches to describing the inelastic responses of structural aerospace adhesives from time-independent plasticity and time-dependent inelasticity.

Time-independent plasticity is considered for adhesives subjected to short-term loading and stress analysis for preliminary study and design. The performances of bulk adhesives and bonded joints in uniaxial and multiaxial tests to obtain yield stress are compared. Commonly used yield criteria for adhesives are also introduced. Due to the chain entanglement and nucleation of cavities, many adhesives demonstrate a hydrostatic stress-dependent yield criterion.

Time-dependent inelasticity includes viscoelasticity and viscoplasticity which are usually manifested under creep and cyclic loading. Nonlinear viscoelastic and viscoplastic behavior can be observed at high creep stress levels. Creep behavior can be sensitive to hydrostatic stress. Ratcheting under cyclic loading is influenced by stress amplitude, mean stress and frequency. Even under a cyclic load with zero mean stress, ratcheting of adhesives may occur with larger strain accumulation than under tensile cyclic load. Under reversed cyclic loading, some adhesives' plastic deformation is accompanied by the Bauschinger effect. Creep behavior can be described by linear viscoelastic models at low stress levels while nonlinear viscoelastic and viscoplastic models should be considered at high stress levels. For modeling the ratcheting behavior of adhesives, nonlinear viscoelastic and viscoplastic models considering the hydrostatic effect showed better agreement with experiment.

As temperature and moisture are the major and common hazards in general aerospace adhesives' working environment, their influence on adhesives' inelastic behavior is discussed. The time-temperature/moisture superposition principle is used to predict creep behavior under long-term loading from short-term experiments. To model the influences of temperature and moisture on the time-dependent behavior of adhesives, it is common to modify the viscoelastic models by accommodating these two factors into material parameters.

References

1. R. D. Adams, J. Comyn and W. C. Wake, *Structural Adhesive Joints in Engineering*, second edition, pp. 2-3, Chapman & Hall, London, UK (1997).
2. M. Abdel Wahab, *Joining Composites with Adhesives: Theory and Applications*, pp. 1-3, DEStech Publications, Lancaster, PA (2015).
3. R. E. Politi, Structural adhesives in the aerospace industry, in: *Handbook of Adhesives*, third edition, I. Skeist (Ed.), pp. 713–728, Chapman & Hall, New York (1990).
4. M. Afendi, T. Teramoto and H. B. Bakri, Strength prediction of epoxy adhesively bonded scarf joints of dissimilar adherends. *Intl. J. Adhesion Adhesives* 31, 402–411 (2011).
5. Y. Chen, Modeling of time-dependent behaviors of adhesives under creep and cyclic loadings, Ph. D. thesis, Washington State University, Pullman, WA, USA (2021).
6. P. Czarnocki and K. Piekarski, Yielding of adhesives. *J. Mater. Sci.* 21, 4296–4298 (1986).
7. J. Y. Cognard, M. Bourgeois, R. Créac'hcadec and L. Sohier, Comparative study of the results of various experimental tests used for the analysis of

the mechanical behaviour of an adhesive in a bonded joint. *J. Adhesion Sci. Technol.* 25, 2857–2879 (2012).
8. M. Ignjatovic, P. Chalkley and C. Wang, The yield behaviour of a structural adhesive under complex loading, *Report No. DSTO-TR-0728*, Aeronautical and Maritime Research Laboratory, Melbourne, Australia (1998).
9. G. Scarselli, C. Corcione, F. Nicassio and A. Maffezzoli, Adhesive joints with improved mechanical properties for aerospace applications. *Intl. J. Adhesion Adhesives* 75, 174–180 (2017).
10. Thick-adherend metal lap-shear joints for determination of the stress-strain behavior of adhesives in shear by tension loading, ASTM D5656-10 (2017).
11. J. Y. Cognard, L. Sohier, R. Créac'hcadec, F. Lavelle and N. Lidon, Influence of the geometry of coaxial adhesive joints on the transmitted load under tensile and compression loads. *Intl. J. Adhesion Adhesives* 37, 37–49 (2012).
12. P. C. Mohapatra and L. V. Smith, Characterization of adhesive yield criteria using mixed-mode loading. *J. Adhesion Sci. Technol.* 33, 1248–1260 (2019).
13. M. Imanaka, A. Fujinami and Y. Suzuki, Fracture and yield behavior of adhesively bonded joints under triaxial stress conditions. *J. Mater. Sci.* 35, 2481–2491 (2000).
14. C. H. Wang and P. Chalkley, Plastic yielding of a film adhesive under multiaxial stresses. *Intl. J. Adhesion Adhesives* 20, 155–164 (2000).
15. W. Althof, Creep, recovery and relaxation of shear-loaded adhesive bondlines. *J. Reinforced Plastics Composites* 1, 29–39 (1982).
16. N. Arnaud, R. Créac'hcadec and J. Y. Cognard, A tension/compression-torsion test suited to analyze the mechanical behaviour of adhesives under non-proportional loadings. *Intl. J. Adhesion Adhesives* 53, 3-14 (2014).
17. J. Y. Cognard, Numerical analysis of edge effects in adhesively-bonded assemblies application to the determination of the adhesive behaviour. *Computers & Structures* 86, 1704–1717 (2008).
18. K-S. Kim and N. Aravas, Elastoplastic analysis of the peel test. *Intl. J. Solids Structures* 24, 417–435 (1988).
19. K-S. Kim and J. Kim, Elasto-plastic analysis of the peel test for thin film adhesion. *J. Eng. Mater. Technol, Trans. ASME* 110, 266-273 (1988).
20. C. Yang, H. Huang, J. S. Tomblin and W. Sun, Elastic-plastic model of adhesive-bonded single-lap composite joints. *J. Composite Mater.* 38, 293–309 (2004).
21. O. Sayman, Elasto-plastic stress analysis in an adhesively bonded single-lap joint. *Composites Part B* 43, 204–209 (2012).
22. Y. Jen, Fatigue life evaluation of adhesively bonded scarf joints. *Intl. J. Fatigue* 36, 30–39 (2012).
23. H. Mardani, N. Stein, P. L. Rosendahl and W. Becker, An efficient stress and deformation model for arbitrary elastic-perfectly plastic adhesive lap joints. *Intl. J. Adhesion Adhesives* 103, 102679 (2020).
24. M. N. Charalambides and A. Olusanya, The constitutive models suitable for adhesives in some finite element codes and suggested methods of generating

the appropriate materials data, *National Physical Lab Report CMMT(B)130*, Teddington, UK (1997).
25. C. Creton and H. Lakrout, Micromechanics of flat-probe adhesion tests of soft viscoelastic polymer films. *J. Polym. Sci., Part B* 38, 965–979 (2000).
26. G. Dean, L. Crocker, B. Read and L. Wright, Prediction of deformation and failure of rubber-toughened adhesive joints. *Intl. J. Adhesion Adhesives* 24, 295–306 (2004).
27. G. Dean and L. Crocker, The use of finite element methods for design with adhesives, *Measurement Good Practice Guide No.48*, Teddington, UK (2001).
28. R. Quinson, J. P. Rez, M. Rink and A. Pava, Yield criteria for amorphous glassy polymers. *J. Mater. Sci.* 32, 1371–1379 (1997).
29. C. H. Wang and L. R. F. Rose, Determination of triaxial stresses in bonded joints. *Intl. J. Adhesion Adhesives* 17, 17–25 (1997).
30. J. Y. Cognard, R. Créac'hcadec, J. Maurice, P. Davies, M. Peleau and L. F. M. da Silva, Analysis of the influence of hydrostatic stress on the behaviour of an adhesive in a bonded assembly. *J. Adhesion Sci. Technol.* 24, 1977–1994 (2010).
31. A. Deb, I. Malvade, P. Biswas and J. Schroeder, An experimental and analytical study of the mechanical behaviour of adhesively bonded joints for variable extension rates and temperatures. *Intl. J. Adhesion Adhesives* 28, 1–15 (2008).
32. G. Dean, Modelling non-linear creep behaviour of an epoxy adhesive. *Intl. J. Adhesion Adhesives* 27, 636–646 (2007).
33. D. Lemme and L. Smith, Ratcheting in a nonlinear viscoelastic adhesive. *Mech. Time-Dependent Mater.* 22, 519–532 (2018).
34. X. X. Yu, A. D. Crocombe and G. Richardson, Material modelling for rate-dependent adhesives. *Intl. J. Adhesion Adhesives* 21, 197–210 (2001).
35. N. Su and R. I. Mackie, Two-dimensional creep analysis of structural adhesive joints. *Intl. J. Adhesion Adhesives* 13, 33–40 (1993).
36. Y. Chen and L. V. Smith, A nonlinear viscoelastic-viscoplastic constitutive model for adhesives under creep. *Mech. Time-Dependent Mater.,* 1–19 (2021). https://doi.org/10.1007/s11043-021-09506-z
37. P. Bidaud, R. Créac'hcadec, P. Jousset and D. Thévenet, A fatigue life prediction method of adhesively bonded joints based on visco-elastic and visco-plastic behavior: Application under cyclic shear loading. *J. Adhesion Sci. Technol.* 30, 1641–1661 (2016).
38. J. Zhang, H. Li, H. Y. Li and X. L. Wei, Uniaxial ratchetting and low-cycle fatigue failure behaviors of adhesively bonded butt-joints under cyclic tension deformation. *Intl. J. Adhesion Adhesives* 95, 102399 (2019).
39. M. Krause and L. V. Smith, Ratcheting in structural adhesives. *Polymer Testing* 97, 107154 (2021).
40. Y. Chen and L. V. Smith, Ratcheting and recovery of adhesively bonded joints under tensile cyclic loading. *Mech. Time-Dependent Mater.,* 1–20 (2021). https://doi.org/10.1007/s11043-021-09532-x

41. D. Thevenet, R. Créac'hcadec, L. Sohier and J. Y. Cognard, Experimental analysis of the behavior of adhesively bonded joints under tensile/compression-shear cyclic loadings. *Intl. J. Adhesion Adhesives* 44, 15–25 (2013).
42. P. D. Chalkley and W. K. Chiu, An improved method for testing the shear stress/strain behaviour of adhesives. *Intl. J. Adhesion Adhesives* 13, 237–242 (1993).
43. W. K. Chiu and R. Jones, Unified constitutive model for thermoset adhesive, FM73. *Intl. J. Adhesion Adhesives* 15, 131–136 (1995).
44. P. Davies, L. Sohier, J. Y. Cognard, A. Bourmaud, D. Choqueuse, E. Rinnert and R. Créac'hcadec, Influence of adhesive bond line thickness on joint strength. *Intl. J. Adhesion Adhesives* 29, 724–736 (2009).
45. J. Y. Cognard, P. Davies, L. Sohier and R. Créac'hcadec, A study of the non-linear behaviour of adhesively-bonded composite assemblies. *Composite Structures* 76, 34–46 (2006).
46. M. Abouhamzeh, J. Sinke, K. M. B. Jansen and R. Benedictus, Kinetic and thermo-viscoelastic characterisation of the epoxy adhesive in GLARE. *Composite Structures* 124, 19–28 (2015).
47. H. Yetgin, Enhancement of performance characteristics of epoxy adhesives through particle reinforcement, Master's thesis, Middel East Technical University, Ankara, Turkey (2019).
48. N. Causse, L. Quiroga Cortes, E. Dantras, C. Tonon, M. Chevalier, H. Combes, P. Guigue and C. Lacabanne, New bonded assembly configuration for dynamic mechanical analysis of adhesives. *Intl. J. Adhesion Adhesives* 46, 1–6 (2013).
49. E. Savonnet, E. Grau, S. Grelier, B. Defoort and H. Cramail, Divanillin-based epoxy precursors as DGEBA substitutes for biobased epoxy thermosets. *ACS Sustainable Chem. Eng.* 6, 11008–11017 (2018).
50. J-S. Pap, M. Kästner, S. Müller and I. Jansen, Experimental characterization and simulation of the mechanical behavior of an epoxy adhesive. *Procedia Mater. Sci.* 2, 234–242 (2013).
51. Z. Qi, N. Hu, G. Li, D. Zeng and X. Su, Constitutive modeling for the elastic-viscoplastic behavior of high density polyethylene under cyclic loading. *Intl. J. Plasticity* 113, 125–144 (2019).
52. H. S. da Costa Mattos, J. M. L. Reis, L. G. M. O. de Medeiros, A. H. Monteiro, S. C. S. Teixeira and E. G. Chaves, Analysis of the cyclic tensile behaviour of an elasto-viscoplastic polyamide. *Polymer Testing* 58, 40–47 (2017).
53. G. K. Hu, F. Schmit, D. Baptiste and D. Francois, Viscoplastic analysis of adhesive joints. *J. Appl. Mech.* 63, 21–26 (1996).
54. C. W. Feng, C. W. Keong, Y. P. Hsueh, Y. Y. Wang, and H. J. Sue, Modeling of long-term creep behavior of structural epoxy adhesives. *Intl. J. Adhesion Adhesives* 25, 427–436 (2005).
55. P. Meshgin, K. K. Choi and M. M. Reda Taha, Experimental and analytical investigations of creep of epoxy adhesive at the concrete-FRP interfaces. *Intl. J. Adhesion Adhesives* 29, 56–66 (2009).

56. S. Vleeshouwers, A. M. Jamieson and R. Simha, Effect of physical aging on tensile stress relaxation and tensile creep of cured EPON 828/epoxy adhesives in the linear viscoelastic region. *Polym. Eng. Sci.* 29, 662–670 (1989).
57. M. E. Tuttle and H. F. Brinson, Prediction of the long-term creep compliance of general composite laminates. *Experimental Mech.* 26, 89–102 (1986).
58. Z. Shuangyin and L. W. Tsai, Computer simulation of creep damage at crack tip in short fibre composites. *Acta Mechanica Sinica* 10, 282–288 (1994).
59. T-R. M. Alejandro and L-U. L. Edmundo, Prony series calculation for viscoelastic behavior modeling of srtuctural adhesives from DMA data. *Ingeniería, Investigación Tecnología* 21, 1-10 (2020).
60. H. F. Brinson and L. C. Brinson, *Polymer Engineering Science and Viscoelasticity: An introduction,*second edition, pp. 231-285, Springer, Boston (2008).
61. O. C. Zienkiewicz, M. Watson and I. P. King, A numerical method of visco-elastic stress analysis. *Intl. J. Mech. Sci.* 10, 807–827 (1968).
62. Y. R. Nagaraja and R. S. Alwar, Viscoelastic analysis of an adhesive-bonded plane lap joint. *Computers & Structures* 11, 621–627 (1980).
63. K. C. Gramoll, D. A. Dillard and H. F. Brinson, A stable numerical solution method for in-plane loading of nonlinear viscoelastic laminated orthotropic materials. *Composite Structures* 13, 251-274 (1989).
64. S. Yadagiri, C. P. Reddy and T. S. Reddy, Viscoelastic analysis of adhesively bonded joints. *Computers & Structures* 27, 445–454 (1987).
65. H. F. Brinson, The viscoelastic-plastic behavior of a ductile polymer, in: *Deformation and Fracture of High Polymers*, H. H. Kausch, J. A.Hassell and R. I. Jaffee (Eds.), pp. 397–416, Springer Science + Business Media, New York (1973).
66. E. Sancaktar and H. F. Brinson, The viscoelastic shear behavior of a structural adhesive, in: *Adhesion and Adsorption of Polymers*, Part A, L.-H. Lee (Ed.), pp. 279-299, Plenum Press, New York (1980).
67. H. F. Brinson, The viscoelastic constitutive modelling of adhesives. *Composites* 13, 377–382 (1982).
68. K. W. Chase and W. Goldsmith, Mechanical and optical characterization of an anelastic polymer at large strain rates and large strains. *Experimental Mech.* 14, 10–18 (1974).
69. E. Sancaktar and S. C. Schenck, Material characterization of structural adhesives in the lap shear mode: 2. Temperature-dependent delayed failure, *Ind. Eng. Chem. Product Res. Develop.* 24, 257–263 (1985).
70. E. Sancaktar, S. C. Schenck and S. Padgllwar, Material characterization of structural adhesives in the lap shear mode: 1. The effects of rate. *Ind. Eng. Chem. Product Res. Develop.* 23, 426–434 (1984).
71. K. K. Choi and M. M. Reda Taha, Rheological modeling and finite element simulation of epoxy adhesive creep in FRP-strengthened RC beams. *J. Adhesion Sci. Technol.* 27, 523–535 (2013).

72. N. Houhou, K. Benzarti, M. Quiertant, S. Chataigner, A. Fléty and C. Marty, Analysis of the nonlinear creep behavior of concrete/FRP-bonded assemblies. *J. Adhesion Sci. Technol.* 28, 1345–1366 (2014).
73. C. Badulescu, C. Germain, J. Y. Cognard and N. Carrere, Characterization and modelling of the viscous behaviour of adhesives using the modified Arcan device. *J. Adhesion Sci. Technol.* 29, 443–461 (2015).
74. P. Majda and J. Skrodzewicz, A modified creep model of epoxy adhesive at ambient temperature. *Intl. J. Adhesion Adhesives* 29, 396–404 (2009).
75. J. Godzimirski and M. Rośkowicz, Numerical analysis and experimental tests for long-term strength of adhesive bonds. *J. Adhesion Sci. Technol.* 28, 1334–1344 (2014).
76. R. A. Schapery, Further development of a thermodynamic constitutive theory: stress formulation, *Purdue University Report No. AA & ES. 69-2*, Lafayette, Indiana (1969).
77. D. Gamby and L. Blugeon, On the characterization by Schapery's model of non-linear viscoelastic materials. *Polymer Testing* 7, 137–147 (1987).
78. M. Henriksen, Nonlinear viscoelastic stress analysis- a finite element approach. *Computers & Structures* 18, 133–139 (1984).
79. J. Lai and A. Bakker, 3-D Schapery representation for non-linear viscoelasticity and finite element implementation. *Computational Mech.* 18, 182–191 (1996).
80. A. D. Crocombe, Modelling and predicting the effects of test speed on the strength of joints made with FM73 adhesive. *Intl. J. Adhesion Adhesives* 15, 21–27 (1995).
81. D. Tscharnuter, S. Gastl and G. Pinter, Modeling of the nonlinear viscoelasticity of polyoxymethylene in tension and compression. *Intl. J. Eng. Sci.* 60, 37–52 (2012).
82. R. W. Bailey, Creep of steel under simple and compound stresses and the use of high initial temperature in steam power plants, in: *Transactions of the World Power Conference* 3, pp. 1089 (1929).
83. G. W. Scott Blair and B. C. Veinoglou, A study of the firmness of soft materials based on Nutting's equation. *J. Sci. Instruments* 21, 149-154 (1944).
84. G. S. Baş and E. Sancaktar, Mechanical behavior of toughened epoxy structural adhesives for impact applications. *ChemEngineering* 4, 38 (2020).
85. J. M. Ting, Tertiary creep model for frozen sands. *J. Geotech. Eng. 109*, 932–945 (1983).
86. A. Nadai, On the creep of solids at elevated temperature. *J. Appl. Phys.* 8, 418–432 (1937).
87. M. Zehsaz, F. Vakili-Tahami and M. A. Saeimi-Sadigh, Modified creep constitutive equation for an epoxy-based adhesive with nonlinear viscoelastic behavior. *J. Strain Analysis for Engineering Design* 50, 4–14 (2014).
88. X. X. Yu, A. D. Crocombe and G. Richardson, Material modelling for rate-dependent adhesives. *Intl. J. Adhesion Adhesives* 21, 197-210 (2001).

89. L. E. Malvern, The propagation of longitudinal waves of plastic deformation in a bar of material exhibiting a strain-rate effect. *J. Appl. Mech.* 18, 203–208 (1951).
90. P. Perzyna, The constitutive equations for rate sensitive plastic materials. *Quarterly Applied Mathematics* 20, 321–332 (1963).
91. R. A. LeMaster, Finite deformation-finite element formulations for elastic-viscoplastic materials, Ph. D. thesis, University of Tennessee, Knoxville, TN, USA (1984).
92. A. Ilioni, C. Badulescu, N. Carrere, P. Davies and D. Thévenet, A visco-elastic-viscoplastic model to describe creep and strain rate effects on the mechanical behaviour of adhesively-bonded assemblies. *Intl. J. Adhesion Adhesives* 82, 184–195 (2018).
93. Y. Chen and L. V Smith, A nonlinear viscoelastic–viscoplastic model for adhesives. *Mech. Time-Dependent Mater.* 25, 565–579 (2021).
94. P. Bidaud, R. Creachcadec, P. Jousset and D. Thevenet, A fatigue life prediction method of adhesively bonded joints based on visco-elastic and visco-plastic behavior: application under cyclic shear loading. *J. Adhesion Sci. Technol.* 30, 1641–1661 (2016).
95. A. Krairi and I. Doghri, A thermodynamically-based constitutive model for thermoplastic polymers coupling viscoelasticity, viscoplasticity and ductile damage. *Intl. J. Plasticity* 60, 163–181 (2014).
96. I. B. C. M. Rocha, F. P. van der Meer, S. Raijmaekers, F. Lahuerta, R. P. L. Nijssen and L. J. Sluys, Numerical/experimental study of the monotonic and cyclic viscoelastic/viscoplastic/fracture behavior of an epoxy resin. *Intl. J. Solids Structures* 168, 153–165 (2019).
97. Z. Xia, X. Shen and F. Ellyin, Cyclic deformation behavior of an epoxy polymer. Part II : predictions of viscoelastic constitutive models. *Polym. Eng. Sci.* 45, 103–113 (2005).
98. F. Ellyin and Z. Xia, Nonlinear viscoelastic constitutive model for thermoset polymers. *J. Eng. Mater. Technol.* 128, 579 585 (2006).
99. Z. Xia, X. Shen and F. Ellyin, An assessment of nonlinearly viscoelastic constitutive models for cyclic loading: the effect of a general loading/unloading rule. *Mech. Time-Dependent Mater.* 9, 281–300 (2006).
100. C. F. Popelar and K. M. Liechti, A distortion-modified free volume theory for nonlinear viscoelastic behavior. *Mech. Time-Dependent Mater.* 7, 89–141 (2003).
101. E. P. Motta, J. M. L. Reis and H. S. da Costa Mattos, Analysis of the cyclic tensile behaviour of an elasto-viscoplastic polyvinylidene fluoride (PVDF). *Polymer Testing* 67, 503–512 (2018).
102. E. P. Motta, J. M. L. Reis and H. S. da Costa Mattos, Modelling the cyclic elasto-viscoplastic behaviour of polymers. *Polymer Testing* 78, 105991 (2019).
103. J. Lemaitre and J.-L. Chaboche, *Mechanics of Solid Materials*, pp. 340-343, Cambridge University Press, Cambridge, UK(1994).

120. H. S. Panda, R. Samant, K. L. Mittal and S. K. Panigrahi, Durability aspects of structural adhesive joints, in: *Structural Adhesive Joints: Design, Analysis and Testing*, K. L. Mittal and S. K. Panigrahi (Eds.), pp. 97–134, Wiley-Scrivener, Beverly, MA (2020).
121. H. Walter, E. Dermitzaki, B. Wunderle and B. Michel, Influence of moisture on humidity sensitive material parameters of polymers used in microelectronic applications, in: Proceedings of the 3rd Electronics System Integration Technology Conference, pp. 1-5 (2010).
122. I. Costa and J. Barros, Tensile creep of a structural epoxy adhesive: Experimental and analytical characterization. *Intl. J. Adhesion Adhesives* 59, 115–124 (2015).
123. R. A. Schapery, Nonlinear viscoelastic and viscoplastic constitutive equations based on thermodynamics, *Mech. Time-Dependent Mater.* 1, 209-240 (1997).
124. W. G. Knauss, Non-linear viscoelasticity based on free volume consideration. *Computers & Structures* 13, 123–128 (1981).
125. W. G. Knauss and I. Emri, Volume change and the nonlinearly thermo-viscoelastic constitution of polymers. *Polym. Eng. Sci.* 27, 86–100 (1987).
126. C. F. Popelar and K. M. Liechti, Multiaxial nonlinear viscoelastic characterization and modeling of a structural adhesive. *J. Eng. Mater. Technol.* 119, 205–210 (1997).
127. D. F. Vieira de Mattos, R. Huang and K. M. Liechti, The effect of moisture on the nonlinearly viscoelastic behavior of an epoxy. *Mech. Time-Dependent Mater.* 24, 435-461 (2020).
128. A. Ameli, N. V. Datla, M. Papini and J. K. Spelt, Hygrothermal properties of highly toughened epoxy adhesives. *J. Adhesion* 86, 698–725 (2010).
129. S. Roy and J. . Reddy, A finite element analysis of adhesively bonded composite joints with moisture diffusion and delayed failure. *Computers & Structures* 29, 1011–1031 (1988).
130. R. Mahadevan and L. Smith, A mechanistic model describing the degradation of polymers. *J. Polymers Environment* 15, 75–80 (2007).
131. A. D. Crocombe, Incorporating environmental degradation in closed form adhesive joint stress analyses. *J. Adhesion* 84, 212–230 (2008).
132. S. Popineau and M. E. R. Shanahan, Simple model to estimate adhesion of structural bonding during humid ageing. *Intl. J. Adhesion Adhesives* 26, 363–370 (2006).
133. W. K. Loh, A. D. Crocombe, M. M. A. Wahab and I. A. Ashcroft, Modelling anomalous moisture uptake, swelling and thermal characteristics of a rubber toughened epoxy adhesive. *Intl. J. Adhesion Adhesives* 25, 1–12 (2005).
134. O. R. Patil, A. Ameli and N. V. Datla, Predicting environmental degradation of adhesive joints using a cohesive zone finite element model based on accelerated fracture tests. *Intl. J. Adhesion Adhesives* 76, 54–60 (2017).

135. A. Mubashar, I. A. Ashcroft, G. W. Critchlow and A. D. Crocombe, Moisture absorption-desorption effects in adhesive joints. *Intl. J. Adhesion Adhesives* 29, 751–760 (2009).
136. S. Popineau, C. Rondeau-Mouro, C. Sulpice-Gaillet and M. E. R. Shanahan, Free/bound water absorption in an epoxy adhesive. *Polymer* 46, 10733–10740 (2005).
137. V. La Saponara, Environmental and chemical degradation of carbon/epoxy and structural adhesive for aerospace applications: Fickian and anomalous diffusion, Arrhenius kinetics. *Composite Structures* 93, 2180–2195 (2011).

Part 2
APPLICATIONS

7

Structural Reactive Acrylic Adhesives: Preparation, Characterization, Properties and Applications

D.A. Aronovich[1]* and L.B. Boinovich[2]

[1]*V.A. Kargin Polymer Research Institute, Dzerzhinsk, Nizhny Novgorod region, Russia*
[2]*A.N. Frumkin Institute of Physical Chemistry and Electrochemistry, Russian Academy of Sciences, Moscow, Russia*

Abstract

This chapter focuses on structural acrylic adhesives (SAAs), which belong to second-generation acrylic adhesives. The complex composition of such adhesives is shown, and the role of various SAA components in achieving the target physicochemical, mechanical, and thermal properties of joints based on multicomponent acrylic adhesives is discussed. Particular attention is paid to the analysis of possibilities of using SAAs with borane initiating curing systems for bonding substrates with low surface energy and the effect of the chemical structure of the components used in formulation on the durability and properties of the resulting adhesive. A comparison of the properties of SAAs and other types of reactive adhesives showed that structural acrylic adhesives have a set of properties that are in many ways superior to the known structural epoxy and polyurethane adhesives. The areas of application of these adhesives are briefly discussed.

Keywords: Adhesion strength, low surface energy substrates, peel strength, shear strength, structural acrylic adhesive

7.1 Introduction

Structural acrylic adhesives (SAAs) belong to the large family of acrylic adhesives: anaerobic, curing in narrow gaps between metal surfaces in the

**Corresponding author*: dovid.aronovich@yandex.ru

K.L. Mittal and S.K. Panigrahi (eds.) Structural Adhesives: Properties, Characterization and Applications, (293–374) © 2023 Scrivener Publishing LLC

absence of atmospheric oxygen; UV curing, when irradiated with UV light; thermosetting, two-part adhesives; mixed curing, aerobic curing adhesives; and cyanoacrylates. The two-part structural acrylic adhesives discussed in this chapter are reactive adhesives, free-radical curing under the influence of a redox system in which the oxidant is in one part of the adhesive and the reducing agent is in the other. In what follows, we will denote such adhesives as SAAs. Historically, in the literature different terms have been used for SAAs, such as acrylics, first-generation acrylics, second-generation acrylics, modified acrylics, surface-activated acrylics, etc. [1].

Reactive acrylics based on methyl methacrylate (MMA) and redox system containing benzoyl peroxide (PB) and tertiary aromatic amine have been known since the 1950s. However, only the development of the so-called second-generation acrylics (SGAs), which were first presented by Du Pont [2], made it possible to widely apply acrylics for bonding in automotive, shipbuilding, transport engineering, and other industrial applications. The term SGA applies to acrylic adhesives containing monofunctional methacrylic monomer or mixtures of (meth) acrylic monomers, elastomeric and polymeric thickeners, curing systems, and various modifying additives.

It should be noted that compositions based on an acrylate, cured by a free radical mechanism, are used not only in the production of adhesives appropriate to create load-bearing joints, but also in many other areas such as composites for anchor bolts, repair composites and putties, "cold" plastic for road marking, in medicine as filling materials, bone cement, etc. The high adhesion and performance characteristics of acrylics have contributed to their widespread use in various industries. They combine the best properties of known reactive adhesives.

Acrylics firmly bond metals, various polymers and composites; have a high cure rate; strong adhesion, impact, peel and shear strengths; good resistance to various solvents and moisture. A distinctive feature of acrylics is their ability to bond surfaces without their preliminary preparation. The successful expansion of acrylics applications was also facilitated by the design of a two-syringe cartridge package with a mixing head, the use of which ensures accurate dispensing of the adhesive and eliminates prolonged contact of the working personnel with aggressive components.

The use of special boron-containing initiators in acrylics makes it possible to firmly bond such low surface energy polymers as polyethylene, polypropylene, polytetrafluoroethylene, poly(ethylene terephthalate), etc., which do not bond well using other adhesives and require preliminary surface preparation. Conventionally, for many other types of adhesives, such preparation includes treatment by corona discharge, plasma, flame, treatment with chemical reagents, the use of special primers, etc.

Since the subject of structural acrylic adhesives is a field of much industrial activity, the academic literature is scarce and earlier findings were mainly addressed in the patent literature. Nevertheless, the number of publications on the subject of structural acrylic adhesives has been steadily increasing over the years.

Due to a number of important advantages of SAAs over other types of adhesives, which we will discuss below, research in this area is actively continuing, due to the constant emergence of new types of technologies that require the use of new adhesives. In the last 20 years, a number of interesting and fairly complete review articles on acrylic adhesives [3–8] have appeared. However, since the field of design, research, and application of such adhesives is continuously evolving, in this chapter we will focus on the latest most interesting results on structural acrylic adhesives. In section 7.2, we will emphasize the compositions and chemistries of SAAs. Physico-mechanical properties of SAAs will be discussed in section 7.3. Peculiarities and detailed properties of adhesives designed for bonding low surface energy materials will be addressed in section 7.4. Section 7.5 is devoted to a brief comparison of the major properties of different types of widely used adhesives. The final section will briefly summarize the state-of-the-art and the development trends of in the field of SAAs.

7.2 Compositions and Chemistries

SAA formulations usually include the following components:

- *Primary methacrylate monomer* (usually MMA), whose chemical structure, including polar and non-polar groups, provides good dissolution of various polymers and wetting of substrates of different nature.
- *Secondary methacrylate monomer*. Such a monomer either has a higher molecular weight than MMA leading to lower volatility or contains functional groups which improve the functional properties of the final product.
- A *crosslinker*, which is a polyfunctional methacrylate monomer used to increase the rate of curing, and heat and chemical resistance.

- A peroxide or hydroperoxide *initiator* which is used for curing. The type of initiator has a decisive influence on the cure rate and stability of the adhesive.
- Another way to affect the cure rate and stability of the adhesive is to utilize curing *accelerators* or *activators*. Metal-organic compounds can serve as accelerators.
- To further enhance the bonding of an adhesive to the substrates, an *adhesion promoter*, usually a monomer with acidic (methacrylic acid) or phosphorus-containing groups is incorporated.
- Enhancement of the toughness and peel strength of an adhesive is obtained by the addition of an *elastomeric component*.
- A core-shell copolymer is most often used as a polymer *impact strength modifier*. The inclusion of a plasticizer provides another avenue to simultaneously increase the elasticity and the impact strength of a structural acrylic adhesive.
- Polymeric *thickeners* or inorganic *fillers* allow increasing the viscosity, reducing shrinkage, and enhancing the crack resistance.
- *Thixotropic additives* are used to improve rheological properties, *stabilizers* increase the shelf-life of adhesives, while *pigments and dyes* provide a simple way to visualize the evenness of the components mixing.

Such a complex formulation of SAAs allows fine-tuning the desired properties for structural acrylic adhesives.

7.2.1 Base Monomer

First of all, it is important to note that MMA is a large-tonnage product with a fairly low price. Many different SAA formulations have been developed using MMA as the main monomer due to a set of its practically attractive features.

The presence of non-polar methyl groups in MMA, and, at the same time, the high dipole moment of the monomer lead to the fact that the use of MMA as the base monomer in SAA contributes to wettability of a wide variety of adherend surfaces, sufficient solubility of most elastomers and polymers, and dissolution of oily contaminants present on the surfaces. Besides, the high volatility of MMA resulting from its structure helps to reduce the inhibitory effect of atmospheric oxygen during curing.

At the same time, the main drawbacks of MMA-based SAAs are related to such properties of the MMA monomer as toxicity, high flammability (a flashpoint of 11 °C), and offensive odor. The low, about 100 °C, glass transition temperature of the poly(methyl methacrylate) (PMMA) matrix does not allow achieving good heat resistance of the adhesive joint. In addition, products containing MMA are hazardous in terms of flammability and require special storage and transportation conditions.

Numerous studies were aimed to replace MMA with other methacrylic monomers, while retaining most of the properties inherent in MMA-based acrylics. These studies have considered such monomers as tetrahydrofurfuryl methacrylate (THFMA), isobornyl methacrylate (IBMA), cyclohexyl methacrylate, isodecyl methacrylate, hydroxyethyl methacrylate (HEMA), hydroxypropyl methacrylate (HPMA), phenoxyethyl methacrylate (PhEMA), hydroxymethyl methacrylate, etc. In these formulations, either a complete or partial replacement of MMA was proposed. The effect of some monomeric methacrylates on the strength properties of adhesive joints is shown in Table 7.1 [9].

The data presented in Table 7.1 show that application of different monomers and their mixtures provides the possibility to control the adhesion strength of SAAs, including the joining of oily surfaces.

The possibility of replacing MMA with a mixture of PhEMA and HPMA monomers was shown in [10]. However, in order to maintain properties such as adhesion to oily surfaces along with high strength characteristics, it was necessary to additionally introduce in formulation a mixture of two elastomers (nitrile-butadiene rubber and methyl methacrylate-butadiene-styrene copolymer) and an acrylic oligomer. The developed SAA was shown to be a non-hazardous substance, which had a high flash point, low odor, sufficient viscosity and thixotropy, and low shrinkage during curing. For this adhesive, high adhesive properties were retained after heat treatment for 1 h at 200 °C, indicating the thermal stability of adhesive [10].

Modification of SAA based on MMA with the acetoacetoxyethyl methacrylate monomer (AAEMA) [11] led to a decrease in the setting time (from 30 to 9 min) due to an increase in the copolymerization rate. The observed variation in the mechanical and thermal properties of SAA with a decrease in MMA/AAEMA ratio was attributed to the formation of hydrogen bonds and crosslinking between the AAEMA-co-MMA chains (Table 7.2).

The authors also associated an increase in the adhesive properties with the possible chelation of the substrate metal ions by the keto-groups of the AAEMA monomer [11].

Table 7.1 Strength of joints of stainless steel and D16AT aluminum alloys using methacrylates based SAAs [9].

Monomer*	Tensile strength, MPa		T-peel strength, N/mm Aluminum alloy**
	Aluminum alloy**	Stainless steel/ Oily stainless steel	
MMA	27	31/20	2.5
BMA	24	31/21	2.1
MMA:BMA (1:1)	16	26/23	2.1
MMA:BMA (2:1)	28	35/24	3.3
THFMA	19	23	3.0
MMA:THFMA (1:1)	25	30/28	4.0
MMA:THFMA (3:1)	19	30	3.0
HCPM	24	28	0.4
GMA	35	36	0.8
MMA:GMA (1:1)	29	37	0.9
GMA:THFMA (1:1)	24	30	3.5
IBMA	23	23	1.4
IBA	21	21	0.7
EGDCPMA	21	22	2.1
EGDCPA	16	21	1.5

*BMA – butyl methacrylate, THFMA – tetrahydrofurfuryl methacrylate, HCPM – 2-hydroxy-3-chloropropyl methacrylate, GMA – glycidyl methacrylate, IBA – isobornyl acrylate, EGDCPMA – ethylene glycol dicyclopentyl methacrylate, EGDCPA – ethylene glycol dicyclopentyl acrylate

** Aluminum alloy of Russian brand D16AT.

Use of methacrylic monomers of the type 3,3,5-trimethylcyclohexylmethacrylate, 4-tert-butylcyclohexylmethacrylate, 3,3,5,5-tetramethylcyclohexylmethacrylate, 3,4,5-trimethylcyclohexylmethacrylate as a monomer base in one of the adhesive parts and the inclusion of epoxy resin into the second part allowed increasing the peel strength on galvanized steel [12].

Table 7.2 Properties of cured SAA depending on the ratio of MMA and AAEMA comonomers [11].

MMA/ AAEMA ratio, %	Tensile strength, MPa	Flexural strength, MPa	Lap shear strength, MPa	Temperature of 10% mass loss, $T_{10\%}$ (°C)	Temperature of maximum decomposition rate T_{max} (°C)
100/0	22	30	2.2	270	377
90/10	32	49	6.4	281	400
80/20	31	45	7.2	278	391
70/30	28	22	4.8	280	397

Although a number of monomer compositions, which successfully replaced MMA, were selected and presented in the literature, many of the SAAs produced in the industry still contain MMA as the main monomer.

7.2.2 Thickeners and Elastomeric Components

Elastomers and thickeners are added to an SAA to ensure high impact strength and elasticity. For example, in the first SAA, PMMA, acrylic copolymers, polystyrene, and polychloroprene were used as a thickener. Later, chlorosulfonated polyethylene,

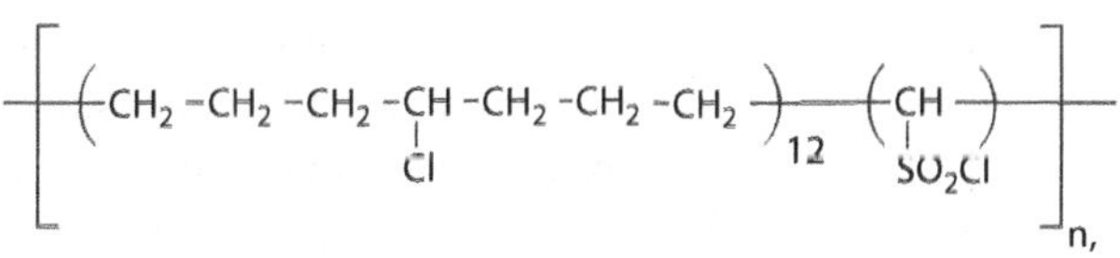

proposed in the DuPont patent [2], found wide applications due to the fact that in the presence of an amine activator, graft polymerization is possible according to the following reactions:

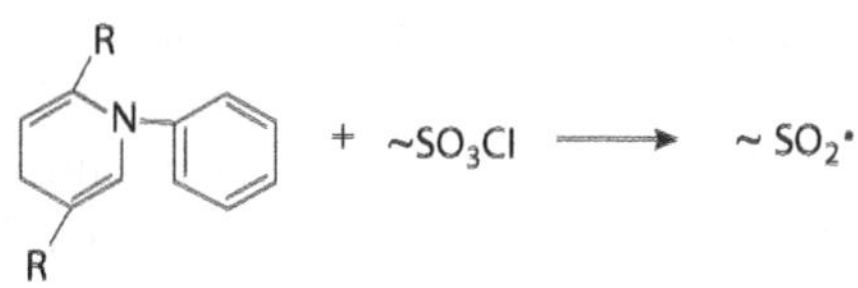

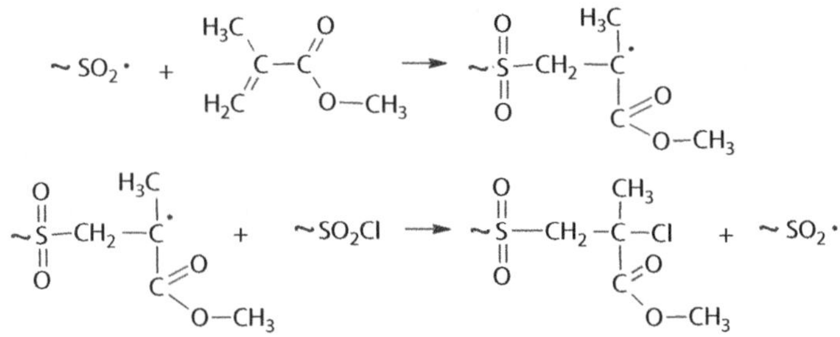

The formation of the graft copolymer in essence provides the unique properties of SAA. In modern SAAs, in addition to the above, chlorosulfonated polypropylene, chlorinated polyethylene, polybutadiene, copolymers of polybutadiene and acrylonitrile, various block copolymers, urethane elastomers, acrylic elastomers, adducts of polybutadienediol with isocyanate, butadiene-acrylonitrile copolymers with end reaction groups, polymers with a core-shell structure, polyurethane methacrylates, etc. are used as thickeners and elastomers.

To illustrate the effect of some methacrylic monomers and polymer thickeners on the strength properties of SAAs, Table 7.3 [9] shows the dependence of the strength properties of adhesive joints of stainless steel and D16AT aluminum alloy on the types of modifiers used in SAAs.

The measurements of impact strength of SAAs formulated on the basis of various methacrylates and rubbers showed an important role of compatibility in the monomer/elastomer system, which affects the morphology, the glass transition temperature, as well as strength characteristics of SAAs.

It has been established that in the case of butadiene-acrylonitrile rubber or fluorine-containing rubber, in contrast to chlorosulfonated polyethylene rubber, the greater the tendency for phase separation, the better are the strength properties. The variation of the fracture energy with the nature of the monomer and elastomer are given in Table 7.4 [13].

Adhesives containing a chlorinated elastomeric polymer, core-shell impact modifier, and urethane methacrylates oligomer exhibit increased strength when bonding fiberglass and acrylic substrates [14].

However, chlorosulfonated polymers have relatively high glass transition temperatures T_g (above −27 °C) and their use in SAAs at low temperatures can lead to the formation of hard and brittle joints. To increase the impact strength at low temperatures, it is recommended to introduce into an SAA elastomers with T_g from −80 °C to −100 °C. As an example, the elastomers containing styrene-butadiene-styrene (SBS) block copolymers of the radial type, such as Kraton D1116 or Kraton 1184 with T_g = −91 °C, work as toughening agents [15].

Table 7.3 Influence of methacrylic monomers and polymer thickeners on the strength properties of SAAs [9].

Monomer	Thickener*	Tensile strength		T-peel strength D16AT, N/mm
		Stainless steel, MPa	Aluminum alloy D16AT, MPa	
MMA	Hypalon-20	24	18	2.4
	Baypren - 233	15	13	3.6
	Copolymer VC-VA 65/35	32	18	2.0
	Copolymer MMA-BA 80/20	22	20	1.0
	Desmocoll-400	30	22	2.4
	OUA-BR-1000	34	20	2.9
	OUA-2000	20	19	1.8
	DST 30R-01	12	4.2	1.6
Isobornyl methacrylate (IBMA)	Hypalon-20	12	11	2.8
	Baypren-233	10	6	5.2
	Copolymer VC-VA 65/35	11	8	3.4
	DST 30R-01	11	10	3.2
	OUA-1000	21	24	1.4
	OUA-BR-1000	21	15	2.8

(Continued)

Table 7.3 Influence of methacrylic monomers and polymer thickeners on the strength properties of SAAs [9]. (*Continued*)

Monomer	Thickener*	Tensile strength: Stainless steel, MPa	Tensile strength: Aluminum alloy D16AT, MPa	T-peel strength D16AT, N/mm
Tetrahydrofurfuryl methacrylate (THFMA)	Hypalon-20	13	6	2.5
	Baypren-233	14	7	4.0
	Copolymer VC-VA 65/35	21	16	2.0
	DST 30R-01	12	6	3.5
	Desmocoll-400	25	19	3.2
	OUA-1000	12	10	3.3
Glycidyl methacrylate (GMA)	Desmocoll-400	38	24	0.6
	Copolymer VC-VA 65/35	23	25	2.2
	OUA-BR-1000	30	29	1.8
	OUA-1000	31	31	2.3
2-Hydroxy-3-chloropropyl methacrylate (HCPM)	Desmocoll-400	28	24	0.4
	OUA-1000	31	25	0.2
Butyl methacrylate (BMA)	Desmocoll-400	31	24	2.1
	Copolymer VC-VA 65/35	35	28	3.4

*Hypalon-20 – chlorosulfonated polyethylene; Baypren-233 – polychloroprene; Copolymer VC-VA 65/35 – copolymer of vinyl chloride and vinyl acetate 65:35; Copolymer MMA-BA 80/20 - copolymer of MMA and butyl acrylate 80:20; Desmocoll-400 – linear hydroxyl polyurethane; OUA-BR-1000 poly(urethane methacrylate) based on 2,4-toluene diisocyanate and butadiene rubber with terminal hydroxyl groups (MM1000); OUA-2000 – poly(urethane methacrylate) based on 2,4-toluene diisocyanate and poly(propylene glycol) (MM1000 or 2000); DST 30R-01 – linear block copolymer based on butadiene and styrene.

Table 7.4 Influence of the types of monomer CH_2=C(CH_3)COOR and elastomer on the fracture energy of rubbers [13].

R	Elastomer	Fracture energy, J/cm^2
C_6H_{11}	chlorosulfonated polyethylene	0.12±0.04
C_6H_{11}	butadiene-acrylonitrile rubber	0.26±0.06
C_6H_{11}	fluorinated rubber	0.18±0.07
C_2H_5	chlorosulfonated polyethylene	2.20±0.55
$_nC_4H_9$	chlorosulfonated polyethylene	1.84±0.52
$_{iso}C_4H_9$	chlorosulfonated polyethylene	0.16±0.07
CH_3	chlorosulfonated polyethylene	2.24±0.31

The effect of elastomer and monomer concentration on the properties of methacrylate/epoxy-based structural acrylic adhesives was analyzed in [16] using Fourier transform infrared spectroscopy, thermogravimetry, and differential scanning calorimetry. For part A of the designed two-part adhesives, the authors used different monomers, such as methyl methacrylate as the primary monomer, and the methacrylic acid and isobornyl methacrylate were applied as the comonomers. Also two elastomers, a vinyl terminated polybutadiene and styrene, and isoprene-based triblock copolymer Kraton, were added to increase the toughness of the adhesive. A curing system consisted of N, N-dimethyl p-toluidine as initiator/reducing agent, and benzoyl peroxide as an oxidizing agent was utilized to promote the free-radical polymerization of part A. Zinc dimethacrylates and mono- and di-phosphate esters of 2-hydroxyethyl methacrylate were used to promote the adhesion. It was found that the introduction of methacrylic acid and isobornyl methacrylate comonomers into SAA and elastomers enhanced the shear strength of the adhesive. The authors concluded that the elastomer-induced increase in toughness was realized through different mechanisms. These included the plasticization, the increase in the ductility of the adhesive, and the generation of a two-phase system where the structural adhesive forms a matrix in which the embedded elastomeric particles act as stress concentrators [16].

Another interesting example of the control of SAA adhesion strength was considered in [17], where the hydroxyl-terminated polybutadiene (HTPB) was added into the composition of methacrylate/epoxy-based SAA. It was found that the optimal elastomer content was about 10 wt.%, which was associated with morphological changes at this rubber concentration with the formation of a regular two-phase system. At the same time, the formation of oligobutadiene urethane acrylate (OBDUA) due to reaction between HTPB, diisocyanate, and hydroxyl-containing epoxy acrylate leads to an increase in the adhesion and thermal characteristics of SAA (Table 7.5). This was associated with the formation of a single-phase system having better compatibility with the main matrix and with the copolymerization of OBDUA with acrylate [17].

The properties of MMA-based elastomeric compositions are influenced by the structure of HTPB. With an increase in the content of double bonds in position 1,2, their ability to copolymerize with MMA increases, which leads to crosslinking, grafting of acrylate to oligomeric rubber, and, as a result, to an increase in the performance characteristics of compositions (Table 7.6) [18].

The study of the polymerization of solutions of fluoroelastomers, polyurethane rubbers, epichlorohydrin rubbers in MMA, activated by the redox system benzoyl peroxide - tertiary aromatic amine showed an acceleration of the polymerization kinetics, while the gel effect was found at a lower degree of monomer conversion. The level of the effect is determined by both the viscosity and physicochemical properties of high molecular weight component. An increase in the content of benzoyl peroxide in the redox system leads to an increase in the number of grafted units. In the presence of even small amounts of double bonds in polymers, phase separation occurs during polymerization, accompanied by an increase in the impact strength of the composites [19].

The optimal composition of SAA based on MMA, 2-hydroxyethyl methacrylate, and methacrylic acid as the main monomers was selected in [20] for obtaining enhanced shear strength. For the preparation of the part A of this adhesive, nitrile-butadiene rubber and cumyl hydroperoxide

Table 7.5 The adhesion and thermal characteristics of SAAs containing hydroxyl-terminated polybutadiene (HTPB) and oligobutadiene urethane acrylate (OBDUA) [17].

Elastomer (10 wt.%)	Lap shear strength, MPa	Peel strength, N/mm	Tensile strength, MPa	Elongation, %	TGA data	
					T_i (°C)*	T_{max} (°C)**
HTPB	12	0.11	30.8	2.9	410	455
OBDUA	15	0.14	40.2	4.8	426	490

*Initial decomposition temperature.
**Temperature at which rate of decomposition is maximum.

Table 7.6 Properties of block cured compositions depending on the structure of the elastomer [18].

Parameter	Elastomer			
	Copolymer of butadiene*	Krasol LBH 3000	NISSO PB G2000	NISSO PB G3000
The content of 1,2 C=C linkages	20	65	85	90
The content of 1,4 C=C linkages	80	35	15	10
Degree of crosslinking, %	2.3	2.7	2.9	3.1
Tensile strength, MPa	2.1	2.9	4.3	5.1
Coefficient of resistance to thermo-oxidative aging by changes in tensile strength/ elongation	1.4/0.4	1.5/0.5	-	1.7/0.6

*Copolymer of butadiene and isoprene (Russian brand PDI-1K).

were used as initiators in addition to monomers. Nitrile-butadiene rubber and hydroquinone were used jointly with triethylamine as an accelerator in part B. The effect of the mass ratio of parts A and B, amount of methacrylic acid, and nitrile-butadiene on the variation of shear strength was discussed.

Testing the effect of graft copolymers of MMA, styrene, and polyvinylpyrrolidone used as thickeners on the properties of MMA-based SAA showed that they affect the rate of curing, shrinkage, water absorption, thermal and chemical resistance [21]. The highest rate of polymerization and the highest degree of conversion were achieved when using a graft copolymer of polyvinylpyrrolidone and poly(methyl methacrylate). The possibility of obtaining SAAs based on copolymers synthesized without the traditional initiators of radical polymerization has been shown. According to the authors, the initiator omission becomes possible due to the intermolecular donor-acceptor interaction between the carbamate group of polyvinylpyrrolidone and the double bond of the methacrylate monomer. It is likely that curing becomes possible due to residual peroxide compounds in the copolymers.

The introduction of comonomers 2-hydroxyethyl methacrylate, glycidyl methacrylate, and vinylpyrrolidone into SAA, as well as Al_2O_3 and SiO_2 used as fillers, leads to a decrease in the rate of polymerization, but such additives make it possible to obtain SAA with improved physical and mechanical characteristics (adhesion strength, hardness, heat resistance) [21].

In [22], high strength properties were achieved by partially replacing MMA with ethoxylated bisphenol A diacrylate and 2-hydroxyethyl methacrylate using chloroprene rubber and acrylonitrile-butadiene-styrene (ABS) copolymer as a thickener. For the obtained adhesive composition, the lap shear, tensile, and peel strengths achieved values of 29.7 MPa, 39 MPa, and 2 N/mm, respectively. The experimental data showed that during heat and humidity aging (63 °C, humidity 90%) for 21 days, the strength of the adhesive practically did not change [22].

The possibility of increasing the thermal properties of SAAs by using acrylonitrile-butadiene rubbers with terminal methacrylates groups in combination with glycidyl methacrylate (GMA) was shown in [23].

In order to develop an SAA, which simultaneously has good elasticity with elongation of 80-120%, an acceptable decrease in strength at elevated temperatures, up to 80 °C, and high durability, it was proposed to introduce two elastomers into the adhesive formulation. For example, in [24], a block copolymer of styrene and isoprene (butadiene) and cross-linked acrylonitrile-butadiene rubber were used. In this formulation, the authors used isophthalic acid as an acidic component in combination with methacrylic phosphate ester. To enhance the adhesion strength and to ensure the transition from adhesional to cohesive failure, the authors additionally used rosin esters and talc [24].

The SAA proposed in [25] forms interpenetrating networks of two continuous phases upon curing. This SAA is characterized by increased fracture toughness, increased Notched Izod Impact Strength, and high shear strength. Such properties are provided by simultaneous introduction of oligourethane acrylate, an impact modifier of the core-shell type, and carboxylated butadiene-nitrile rubber into the initial methacrylate composition. When this SAA was used for bonding polycarbonate, poly(vinyl chloride) (PVC) and copolymer of acrylonitrile, butadiene and styrene (ABS), either mixed failure or material failure occurred during testing. When aluminum was bonded, cohesive failure in adhesive occurred [25].

An SAA with high elasticity characterized by high elongation at contact rupture and with elastic deformation well described by Hooke's law was presented in [26]. The studied adhesive composition contained a mixture of two methacrylic monomers, MMA and tetrahydrofurfuryl methacrylate (THFMA), and a liquid elastomer with a molecular weight of 1000-9000 containing methacrylate groups. In this work, poly(urethane methacrylate) based on isophorone diisocyanate and poly(propylene glycol) (MW of 4000) was used. For the proposed formulation, elongation at rupture of more than 200%, a tensile strength of 8-10 MPa, and the possibility of adhesive joining of the materials with different coefficients of thermal expansion were achieved. It is noted in [27] that the proposed SAA formulation can be used as an adhesive for joints such as aluminum with poly(vinyl chloride) (PVC), and glass with aluminum.

Also, high elongation at rupture, high resistance of the cured polymer at short-term exposure to 200 °C, and an improvement in the thixotropy index of SAA can be achieved by using a mixture of monomer-insoluble core-shell particles with a styrene-free block copolymer as a thickener. For example, in [28] block copolymers PMMA – polybutadiene - PMMA or PMMA – poly(butyl acrylate) - PMMA were used.

In [29], block acrylic copolymers, vinyl-terminated acrylonitrile-butadiene rubber, silica treated with silane, and other additives were used as thickeners and fillers to modify the composition of an MMA-based adhesive together with a phosphorus-containing adhesion promoter. For the compositions obtained, when bonding aluminum, shear strength of 27.5-31.5 MPa and peel strength of 3.7-7.2 N/mm were achieved [29].

A copolymer of styrene with acrylonitrile as a toughening agent for the SAA was used in the assembly of wind turbine blades. The proposed composition also contained MMA, a polyfunctional monomer, a thickener, stabilizers, initiation systems, an adhesion promoter, and a diluent. It was noted in [30] that such a copolymer increases the adhesion to fiberglass, flexibility at low temperatures, decreases volatility, and increases the composite glass transition temperature without increasing its brittleness. This variation in properties of composite allows avoiding microcracks formation when the cured adhesive is exposed to thermal cycles and thermal shock. The adhesive is characterized by a fast attaining of strength and high strength characteristics when bonding carbon fiber (shear strength 21.4 MPa) with 100% cohesive rupture, improved creep resistance and fatigue properties. This SAA is well suited for forming a wind turbine blade [30].

7.2.3 Adhesive Additives

An additional introduction of epoxy resin into the acrylate part, and the amine hardener into the other part makes it possible to obtain hybrid adhesives that combine the advantages of SAAs and epoxy adhesives [31].

To increase the adhesion strength of SAA based on MMA when bonding aluminum, it was proposed to use in its composition an adhesion promoter based on a diol which contains acetylene bonds, with a chemical structure such as [32]:

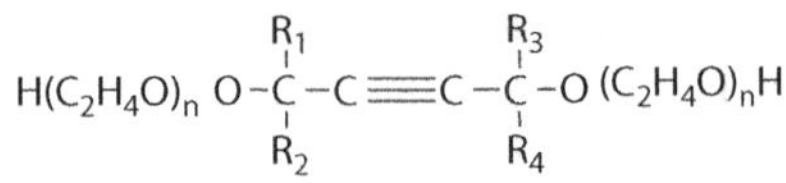

The use of phosphorus-containing adhesion promoters like the ones described in the patent [33], for example, those presented in Figure 7.1 makes it possible to increase the strength properties of SAA when bonding galvanized steel.

To ensure sufficient strength when bonding galvanized steel surfaces, it was proposed to introduce etching acid components (such as phosphoric, acetic, formic, etc.) into the SAA, along with phosphorus-containing adhesion promoters [34]. The strength of adhesive joints made of galvanized steel with such an adhesive reaches 10.3-12.4 at 23 °C, 11.0-13.8 at 82 °C, and 6.2-9.0 MPa at 121 °C.

The introduction of chelating components into the composition of SAA (for example, tetrasodium salt of ethylenediaminetetraacetic acid, 1-2 wt.%) increases the stability of the adhesive during storage. The addition of wax or poly(vinyl acetate) reduces the weight loss during the exothermic curing reaction by reducing MMA emission. Block copolymers with a glass transition temperature from −50 °C to −110 °C (for example, styrene-butadiene-styrene block copolymer Kraton®1116) [35] were used to increase the impact toughness at low temperatures (−40 °C), while maintaining it at elevated temperatures.

The addition of small amounts (5 wt.%) of acetal-containing methacrylates to SAAs, for example, of the structures presented in Figure 7.2, makes

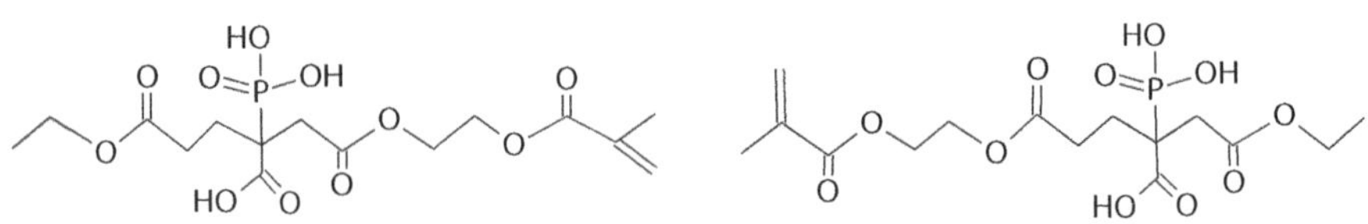

Figure 7.1 Examples of adhesion promoters based on acrylic phosphonic acid esters [33].

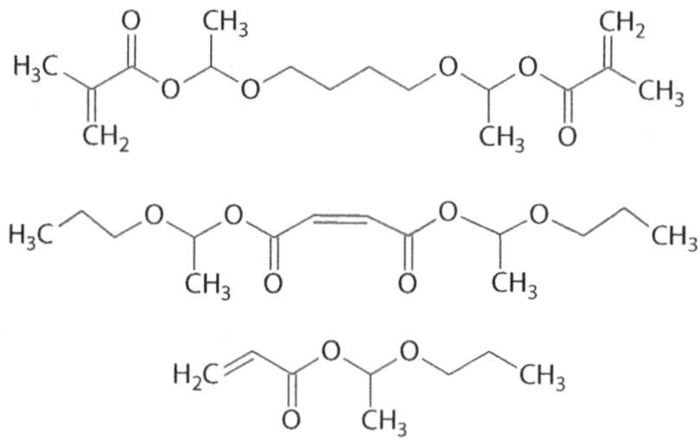

Figure 7.2 Examples of acetal-containing free radically curable compounds [36].

it possible to obtain an adhesive joint that can be easily dismantled if necessary when heated to 100 °C without damaging the substrate. Moreover, such SAAs are sufficiently stable when the adhesive joints are kept in a humidity chamber with cyclic temperature changes from 25 to 65 °C and humidity of 95% [36].

When SAAs are modified with polyimides with terminal maleimide groups, along with a rubber modifier and an adhesion promoter, the adhesion characteristics also increase both at room temperature and at T=80 °C (Table 7.7). Besides, the subsequent heating of such compounds at 150-190 °C results in an additional increase in the strength characteristics [37].

Table 7.7 Comparison of the strength characteristics of commercial SAA and SAA modified by polyimide [37].

	Shear strength, MPa (galvanized steel)				Impact strength, N/mm (wedge test)		
Adhesive	**initial**	**−30 °C**	**+80 °C**	**oiled**	**initial**	**−30 °C**	**+80 °C**
Adhesive containing maleimide-terminated polyimide [32]	15.8	19.8	8.6	15.1	26.8	13.8	23.9
Commercial Versilok®	13.3	18.2	6.3	12.9	18.8		20.3

7.2.4 Initiators

As was mentioned above, the cure of structural acrylic adhesives is initiated by a two-part redox system. For the purpose of distinction, that part of the redox system present in the base component of the adhesive which triggers chemical reactions will be defined as the initiator [1]. In contrast, the second part of the redox system is present in the accelerator or curative component. Reaction accelerators serve as catalysts that accelerate the progress of the chemical reaction.

The well-known redox initiation system benzoyl peroxide (BPO) - tertiary aromatic amine contains aminoalkyl radical which forms in the course of reaction [38]:

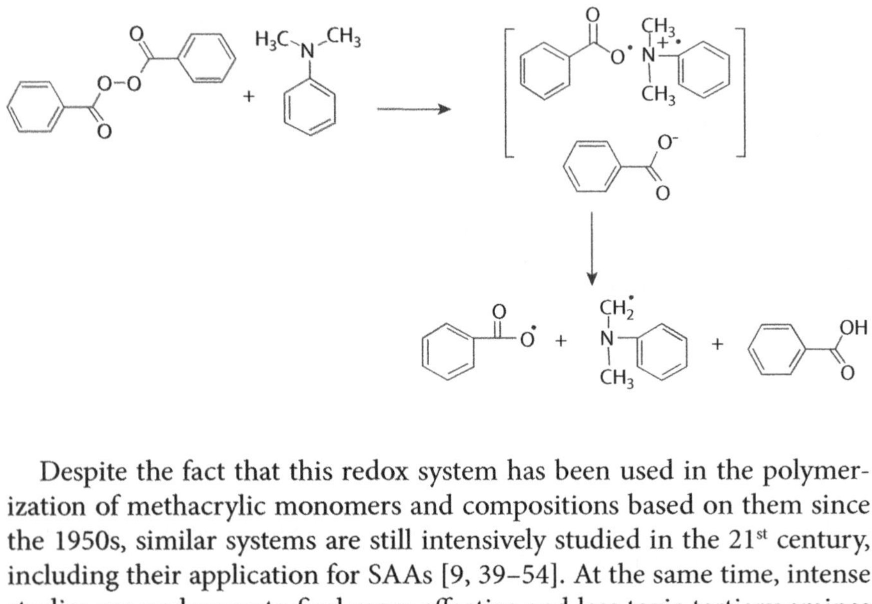

Despite the fact that this redox system has been used in the polymerization of methacrylic monomers and compositions based on them since the 1950s, similar systems are still intensively studied in the 21st century, including their application for SAAs [9, 39–54]. At the same time, intense studies are underway to find more effective and less toxic tertiary amines to replace traditional dimethylaniline (DMA) and dimethyl-p-toluidine (DMPT). For example, it was shown that the replacement of DMPT/BPO with dihydroxyethyl-p-toluidine/BPO system provides a higher rate of MMA curing at room temperature [44].

Comparison of the efficiency of initiation systems based on BPO with reducing agents N,N-di-isopropanol-p-toluidine (DIIPT) or mixtures thereof with either dimethylaniline (DMA), dimethylpiperazine (DMP), or co-accelerator 1,4-diazabicyclo [2.2 .2] octane (DABCO) showed the highest efficiency of the latter mixtures [55]. The authors of this patent associated the acceleration of curing with the formation of a stronger complex with acidic adhesion promoters. An increase in the strength of such complexes is stimulated by the high basicity of the co-accelerator. The formation of strong complexes, in turn, suppresses the inhibition of curing. Data on the effect of the composition of the initiation systems on the time to reach the maximum curing temperature (τ_t), the time to handling strength (time is defined as the time required after the adhesive bead has been applied and mated for the bond strength to exceed 100 psi) (τ_s) and the open time (τ_o) for mixtures investigated in [55] are given in Table 7.8.

When curing at low positive and negative temperatures (up to −25 °C), it is preferable to use 2-[4-(dimethylamino)-phenyl]ethanol as a reducing agent. Such a reducing agent makes it possible to obtain adhesives of higher strength than using traditional tertiary amines DMA and DMPT [45].

The effect of various tertiary amines as components of initiator systems together with BPO was studied in [46, 47]. Comparison of the activities of some of the donor-acceptor curing systems according to DSC data is given in Table 7.9.

In [48], the high efficiency of N-(4-methoxyphenyl)pyrrolidine as a reducing amine agent was established.

The combination of traditional monomers, monomers with acetoacetoxy groups, and silane-containing monomers in the presence of a curing system BPO/tertiary aromatic amine (N,N-diisopropanol-p-toluidine, N,N-dihydroxyethyl-p-toluidine, N,N-methylhydroxyethyl-p-toluidine) ensures complete curing of the SAA at −10 °C. In contrast, a commercially available SAA does not achieve full strength at this temperature [56].

The synthesis of new amine curing accelerators that can be used as reducing agents was described in [57]. Some of proposed compounds - derivatives of indoline and tetrahydroquinoline are shown in Figure 7.3. The strength properties of SAAs containing such accelerators are shown in Table 7.10.

Table 7.8 The influence of accelerator mixture composition on the curing parameters [55].

Mixture of amines	τ_t, min-sec	τ_s, min	τ_o, min	τ_s/τ_o
DIIPT/DMA	17-20	45	16	2.81
DIIPT/DMP	15-34	34	19	1.80
DIIPT/DABCO	15-09	27	18	1.50

Table 7.9 The efficiency of the tertiary amine - BPO systems in the curing of acrylate compositions [46, 47].

Tertiary amine	Decomposition temperature of complexes, °C	Curing start temperature, °C	Maximum curing temperature, °C
$C_6H_5N(CH_3)_2$	–5.05	3.31	16.17
$C_6H_5N(CH_2CH\text{-}CH_3)_2$ \ / O	–4.18	6.19	22.14
$H_5C_6Si(OCH_3)_2[OCH_2CH_2N(CH_3)_2]$	–3.15	6.30	23.12
$(H_9C_4O_2)Ti[OCH_2CH_2N(CH_3)_2]$	–3.53	6.10	23.40

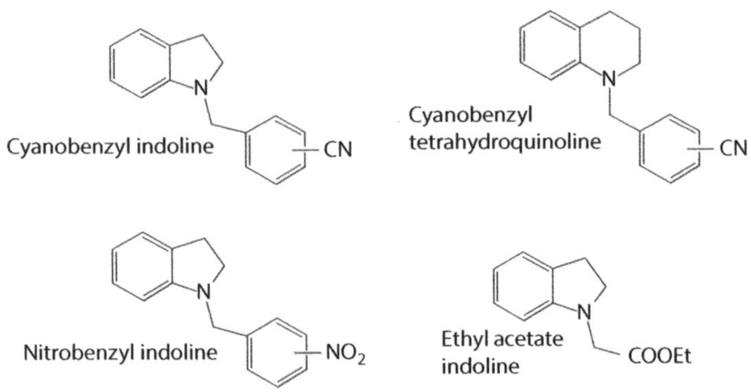

Figure 7.3 Examples of tertiary amines from [57].

Table 7.10 Strength properties of SAAs containing amine curing accelerators [57].

Accelerator	Lap shear strength, MPa			T-peel strength, N/m (steel)	Impact strength, kN/m (wedge test)
	Steel	Stainless steel	Al		
Cyanobenzyl indoline	6.2	11.2	10.0	6.5	21.0
Cyanobenzyl tetrahydroquinoline	10.8	11.9	11.3	5.8	23.7
Nitrobenzyl indoline	9.4	9.4	10.3	4.3	20.3
Ethyl acetate indoline	12.8	12.6	12.4	7.3	20.8

Table 7.11 Properties of triphenylphosphine containing SAAs [61].

Triphenylphosphine content, wt.%	Open time, s	Fixture time at 40 °C, s	Stability of SAA at 82 °C, h
0	255	60	4
1.4	135	30	15
2	105	30	15

A combination of a prolonged open time and fast curing can be achieved using an initiator system containing fatty acid peroxide such as lauryl peroxide and tertiary amines [58].

Combination of dilauroyl peroxide and 3,5-diethyl-1,2-dihydro-1-phenyl-2-propylpyridine suggested in [59] allows obtaining higher stability, fast curing and high adhesion properties, when joining both metals and plastics.

Use of 4-dimethylaminobenzylmethacrylate as a reducing agent allows one to reduce the toxicity of the material due to binding of amine with MMA during free radical copolymerization [60].

BPO together with ferrocene can also initiate the curing of SAA as follows [4]:

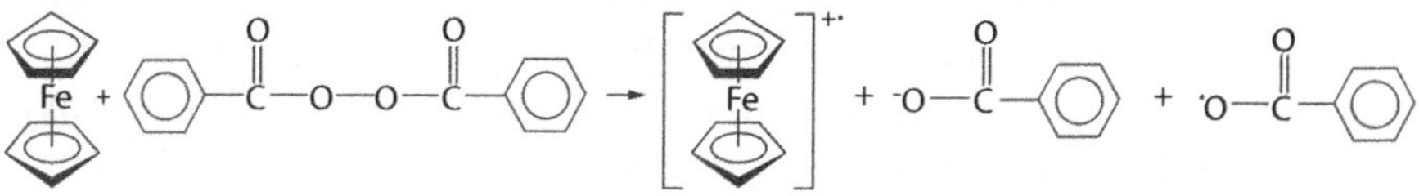

The additional introduction of the reducing agent triphenylphosphine into SAAs containing amine and quinone can shorten the open time and fixture time and, at the same time, increases the stability of the SAAs (Table 7.11) [61].

Here it is worth mentioning the hydroperoxide - dihydropyridine initiation system, the design of which is associated with a very significant milestone in the development of SAAs. In the presence of this initiation system, which is a product of the interaction of a primary amine and an aldehyde, initiation is possible according to the following scheme:

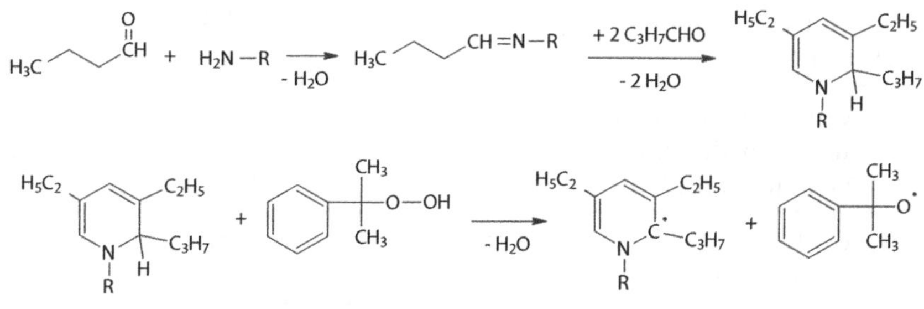

The additional inclusion of metals of variable valence makes possible the following well-known redox reactions:

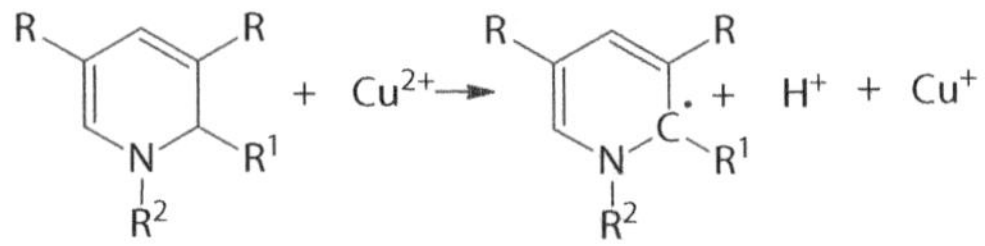

$$Cu^{+} + R\text{-}O\text{-}OH \rightarrow Cu^{2+} + R\text{—}O^{\bullet} + HO^{-}$$

$$Cu^{2+} + R\text{-}O\text{-}OH \rightarrow Cu^{+} + R\text{—}O_2^{\bullet} + H^{+}$$

Nowadays, SAAs are mostly produced with this initiation system, since it provides greater stability of the compositions than BPO-amine system, high strength, and very high cure rate.

The use of hydroperoxide systems together with aldimines, dihydropyridines, acetylphenylhydrazine, substituted thioureas, etc., can significantly alter the key characteristics of SAAs. This is illustrated in Tables 7.12 and 7.13 which present a comparison of some initiation systems on their effects on curing time and strength properties of SAAs.

As can be seen from Table 7.12, it is possible to obtain an SAA free from peroxide initiator. Compositions containing compounds with sulfonyl chloride groups, aldehyde-amine condensate, and metal complex with 8-hydroxyquinoline, as initiating components, give an example of such SAAs [63].

Table 7.12 The influence of initiators on the properties of SAAs [62].

Initiation system	Setting time, min	Shear strength, MPa
1% ferrocene 1% toluenesulfonic acid 2 % cumene hydroperoxide (CHP)	4	3.5
2% DMA 2% CHP	60	1.4
1% condensation product of butylaldehyde with butylamine 0.025% Cu (II) naphthenate	4	4.9
1% Cu(I) 1% ascorbic acid	4	9.8

Table 7.13 SAA* properties with different initiation systems [9].

Parameter	Initiation system				
	BPO (2%), DMPT (1%)	CHP (2%), Accelerator 833 (2%)	CHP (3%), Accelerator 808 (2%)	CHP (2%), TMTU (1%)	TEB·HMDA (2%), BPO (0.1%)
Setting time, min	4	3	14	5	8
Tensile strength, MPa: - stainless steel - oiled stainless steel	 20.8 20.6	 28.2 27.0	 19.6 18.4	 11.5 11.2	 12.5 11.6
Shear strength, MPa, -stainless steel	23.8	23.0	24.6	17.0	14.8
Tensile strength, MPa, Al	19.8	24.4	18.1	11.0	11.8
Shear strength, MPa, Al	8.8	8.5	4.8	6.8	10.1
T-peel strength, N/mm,Al	2.4	2.5	2.4	2.5	1.8

*SAA contains Hypalon-20; BPO – benzoyl peroxide; DMPT – dimethyl-p-toluidine; CHP – cumene hydroperoxide; Accelerator 833 – condensation product of butylamine and butyl aldehyde; Accelerator 808 – condensation product of aniline and butyraldehyde; TMTU – tetramethylthiourea; TEB·HMDA – complex of triethylborane with hexamethylenediamine.

Another approach to shorten the fixture time (time - the ability of the adhesive to withstand 100 psi lap-shear stress) on the copper substrate for the commercial 3M Scotch-Weld® Low-Odor Acrylic Adhesive DP810 was proposed in [50]. It was shown that at room temperature, a significant reduction of fixture time (from 10 min to 2.5 min) is achieved by the introduction of saccharin-DMPT accelerators. Such systems were previously widely used in anaerobic adhesives. It is worth noting that such modified adhesives have a lower degree of crosslinking, but the achieved final strength practically remains equal to the strength of the unmodified adhesive.

Another type of accelerators, benzoyl thioureas [64], some examples of which are presented in Figure 7.4, were used for curing SAAs with peroxide.

Besides, initiation systems containing inorganic and complex salts of metals of variable valence, along with a peroxide initiator, are widely used. For example, an SAA having a cure time of 10-20 seconds at room temperature or 60 seconds at −30 °C, while gaining a full strength in 30 min, was described in [65]. The initiation system formulation in [65] contained CHP as an initiator. As a reducing agent, it was suggested to use any of the following components: vanadium or copper salts (vanadyl acetylacetonate, copper naphthenate) or ethylene thiourea and an amine (for example, polyethylenimine).

It is interesting to note here that an SAA with an initiator system comprising CHP, DMPT, saccharin, and copper (II) ethylhexanoate is shown to be effective to join chemically inactive substrates with low surface energy, such as polyethylene [66].

It was shown in [67] that the introduction into MMA-based SAA of an acrylic polymer and a hydroperoxide initiator in Part A, and a composition

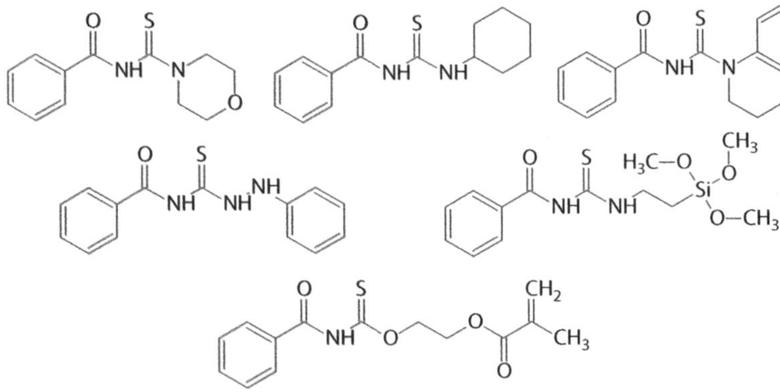

Figure 7.4 Examples of benzoyl thioureas used as accelerators for SAAs [64].

of a dihydropyridine reductant, an organometallic curing promoter (for example, copper acetylacetonate), and an accelerator (ethylene thiourea) into Part B, allows increasing the maximum temperature and completeness of exothermic curing, and reducing the curing time. This also leads to an increase in the heat and moisture resistance of adhesive joints at elevated temperatures (80 °C). Besides, a decrease in the concentration of unreacted components of SAA was detected for the joints of thermoplastic pipes and fittings made of poly(vinyl chloride) (PVC), chlorinated PVC, acrylonitrile-butadiene-styrene (ABS) plastic. Similar properties were demonstrated by SAAs additionally containing p-toluenesulfochloride in Part A [67].

In order to reduce the inhibiting effect of air oxygen during SAA curing, it was proposed in [68] to use a metal salt or a metal complex of a transition metal, hydroperoxide, and a β-dicarbonyl compound as an initiation system. In the last compound, the carbon atom between the two carbonyl functions is bonded with only one hydrogen atom as, for example, in structures of the type presented in Figure 7.5. The time for complete curing of such SAAs is 40-320 minutes.

A higher curing rate can be achieved by including weak acids (citric, sorbic, phthalic, etc.) into the formulation of SAAs [69]. It is worth noting that the replacement of the odorous adhesive additives – acrylic and methacrylic acids which are actively used in SAAs – by the notably less odorous ones significantly enhances the attractiveness of SAAs for wide industrial applications.

In this case, it is possible to use monomers with carboxyl groups of higher molecular weight, for example, mono-2-(methacryloxy)ethyl succinate. Another approach to odor suppression is based on the replacement of MMA with a mixture of methacrylic acid and a reaction product of a (hydroxy) acrylate with an anhydride, predominantly mono-2-(methacryloxy)ethyl phthalate [70].

The use of a redox curing system containing n-halogen-substituted derivatives of DMA as a reducing agent [71, 72] eliminates stickiness at the edges of the joints.

The differences in the electrochemical activities of metals allow free radical polymerization and curing of adhesives on metal surfaces.

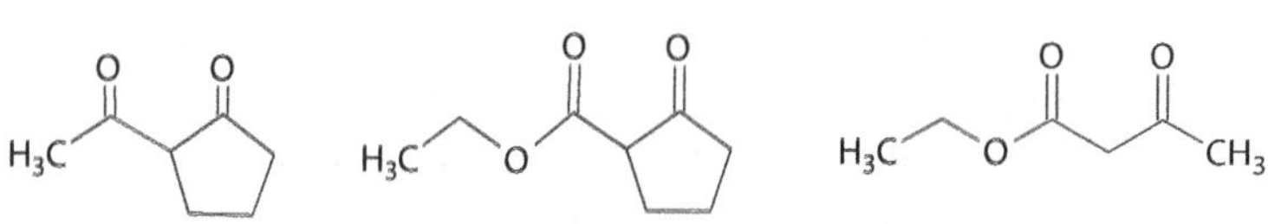

Figure 7.5 Examples of dicarbonyl compounds [69].

The capability of initiating polymerization is based on the fact that the metal salt added to the composition (for example, $Cu(BF_4)_2$, $Fe(ClO)_4$) has a standard reduction potential greater than the standard reduction potential of the substrate [73]. Upon contact of the composition with the metal surface, the metal salt contained in it is reduced on the surface and interacts with the peroxide initiator, causing curing of the methacrylate derivative [73].

To stabilize the exothermic reaction of the SAAs curing, it is recommended to add to their formulation the regulators such as cycloheteroatomic zirconates and titanates, as well as water and chelating agents. These additives can extend the time to reach the maximum exothermic effect during adhesive curing. Such extended exothermic reaction time is desirable when bonding large objects or panels where significant working time is required. So, adding 1 wt. % of water or an aqueous solution of tetrasodium salt of ethylenediaminetetraacetic acid increases the time to reach the maximum temperature of the reaction products nearly twice [15].

In contrast, the addition of a tertiary polyamine, for example, 4,4',4"-methylidyne-tris (N, N-dimethylaniline), make it possible to accelerate curing when using an initiating system with BPO [74].

It is worth noting that the ratio of the adhesive components affects the physical and mechanical properties of the adhesive, which was demonstrated in the literature. For example, in [75] it was shown that with an increase in the hardener: resin ratio in the My-T-Bond 1001AB adhesive, an increase in tensile elongation and ultimate deformation was found.

An approach for achieving a combination of improved shear strength and peel strength when bonding aluminum with an SAA is described in [29]. The effect in this work was achieved by using in the SAA an adhesion promoter, two types of reinforcing elastomers as block copolymer and a methacrylate-terminated liquid rubber, fillers, and the cure initiator system, which comprised an organic peroxide, a barbituric acid derivative, a metal salt and ammonium chloride salt.

In order to increase the SAA open time without reducing the strength properties after curing, it was proposed to add into adhesive formulation the secondary aromatic or cyclic amines in an amount of 1-5 wt.%, which practically does not affect the stability of the adhesive during storage [76].

2-nitrobenzoic acid can be used to increase the adhesive stability and to control the curing time for SAAs containing an organic peroxide, salt or oxide of a polyvalent metal, non-metallic halide salt and beta-dicarbonyl compound as components of the cure initiation system [77].

Thus, as was shown above, the SAA cure initiation systems are important components and can affect cure rate, adhesion, and performance characteristics.

7.2.5 Aerobically Curable Systems

The use of compounds that are self-oxidizing in air as initiators of polymerization provides an alternative way to obtain SAAs (possibly containing no peroxide compounds).

The seminal paper of Horner and Knapp, where studies on spontaneous autooxidation of N-dialkyl substituted arylamine compounds activated in the presence of metal salts Co (II) and Fe (III), acetic acid, and benzyl chloride were performed, substantiates such mechanisms [78].

Examples of such compounds are tertiary aromatic amines, dihydropyridines, Schiff bases, β-diketones together with salts of metals with variable valence, and weak acids that accelerate autooxidation, can also be used as such initiators to obtain SAAs [79, 80]. When oxidized with atmospheric oxygen, these components form peroxide compounds initiating the polymerization of methacrylic derivatives. The advantage of the proposed aerobic-curing SAA, which can be either one-component or two-component, is the independence of curing on the nature of the substrate and the possibility of curing in thick gaps.

Using the aforementioned initiating systems, adhesives with a dual curing mechanism were developed: a radical one and the mechanism based on the interaction of air moisture with oligomers containing terminal isocyanate or silanol groups. This allows obtaining high initial and final strength characteristics [81].

To stabilize such auto-oxidizing SAAs, it was proposed to use stable nitroxyl radicals. The use of 1,4-dihydropyridine and onium salts (diazonium, sulfonic, iodonium) together with ionic salts has been described in [82]. The study of such adhesives by dynamic thermal-mechanical analysis showed the stability of their properties during storage for a year [83].

7.2.6 Fillers

As discussed in a recent review [84], the role of fillers is very important for adhesives of different nature. Fillers used in SAAs promote a reduction in shrinkage stresses and affect the physical and mechanical properties of the cured adhesive. The effect of quartz sand on the properties of SAA based on MMA and containing a styrene-MMA copolymer as a polymer modifier was considered in [85]. It was shown that both the glass transition temperature T_g and the calculated value of the Simcha-Boyer constant, which characterizes the fraction of free volume during glass transition, depend on the filler concentration C (Figure 7.6). Besides, it was found that a decrease

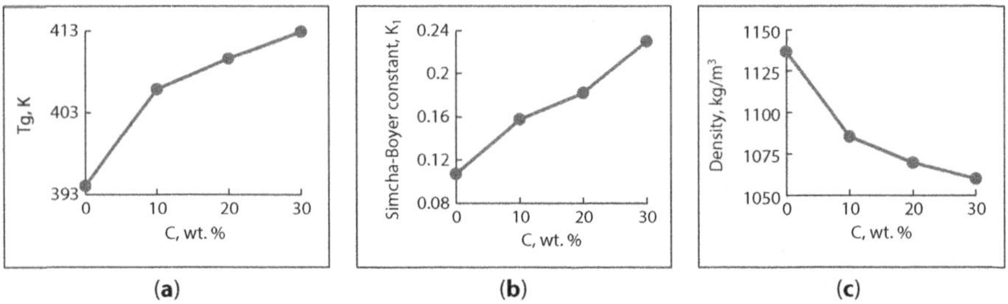

Figure 7.6 Dependence of glass transition temperature (a), Simcha-Boyer constant (b) and the density at 25 °C (c) on the concentration of quartz filler C in an acrylic adhesive [85].

in density in the bulk polymer phase of the filled system leads to some decrease in adhesive strength.

The variation of the dynamic shear modulus of filled systems with an increase in temperature is more intense than that of an unfilled polymer resin, which in [85] was attributed to the adsorption and curing of the resin on the surface of the quartz filler.

The application of finely dispersed zinc oxide and mica as modifying additives together with methacrylic acid increases the mechanical characteristics of acrylic adhesive: for compression strength from 83.6 to 98.6 MPa, the tensile strength from 18 to 28 MPa, and shear strength from 32 to 41 MPa [86].

Analysis of modern SAA compositions, in which dispersed particles were embedded in a polymer matrix showed significantly increased fracture toughness and impact toughness, as well as shear and peel strengths [87].

The effect of TiO_2 and Al_2O_3 nanoparticles on the adhesion strength of steel-glass/epoxy composite joints bonded with a commercial two-part structural acrylic adhesive Macroplexx 3295 (Holdtite Company, UK) was studied in [88, 89]. It was found that the shear and tensile strengths of the adhesive joints increased with the filler content up to 3 wt.% of TiO_2 and 1.5 wt.% of Al_2O_3. Besides, the presence of nanoparticles leads to improvement of the glass/epoxy substrate wettability by the adhesive and alters the fracture morphology from smooth to rough fracture surfaces. The DSC curves revealed that the glass transition temperature of the nanocomposites increased by increasing the filler content.

A similar effect on the adhesion strength of steel–glass/epoxy composite joints bonded with two-part structural acrylic adhesives was observed for silicon carbide nanoparticles of 20-30 nm in diameter [90]. It was found that adding SiC nanoparticles to acrylic adhesive caused nonmonotonous variations of the shear and tensile strengths of the composite joints with the maximum achieved when the nano-SiC content was 1.5 wt. %. However, the monotonous reduction in the peel strength of the joints and the rise of T_g were also found. SEM images revealed that the addition of nanoparticles altered the fracture morphology from smooth to rough fracture surfaces as was discussed above for TiO_2 nanoparticles.

A comparison of the adhesion of SAAs filled with different nanoparticles, such as SiO_2, TiO_2, and Al_2O_3, was given in [91] and is shown in Table 7.14. As follows from the presented data, nano-Al_2O_3 has the greatest effect on the shear and tensile strengths: the shear strength increases from 22.6 to 32.3 MPa (by ~ 43%), and the tensile strength from 27 to 43.9 MPa (by ~ 63%). For all fillers, there was a significant decrease in peel strength, which was explained by an increase in the brittleness of the adhesives.

An opposite effect, i.e., a decrease in strength characteristics, was observed when carbon nanoparticles (1 wt.%) were added to SAA: tensile strength drops from 73.7 to 20.6 MPa, and shear strength - from 7.4 to 2 MPa [92]. Such a decrease, most likely, can be associated with the ability of carbon nanoparticles and nanotubes to inhibit free radical polymerization [93].

At the same time, recent studies showed that a combination of carbon nanotubes (CNTs) and a toughening agent makes it possible to achieve high shear strength (24.5 MPa). In the case of using only CNTs without a toughening agent, a decrease in strength to 1.9 MPa was observed. The significant effect of CNTs morphology on the strength characteristics of the composite was found [94].

For the use of SAA in display devices, it was proposed to introduce polyhedral silsesquioxane (POSS) in an amount of 0.5-2.0% into the part of

Table 7.14 Effect of different nanoparticles (nanofillers) on shear, tensile, and peel strengths of acrylic nanoadhesive joints [91].

Nanofiller* (1.5 wt.%)	Shear strength, MPa	Tensile strength, MPa	Peel strength, N/mm
-	22.6	27	47.5
Al_2O_3	32.3	43.9	31
SiO_2	29.2	35	34
TiO_2	28.2	31.7	26.5

*Nanofillers in the amount of 1.5 wt.% were incorporated into acrylic adhesive Macroplexx 3295.

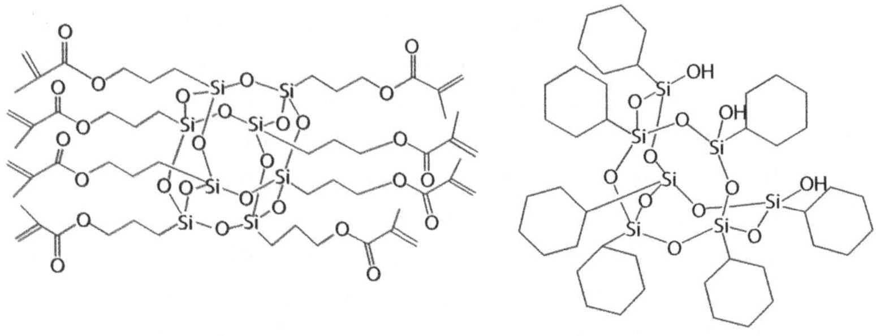

(meth)acryloxypropyl POSS Trisilanolcyclohexyl POSS

Figure 7.7 Examples of polyhedral silsesquioxanes (POSS) [95].

the adhesive containing curing accelerators. The chemical structures of the accelerators used in [95] are shown in Figure 7.7.

Data on lap shear strength for a variety of materials including metals and polymers, presented in Table 7.15, indicate that the introduction of POSS increases the bonding strength, as well as somewhat accelerates the curing of SAA.

The use of a mixture of two fillers in the SAA composition - fibrillated filler with a length-to-diameter ratio of more than 20 and halloysite clay or fumed silica, along with strengthening elastomers-makes it possible to achieve high shear strength and peel strength when bonding aluminum (up to 28 MPa and 7.2 kN/m, respectively) [29]. The simultaneous attainment of such adhesive characteristics is infeasible when using other adhesives.

Table 7.15 Strength characteristics of SAA containing POSS [95].

Substrate	Lap shear strength, MPa for SAA containing		
	methacryloxypropyl POSS	**trisilanolcyclohexyl POSS**	**no POSS (for comparison)**
Mild Steel	10.8	12.1	5.0
Stainless Steel	10.2	10.2	7.1
Aluminium	10.8	8.6	5.1
Nylon	1.0	1.1	1.0
Polystyrene-ABS	4.2	4.3	3.2
Polycarbonate	3.4	3.5	2.6

Since the use of natural fillers such as wood flour in SAAs causes inhibition of curing by tannin-containing products, it was proposed to introduce into SAAs simultaneously natural filler and oxygen scavengers - triphenylphosphine or dihydropyridine [96].

7.3 Physico-Mechanical Properties of SAAs

In recent years, the adhesive joints obtained using SAAs have been studied in detail in order to establish the possibilities of their application in various fields of technology under different operating conditions [97–137].

For example, the authors of [97] studied the effect of silanization parameters (pH, solution concentration and drying temperature, surface preparation before silanization) on the fracture toughness of adhesive joints made of an aluminum alloy during heat and moisture aging.

The strength characteristics and relaxation processes occurring in cured SAA (Plexus-MA420), characterized by adhesive relaxation with non-linear stress-strain dependence, were analyzed in [105]. Based on the quasi-static tensile and relaxation test results for the above adhesive, the complex characteristics of the viscoelastic material parameters of the adhesive for different stress levels were determined [107].

In [111], a study on the mechanical behavior of a viscoelastic ductile SAA (SikaFast 5221NT grade based on tetrahydrofurfuryl methacrylate) under tension and compression and at different strain rates was carried out. A logarithmic dependence on increasing strain rate was found for the behavior of yield stress, elastic modulus, and failure strain, while yield strain and stiffness after yielding were insensitive to strain rate. Based on the high recovery of the adhesive after 48 h, the authors concluded that no residual deformation was caused by damage or plastic flow.

The expediency of using an SAA in joints made of high-strength steel and subjected to static loads was confirmed in [110]. The conclusions of the authors were made based on the results of tests which, to minimize the influence of the size effect, were carried out on elements with dimensions comparable to the actual structural elements.

One of the important tasks that can be solved with the help of an SAA is the bonding of carbon fiber reinforced polymers to steel. Comparison of epoxy adhesive with an SAA under different loading conditions showed a significant advantage of the latter when joining load-bearing structures made of carbon fiber reinforced polymers with high-strength steel, especially during fatigue cyclic tests [85, 108, 111].

It is known that the surface treatment of substrates before bonding leads to a change in the surface free energy. This, in turn, affects the bond strength [138]. The effect of surface treatment on the strength characteristics of the adhesive bond and the surface energy of the material for SAA/glass reinforced polymer (epoxy) composite joints was investigated in [101]. In this study, Loctite Hysol H4800 was used as SAA. The surfaces of glass reinforced polymer (epoxy) composites were modified by degreasing, atmospheric plasma, grit blasting, and grit blasting plus silanization with γ-methacryloxypropyltrimethoxysilane (γ-MPS) treatments. It was shown that all types of surface treatments increased the surface energy of the epoxy composite. Both the grit blasting and grit blasting followed by silanization led to enhancement of the adhesion between the SAA and the composite. In contrast, atmospheric plasma treatment caused adhesion failure (interfacial failure) when mechanical stress was applied to the joint. Crack resistance of adhesive bonds for different studied surface treatments well correlated with shear strength. The fracture energies for SAA/modified epoxy composite system had values near 500 J/m^2 for degreased and atmospheric plasma-treated specimens with interfacial failure for both types of treatments. In contrast, grit blasting and γ-MPS surface treatments showed fracture energies of 1796 J/m^2 for the former and 1570 J/m^2 for the latter and demonstrated an interlaminar cohesive failure of the composite [101].

The effect of plasma treatment on the durability and strength of adhesive joints was also studied for polypropylene (PP) and polycarbonate (PC) joined using an SAA of Permabond Ell grade based on ethyl methacrylate and chlorosulfonated polyethylene. The shear strength for adhesive joints increased after treatment with oxygen plasma and exposure to air. At the same time, the aging of adhesive joints in water at 20 and 90 °C after plasma treatment led to a decrease in shear strength [102]. Analysis of the surface condition showed that plasma treatment induced the formation of an oxygen-enriched layer on the treated surfaces. In the case of PP, this layer did not change its composition when immersed in water, while for PC, leaching of oxidized species with low molecular weights was observed. The decrease in shear strength was accompanied by adhesion (interfacial) failure for untreated PP and cohesive failure for plasma-treated PP [103].

The effect of roughness on the shear strength of an adhesive bond was studied for the joints of steel samples with a glass-reinforced epoxy composite using SAA of Macroplexx 3295 grade based on MMA [104]. A decrease in the r.m.s. roughness of the steel surface from 90.77 ± 1.81 nm to 2.97 ± 0.05 nm led to an increase in shear strength by 20-25% and peel strength by 7.5-8.5%.

Fracture toughness in three different joints of the substrates fabricated from aluminum (Al/Al joint), from glass fiber reinforced polypropylene matrix composite (GF-PP/GF-PP joint) and from carbon fiber-reinforced epoxy matrix composite (CF-EP/CF-EP joint) was studied in [105]. All joints were bonded with SAA 3M Scotch-Weld™ DP8005. The analysis of the fracture surfaces indicated that the fracture toughness of the joints for all studied specimens was closely related to the fracture morphology. Fractures of the aluminum alloy joints occurred mainly between the aluminum alloy and the adhesive. An analysis of the values of fracture toughness of joints for composite substrates and fracture surfaces confirmed relatively poor bonding between the GFs and PP, which caused delamination of the adherends in the GF-PP specimens. The fracture surfaces of the CF-EP specimens indicated cohesive failures in CF-EP composite.

The strength properties of an adhesive joint are influenced by the wetting of the substrate surface which, in turn, depends on the viscosity of the adhesive and its surface tension, since these parameters determine the mode of the contact (homogeneous or heterogeneous wetting) between the substrate and the adhesive. Addition of a thickener increases both the viscosity and the surface tension of the adhesive. Studies on the properties of SAA based on MMA containing a neoprene thickener and aldimine-cumene hydroperoxide as an initiation system performed in [106] showed that there is an optimal viscosity that provides a homogeneous contact between the substrate and the adhesive, low shrinkage during curing, and high shear strength. The low content of the thickener, and hence the low viscosity, helps reducing the impact strength; the toughness above the optimal value leads to a decrease in strength due to the formation of voids at the interface.

A method of fabricating functionally graded adhesive joints when two SAAs with different elastic moduli, i.e., one rigid and other elastic adhesive were mixed just prior to bonding, was developed in [107–109]. It was shown that it is possible to control and widely vary the mechanical properties by varying the mixing ratio. For example, Young's modulus can vary for such adhesives over a wide range from 1.37×10^3 to 3.72×10^2 MPa [109].

The role of aging and degradation of adhesive joints in different media has been widely studied with various substrates and SAAs. For example, in [110] the durability of aluminum alloy adhesive joints obtained with SAA of Loctite Hysol H4800 Speedbonder grade was studied under exposure to moisture, immersion in deionized and salt water. Besides, the effect of surface pretreatment of samples, such as degreasing, abrading, iron-phosphate etching, and silane treatment, on the resistance of joints in a humid atmosphere was analyzed for 90 days at T=50 °C and relative humidity of 95%. The best stability of the adhesive joint properties was observed for the samples which were etched and silane treated.

The variations in the properties of phosphated steel adhesive joints with an SAA based on MMA and chlorosulfonated polyethylene after aging in air and in water at temperatures of 20 and 90 °C were studied in [110] and presented in Table 7.16. XPS data obtained from the analysis of the fracture surfaces showed a significant increase in the amount of chlorine at the surface after adhesive joints immersion in water. The authors attributed the observed effect to the sorption of water, which caused plasticization of the adhesive and diffusion of small domains of chlorine-containing rubber [111].

Test results for adhesive joints immersed in demineralized water at 45 °C were obtained in [112] for aluminum/aluminum and galvanized steel/galvanized steel adhesive joints with a highly elastic SAA SikaFast 5215 NT based on tetrahydrofurfuryl methacrylate. The impact of various substrate treatments on the mechanical properties of the joints subjected to demineralized water for 21 days was analyzed. The data obtained showed a deterioration in mechanical properties, especially for anodized aluminum and galvanized steel. The authors of the study [112] concluded the positive effect of roughening on adhesion and aging resistance for all tested joints using SAAs.

Studies on the water resistance of adhesive joints made using various SAAs allowed the authors of [113] to propose an accelerated method for assessing the water resistance of adhesives. It was shown that for acrylic adhesives, the water resistance at room temperature can be evaluated on the basis of the shear strength retention values after aging in water at 80 °C.

Table 7.16 Shear strength for adhesive joints (steel-steel) at different aging conditions [111].

Aging conditions		**Shear strength, MPa**
Immersion time, h	**Water temperature, °C**	
0	-	22.6
100	20	21.2
200	20	20.1
100	90	17.9
200	90	12.7

Aging of adhesive joints and cured block SAAs in a climatic chamber showed retention of mechanical properties and cohesive nature of fracture under the following cyclic exposures: (1) exposure for 15 h at a temperature of 40 °C and at relative humidity higher than 90%; (2) exposure for 2 h at a temperature of 220 °C; 3) exposure for 4 h at a temperature of 70 °C and a relative humidity value of 50%; (4) return to a temperature of 40 °C and relative humidity not lower than 90%. At the same time, a subsequent immersion of the test specimens in a saline solution for 8 h at 23 °C causes corrosion of metal surfaces, which significantly reduced the strength parameters and changed the nature of failure from cohesive to interfacial. At the same time, the block polymer retained its high deformability after exposure to a saline solution [132].

Comparative analysis of strength characteristics of the welded aluminum joints against adhesive joints based on cured adhesive LORD 406-19GB was performed in [120]. Some characteristics of adhesive joints versus temperature are shown in Figure 7.8. It was concluded that in the manufacture of dynamic message signs, welded parts can be reliably replaced by adhesive joints [120].

The study of the influence of a temperature increase on the strength properties and the nature of the failure of adhesive joints with Plexus MA310 grade SAA [132] showed a decrease in strength by 14% and 26% at 50 °C and 80 °C, respectively, compared to the strength under peel loads at temperature of 20 °C. Simultaneously, with an increase in temperature, the adhesive acquires greater plasticity and its cohesive properties change.

As noted above, an important property of SAAs is their ability to bond oily parts without a significant decrease in strength. This reduces the requirements for surface preparation prior to bonding. The methacrylic monomers and elastomers used in the composition of SAAs are capable of dissolving oily contaminants from the surface and thus eliminate the degreasing stage.

Table 7.17 shows comparative data for bonding fat-free and oily substrates using SAA type HP (High Performance) [4].

The plasticity of acrylic adhesive Sika Fast 5221NT and the restoration of its characteristics after deformation were studied in [133]. The quasi-static characteristics of true strain were determined at high rates of tensile and compressive strain. It was found that the yield point, elastic modulus, and fracture strain depended logarithmically on the strain rate. In contrast, the yield strain and post-strain stiffness were insensitive to strain rate. It was shown that recovery after unloading depended on time, type of loading, strain rate of loading and strain at unloading. The mechanical

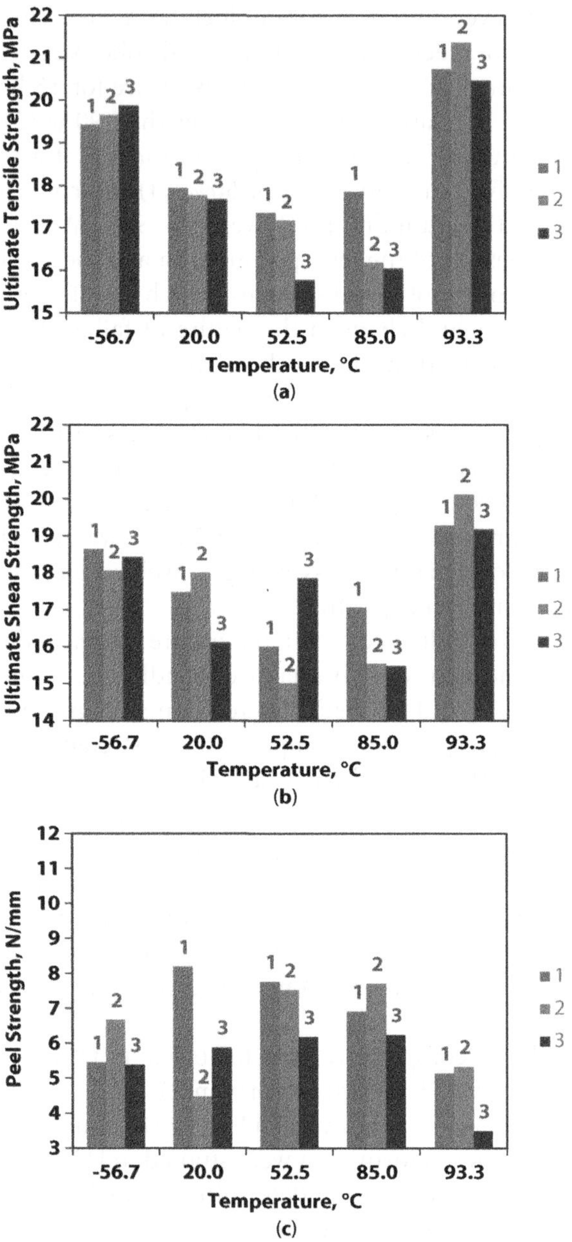

Figure 7.8 Influence of temperature and sample geometry on tensile strength (a), shear strength (b), and T-peel strength (c) of adhesive joints for adhesive thickness of 13 mm (1), 25 mm (2), 38 mm (3) [120].

Table 7.17 Strength properties of SAAs when bonding substrates with various surface preparation methods [4].

Substrate	Lap shear strength, MPa
Steel SAE 1010 (solvent wiped)	42.1
Steel SAE 1010 (grit blasted)	41.0
Steel SAE 1010 (oily)	44.1
Aluminium 6061-T6 (solvent wiped)	33.6
Aluminium 6061-T6 (grit blasted)	34.7
Aluminium 6061-T6 (oily)	35.1
Aluminium 2024-T3 (solvent wiped)	32.5
Aluminium 2024-T3 (oily)	35.1

characteristics of the adhesive 48 h after unloading allowed concluding the absence of the residual deformation.

Studies of the cyclic fatigue behavior of the SAA (Hardloc C355-20 A/B, Denka Co.,) performed using the fracture mechanics approach have shown a significant impact of the adhesive thickness and loading conditions on the crack growth. The authors concluded an advantage of a thicker adhesive layer to increase fatigue resistance when using SAA under conditions of vibration or cyclic loading [109].

7.4 Adhesives for Low Surface Energy Materials

7.4.1 Initiators Based on Trialkylboranes

It is well known now that low surface energy polymers are poorly bonded by conventional adhesives and require surface pretreatment. Such a preatreatment of polymeric materials can be performed by different ways. As a rule, such a process includes surface activation by treatment with chemical reagents, corona discharge, plasma, flame, the application of special primers, etc. Strong adhesion of low surface energy substrates reached using SAAs is based on the graft polymerization of methacrylic monomers onto the substrate in the presence of an initiating system, containing trialkylborane (TAB) R_3B and oxygen.

A large number of works have been devoted to the study of the polymerization of methacrylic under the influence of the TAB - oxygen system, and such studies remain relevant at present. The high oxidizing ability of TAB during polymerization leads to the formation of radicals causing radical polymerization, which proceeds in high yields at room temperature. Note that this process is practically non-inhibited by atmospheric oxygen. The reversible addition – fragmentation chain transfer process makes it possible to obtain block copolymers, graft copolymers, star-polymers, etc. [139–143].

The use of TABs in the usual state turned out to be practically inconvenient, since due to their high reactivity when interacting with atmospheric oxygen, they exhibit pyrophoric activity. At the same time, complexes of trialkylboranes with basic compounds are stable in the air. TABs can be displaced from these complexes by various reagents. It was proposed to use such systems in two-component SAAs [144].

One of the first patents for thermoplastics bonding, including polytetrafluoroethylene and PVC, registered in Russia [145], used a two-component acrylic composition based on MMA and butyl acrylate. In this work, a complex of amine and TAB was used as the initiator of curing, and (meth) acrylic acid as a deblocking agent. Somewhat later, a similar patent was registered by 3M [146].

The mechanism of the adhesive interaction of acrylate-borane systems with low surface energy substrates was described in [147–152]. Initiation of polymerization occurs when the TAB-amine complex is unblocked with an acid:

$$R_3B{\rightarrow}N(R)_3 + R'OOH \rightarrow R_3B + N(R)_3{\bullet}R'OOH,$$

followed by oxidation of the TAB with oxygen dissolved in SAA with the generation of free radicals according to the following scheme:

$$
\begin{aligned}
&R_3B + O_2 \rightarrow R_2BOOR\\
&R_2BOOR \rightarrow R_2BOO{\bullet} + R{\bullet}\\
&R{\bullet} + O_2 \rightarrow ROO{\bullet}\\
&ROO{\bullet} + R_3B \rightarrow R_2BOOR + R{\bullet}\\
&ROO{\bullet} + R{\bullet} \rightarrow ROOR\\
&R{\bullet} + ROOR \rightarrow RO{\bullet} + ROR\\
&R_2BOOR \rightarrow R_2BO{\bullet} + RO{\bullet}
\end{aligned}
$$

$R_2BO\bullet + R_3B \rightarrow R_2BOBR_2 + R\bullet$

$3R_2BO\bullet \rightarrow$ 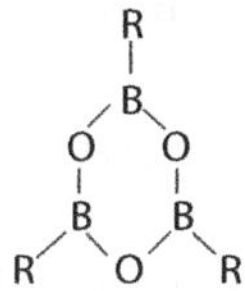 $+ 3\,R\bullet$

$RO\bullet + R_3B \rightarrow R_2BOR + R\bullet$

$R_2BOOR + R_3B \rightarrow R_2BO\bullet + R\bullet + R_2BOR$

$R_2BOOR + R_3B \rightarrow R_2BOBR_2 + RO\bullet + R\bullet$

The interactions of radicals with the substrate and methacrylic monomer lead to the establishment of adhesion contact

$RO\bullet + R'H$ (substrate) $\rightarrow ROH + R'\bullet$

$R'\bullet + nM \rightarrow R'\,nM\bullet$ (grafting)

$R\bullet + nM \rightarrow RMn\bullet$

$RO\bullet + nM \rightarrow RO\,Mn\bullet$

$R'\bullet + R(Mn\bullet)$ or $(RO\,Mn\bullet) \rightarrow R'\,M\,nR$ or $(R'OM\,nR)$ (adhesion)

The measurement of the amount of grafted polymer on the surface of polypropylene by the gravimetric method showed [151] that growth of the polymer layer on the surface occurs in the first 30–40 min (Figure 7.9). The kinetics of grafting well correlates with the increase in the strength of the adhesive joints after the beginning of the adhesive curing (Figure 7.9, inset).

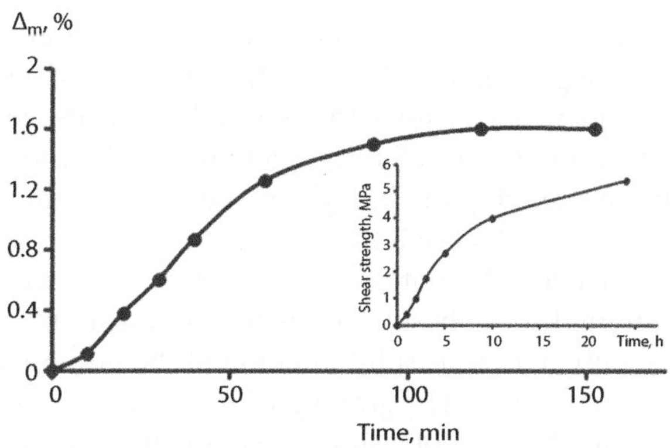

Figure 7.9 The rate of mass growth (Δm, %) of the grafted polymer on polypropylene; inset shows the variation in the shear strength of polypropylene adhesive joints with time [151].

The kinetics and mechanism of initiation and re-initiation of such reactions were considered in [143, 152]. It was shown that the depletion of oxygen in the system occurs very quickly. Therefore, the dissolved oxygen in such systems does not reveal its inhibitory property, but only acts as a co-initiator due to interaction with the TAB.

Subsequent formation of radicals can occur because of re-initiation due to the growth of macromolecular chains with a terminal boroxyl group PMMA–O–BR_2 that originated at the beginning of the reaction, as well as a result of decomposition of organoboron peroxides [143, 152]. An assumption was also made that at the final stage of bonding, the formation of radicals in an oxygen-free environment occurs in viscous and solid media due to the interaction of TAB with organoboron peroxides [151].

It was noted that if the initial concentration of the acid that deblocks the amine complex of the TAB is kept low (<0.1 M), then alkoxy radicals are formed earlier than any other types of radicals. Due to the fact that alkoxy radicals are more reactive when interacting with α-olefins, then the observed grafting and polymerization rates will be higher than in the case of alkyl radicals formed at higher acid concentrations. The authors recommend taking this factor into account when creating adhesives and coatings for low surface energy substrates [153].

The degree of MMA grafting onto polyolefin in the presence of TAB critically depends on the presence of oxygen in the system. Thus, in the case of oxygen deficiency, MMA grafting occurs when the formed alkoxy radicals replace hydrogen atoms on the substrate. With constant oxygen supply, peroxyborane $Bu_2B(OOBu)$ is rapidly decomposed by unreacted Bu_3B [154].

When considering an acrylate compound for adhesion to inert thermoplastics it was suggested in [155] that the formation of boron-centered radicals is possible, which ensures the removal of a hydrogen atom from polyethylene (PE) and PP and a fluorine atom from a fluoroplastic followed by grafting of polymerizing methacrylate. The corresponding scheme for the reaction is shown in Figure 7.10.

The author considers the noticeable effect of substituents at the boron atom on the strength of adhesive bonds in the transition from triethylborane to triisobutylborane as substantiation of the proposed mechanism [155]. In addition, the fact that grafting of monomers onto the substrate does not occur in the case of using other initiation systems, in which carbon- and oxygen-centered radicals are also generated, can also be considered as an indication of the discussed mechanism.

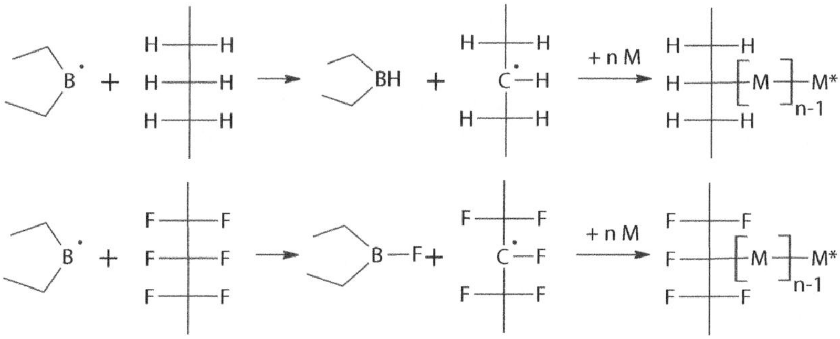

Figure 7.10 A possible scheme for the interaction of a boron-centered radical with polymer substrate: a - polypropylene, b - fluoroplastic [155].

However, to date no direct evidence for the formation of boron-centered radicals has been described in the literature. During the polymerization of MMA and other vinyl monomers in the presence of TAB or boracyclane, the EPR spin trap method detected only alkyl, alkoxy, and propagating radicals. The polymerization of (meth) acrylates proceeds by the mechanism of reversible inhibition, i.e., controlled pseudoliving radical polymerization under reversible termination, in which the resulting boroxyl radicals serve as reversible termination agent [152].

The application of TAB – amine complexes in SAAs has been widely discussed in the literature. It has been shown that the stability of such TAB complexes is determined by the B-N bond energy, which can be measured by various methods, including 11B-NMR, DSC, thermal viscometry, or estimated by numerical approaches [148].

The advantage of using one or another TAB and amine in terms of the stability of the complex can be analyzed using the decomplexation temperature and the stabilization energy of the complex [148, 156]. Substituents at the boron atom in TAB affect both the strength and stability of the complexes (see Table 7.18), and, accordingly, the shelf-life of the adhesives [156].

Upon formulation of an SAA, attention is paid to the appropriate selection of amine complexes of trialkylboranes from the point of view of their initiating activity and stability during storage. Primary and secondary amines, diamines, polyamines, polyoxyalkylenepolyamines, oxyalkylene amides, amino-substituted alicyclic compounds, N, N'-guanidines, conjugated imines, etc. can be used as complexing agents of TAB. The effect of the amine component in the amine-tri-n-butylborane complex on the stability of the complex and the adhesion strength to polypropylene is shown in Table 7.19 [156].

Table 7.18 Stability of the complex and strength property of SAA depending on the alkyl radical in TAB [156].

Trialkylborane*	Complex decomposition temperature, °C	Shear strength, MPa
Tri-n-butylborane	53	4.5**
Tri-iso-butylborane	39	4.2**
Tripropylborane	>100	0
Trioctylborane	>100	3.3**
Diethylmethoxyborane	68	0.25
Diethylisopropylborane	77	0.44

*Complex with dimethylaminopropylamine; **cohesive failure in PP

Table 7.19 Properties of SAAs containing amine complexes with tri-n-butylborane [156].

Amine compound	Complex decomposition temperature, °C	Shear strength, MPa*
Hexamethylenediamine	44	3.1
Dimethylaminopropylamine	52	6.2
Morpholine	20	5.6
Aminopropyl morpholine	50	4.7
Cyclohexylamine	20	6.4
Isophoronediamine	46	6.2
Methoxypropylamine	63	4.4
Piperidine	85	5.3
Pyrrolidine	>100	4.9
Cyclopentylamine	25	6.5
1,4-Diazabicyclo[2.2.2] octane	25	5.3
Hexylamine	45	6.1
2-Methyl-2-imidazoline	35	6.9

* cohesive failure in polypropylene in all cases

Table 7.20 The effect of some amines and aminosilanes on adhesion strength properties of SAAs [159].

Amine compound*	Shear strength, MPa
3- Methoxypropylamine *	2.7
Isophorone diamine*	3.4
3-Aminopropyltriethoxysilane #	2.3
3-Aminopropyltriethoxysilane*	2.9
N-(2-aminoethyl)-3-aminopropyltrimethoxysilane*	3.0
N-(2-aminoethyl)-3-aminopropylmethyldiethylsilane*	3.5

*the deblocking agent is isophorone diisocyanate, # the deblocking agent is 3-isocyanatopropyltriethoxysilane

In order to increase the stability of SAA during storage, it was recommended to introduce radical polymerization inhibitors. Hydroxylamines (bis-(N-dodecyl)-N-hydroxylamine) and nitroxyl radicals (4-hydroxy-2,2,6,6-tetramethylpiperidine-1-oxyl) were mentioned in the literature as the most effective [157, 158].

Comparison of the strength properties of SAAs containing amine and aminosilane complexing agents of TABs when bonding poly(ethylene terephthalate) is shown in Table 7.20 [159].

It was demonstrated in [160] (Figure 7.11), that the efficiency of various amine complexes with TABs as a component of SAA can be characterized by the thermokinetic curves of curing by DSC method.

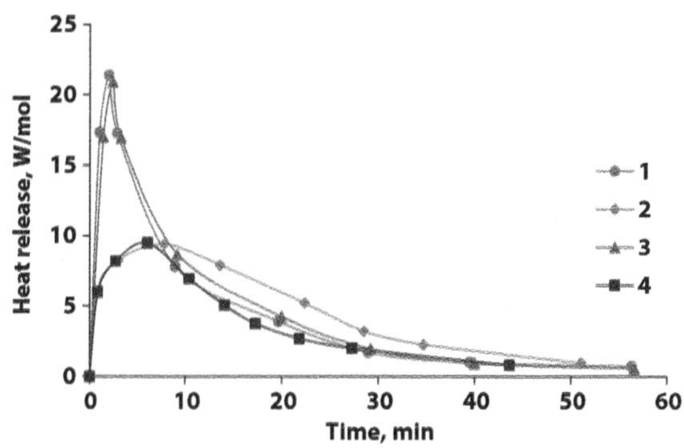

Figure 7.11 Variation in heat release during curing of SAA containing TAB complexes: 1 – $Et_3B{\bullet}0.5NH_2(CH_2)_6NH_2$; 2 – $Et_3B{\bullet}CH_3O(CH_2)_3NH_2$; 3 – $Pr_3B{\bullet}0.5NH_2(CH_2)_6NH_2$; 4 – $Pr_3B{\bullet}CH_3O(CH_2)_3NH_2$ [160].

In a study of more than 50 different complexes of TAB with amines, it was found that the highest strength characteristics of PE joints were obtained with dimethylaminopropylamine, piperidine, methoxypropylamine, hexyl- and cyclohexylamine, morpholine, and 2-methyl-2-imidazoline, upon their deblocking with methacrylic acid [147].

For the formation of stable complexes of TABs, it was proposed in the literature to use amidines as complexing agents with the structure shown below [161]:

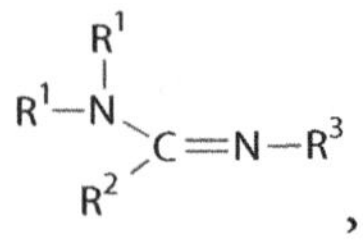

where R^1 – H, $C_{1\text{-}10}$-alkyl or $C_{3\text{-}10}$-cycloalkyl, R^2 – H, $C_{1\text{-}10}$-alkyl, $C_{3\text{-}10}$-cycloalkyl or N $(R^1)_2$, R^3 may be H, CH_3 or part of a cyclic ring, as well as conjugated imines. Examples of conjugated imines – TAB complexing agents [161] – are shown in Figure 7.12.

Complexes of a TAB with an agent containing a bifunctional Lewis base with an amine group and a second functional group having lower basicity make it possible to increase the bond strength to PP and to increase the SAA stability. The compounds presented in Figure 7.13 can be considered as examples of such additional functional groups [162].

It was shown in [162] that the stability of such TAB complexes increases, both due to the formation of a hydrogen bond between the hydrogen of the primary amine group and a heteroatom of another functional group and

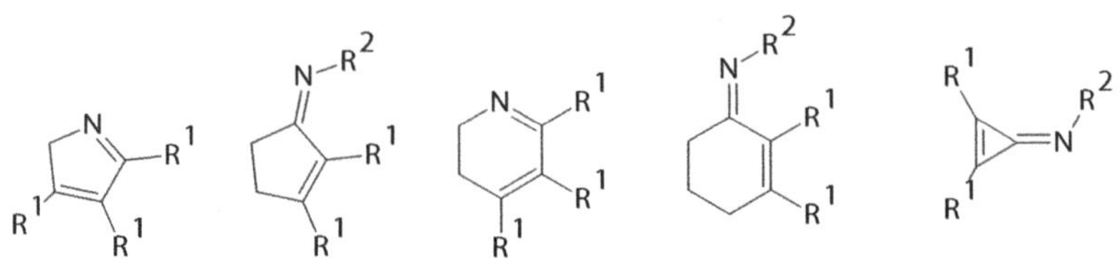

Figure 7.12 Examples of conjugated imines used as TAB complexing agents. Here R^1 is H or CH_3; R^2 is H, $C_{1\text{-}10}$ alkyl, $C_{3\text{-}10}$ cycloalkyl [161].

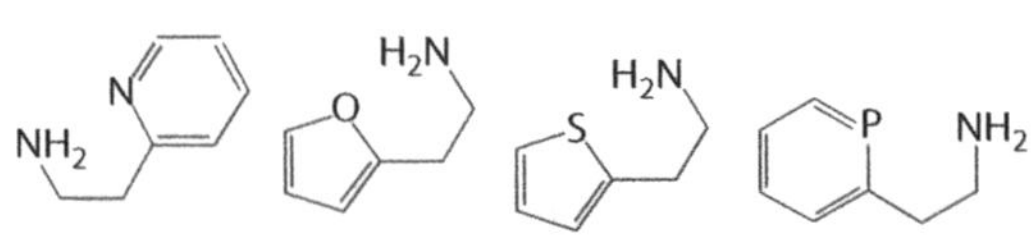

Figure 7.13 Examples of bifunctional amine complexing agents [162].

Table 7.21 Properties of SAAs containing various compositions with TAB [162].

Amine complexing agent	Shear strength of PP substrates, MPa	Change in SAA viscosity on storage at 23 °C for 21 days, %
2-(2-aminoethyl) pyridine	13.1	45
3-(methylamino) propylamine	10.3	140
3-(methoxypropyl) amine	6.9	450

due to the delocalization of the electrons of the aromatic ring. The data presented in Table 7.21 show the advantage of SAA containing a blocking bifunctional complex of TAB based on 2-(2-aminoethyl)pyridine over monofunctional amine complexes based on 3-(methylamino) propylamine and 3-(methoxypropyl)amine.

On the basis of silyl-containing boranes which were synthesized in [163], the following compositions with aminosilanes can be formed:

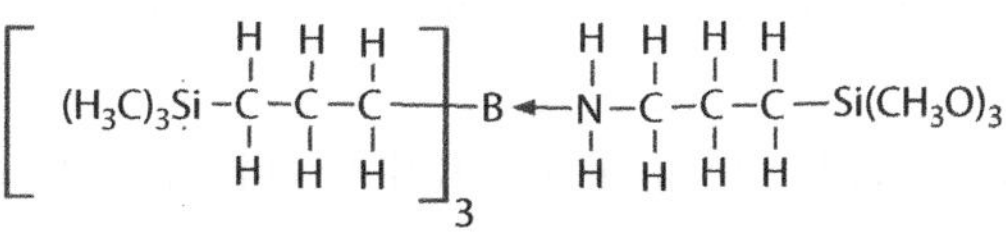

Utilization of such SAA complexes with silicon-containing methacrylates, for example methacryloxypropyldimethylsiloxy-terminated poly(dimethylsiloxane) (PDMS), allows reducing the curing temperature, reaching a single-phase structure of the cured adhesive, and enhancing the thermal properties of adhesive joints. Besides, siloxane-containing amines as complexing agents provide high adhesion, impact resistance, flexibility, chemical resistance, as well as low emission of volatile components when joining elastic roofing materials based on PE and PVC [164].

Siloxane-containing amide compounds, including polymers, have been proposed as complexing agents. For example, the structures shown in Figure 7.14 can be used as such compounds [165]. The complexes of siloxane-containing amides with boranes are quite stable and make it possible to obtain adhesives with a long shelf-life and improved adhesive properties.

In addition, substituted imidazoles and tri(alkyl)arylphosphines (Figure 7.15) can be used as complexing agents for TABs [166].

It is interesting to note that recently it has been proposed to introduce complexes of amines with TAB in the form of functionalized nanoparticles of metal oxides Z – NHR^1 – B $(R^2)_3$, which are obtained by treating

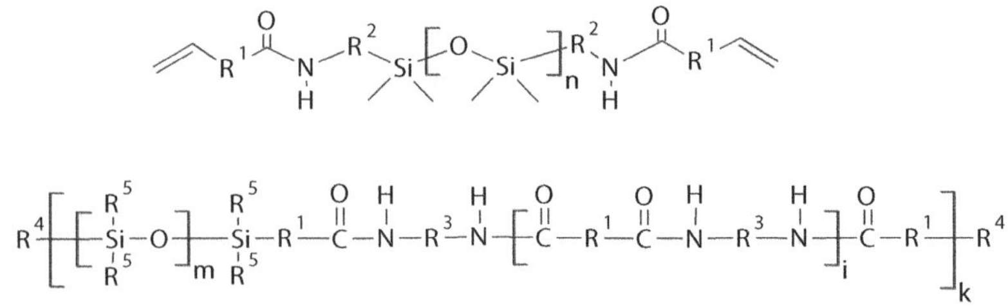

Figure 7.14 Examples of siloxane-containing amide compounds [165].

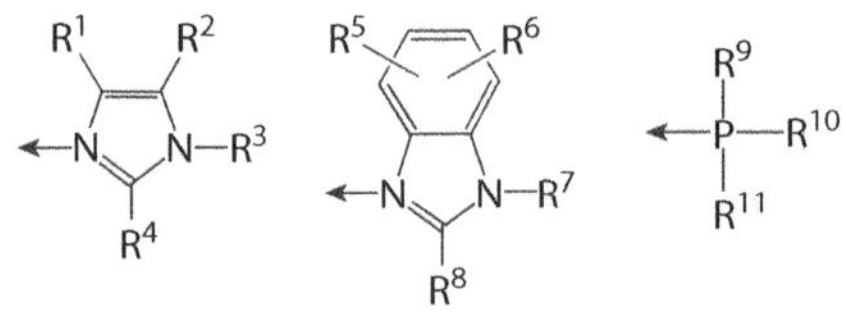

Figure 7.15 Examples of substituted imidazoles and tri (alkyl) arylphosphines proposed as TAB complexing agents [166]. Here R^1, R^2, R^3, R^4, R^5, R^6, R^7, R^8, R^9, R^{10}, R^{11} - hydrogen atom, a halogen atom, an alkyl group, an aryl group, or an alkoxy group.

the particle surface with aminosilanes followed by functionalization using TAB [167].

As for deblocking agents for complexes of TABs, one can use such compounds as acids, including Lewis acids (for example, $SnCl_4$, $TiCl_4$), aldehydes, ketones, anhydrides, epoxides, isocyanates, sulfonyl chlorides, and other compounds that can react with an amine group.

The possibility of increasing the work-life (the maximum time period available for bringing the adhesive composition into contact with the substrate to be bonded) of SAA when using itaconic acid or itaconic anhydride as a deblocker was reported [168]. To deblock amine complexes of TABs, nanoparticles of metal oxides treated with acids can be used as well [169].

Deblocking of complexes of triethylboron with N,N-dimethylaminopyridine and hydrazine in the vapor phase using various acids was carried out in [170]. The result of deblocking was evaluated by the degree of conversion of the acrylic monomer and the thickness of the resulting film on the polyethylene surface. Some of the results of this study are shown in Table 7.22.

A reversible addition–fragmentation chain transfer (RAFT) process was applied to study the polymerization initiated by TAB-amine complex. It is shown that the amount of expended deblocking agent depends on its

Table 7.22 The efficiency of acid deblocking agents for triethylboron complexes with N, N-dimethylaminopyridine (Et_3B • 4DMAP) and hydrazine (Et_3B• HZ)* [170].

Deblocker	(Et_3B • 4DMAP)		(Et_3B • HZ)	
	Double-bond conversion, %	**Film thickness (μm)**	**Double-bond conversion, %**	**Film thickness (μm)**
Formic acid	68	120	77	80
Acetic acid	35	81	82	120
Propionic acid	38	70	35	85
Acrylic acid	59	46	64	6
Hydrochloric acid	71	33	6.5	18
Acetic anhydride	-	-	78	71
Silicon tetrachloride	65	95	37	10
Tin tetrachloride	-	16	67	135
Oxalyl chloride	74	148	65	127
Acetyl chloride	62	73	69	54
Acetic aldehyde	-	-	76	380
Sulfur dioxide	-	-	74	117

* Contact time of an acrylic composition containing a complex with a deblocking agent - 3 min

Table 7.23 The influence of sulfo-containing additives on the shear strength of PE and PP joints [172].

	Shear strength, MPa	
Modifying additive *	**PE-PE**	**PP-PP**
-	4.7	1.8
p-Toluenesulfonil chloride	5.5	5.1
Na-p-toluenesulfinate	5.8	2.5
Na-p-toluenesulfonate	4.8	2.4
Benzene sulfonyl fluoride	4.5	2.8
p-Toluenesulfonyl hydrazide	4.6	2.0
4,4'-diphenylsulfonyl chloride	2.5	2.8

*Part A contains tetrahydrofurfuryl methacrylate, 2-ethylhexyl methacrylate, Kraton SBS, triethylborane diaminopropane complex. Part B contains tetrahydrofurfuryl methacrylate, 2-ethylhexyl methacrylate, mono- (2-methacryloyloxy) ethyl maleate, Kraton SBS, and additive (0.25 wt %)

nature. For example, carboxylic acid is required in greater excess than isocyanate. Besides, the deblocking agent can affect the kinetics of polymerization, molecular weight and dispersity of the resulting polymer [171].

Adding a compound comprising of a sulfonyl moiety or sulfinyl moiety as acidic deblockers in an amount of 0.1-2.5 wt. % also allows increasing the rate of curing and joint strength when bonding PP, PE (Table 7.23) [172].

Currently, the choice of borane compounds for the production of adhesives is quite wide. Organosilicon boranes can be used as borane compounds, for example with the structures shown in Figure 7.16 [163, 165].

7.4.2 Comparison of the Initiation System Containing Trialkylborane with the Redox System Benzoyl Peroxide (BPO) - Tertiary Aromatic Amine

A comparative study of the efficiency of MMA polymerization using 2 types of initiating systems was carried out in [139, 173, 174]. It was shown that the initiation of polymerization with tri-n-butylborane (TnBB) and partially oxidized TnBB (TnBBO) on the one hand, and BPO/tertiary aromatic amine system, on the other hand, showed the advantages of a

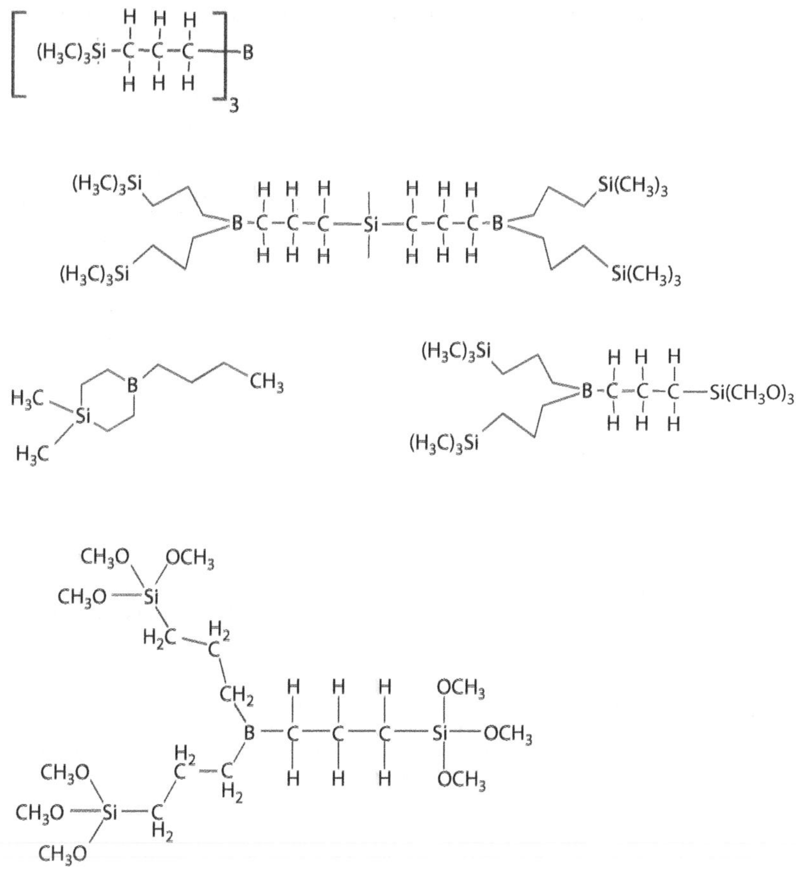

Figure 7.16 Examples of silane-containing boranes [163, 165].

boron-containing initiator. These advantages were associated with a higher conversion of MMA and a higher molecular weight of the resulting polymer. EPR analysis showed that the radicals in the TnBBO-initiated system remained active for a longer period than in the BPO/amine-initiated system.

Gel permeation chromatography was used to compare the properties of poly(methyl acrylate) formed, on the one hand, during the polymerization initiated by the TnBB-methoxypropyl amine complex (TnBB-MOPA) with two deblocking agents - isophorone diisocyanate (IPDI) and propionic acid (PA), and, on the other hand, during polymerization with an initiator (BPO/DMPT) [140]. The data presented in Table 7.24 indicate a higher cure rate, a greater achieved conversion and a larger molecular weight of

Table 7.24 Impact of various initiating systems on methyl acrylate polymerization [140].

Initiating system *	Deblocking agent	Molecular weight M_n (kDa)	Polydispersity	Time to reach the maximum exothermic reaction, s	Conversion of MA, %
TnBB-MOPA	IPDI	91.2	4.7	37	99.6
TnBB-MOPA	PA	191.0	4.6	77	96.3
BPO/DMPT	-	19.7	2.7	-	89.0

*TnBB – tri-n-butylborane, MOPA - methoxypropyl amine, BPO - benzoyl peroxide; DMPT – dimethyl-p-toluidine

Table 7.25 Bonding strength of PTFE depending on initiation system used [9].

Tensile strength of SAA (MPa), containing the initiation system*				
BPO (2%), DMPT 1%)	CHP (2%), Accelerator 833 (2%)	CHP (3%), Accelerator 808 (2%)	CHP (2%), TMTU (1%)	TEB·HMDA (2%), BPO (0.1%)
1.5	1.0	0.7	0.8	4.5

*BPO – benzoyl peroxide; DMPT – dimethyl-p-toluidine; CHP – cumene hydroperoxide; Accelerator 833 – condensation product of butylamine and butyl aldehyde; Accelerator 808 – condensation product of aniline and butyraldehyde; TMTU – tetramethylthiourea; TEB·HMDA – complex of triethylborane with hexamethylenediamine.

poly(methyl acrylate) in the polymerization initiated by TAB. Table 7.24 also allows concluding the advantage of using the isocyanate as a deblocking agent.

The advantages of the organoborane initiation system for bonding polytetrafluoroethylene with SAAs containing chlorosulfonated polyethylene in comparison with the BPO-DMPT system and other initiating systems are shown in the Table 7.25 [9].

7.4.3 Alternative Types of Trialkylborane Initiators

Along with TAB, other boron-containing initiators have been proposed, which are more stable in air as well as in adhesive compositions. Quaternary borate compounds containing a metal cation or a quaternary ammonium cation with the structures shown in Figure 7.17, for example, sodium tetraethyl borate, lithium tetraethyl borate, lithium phenyl triethyl

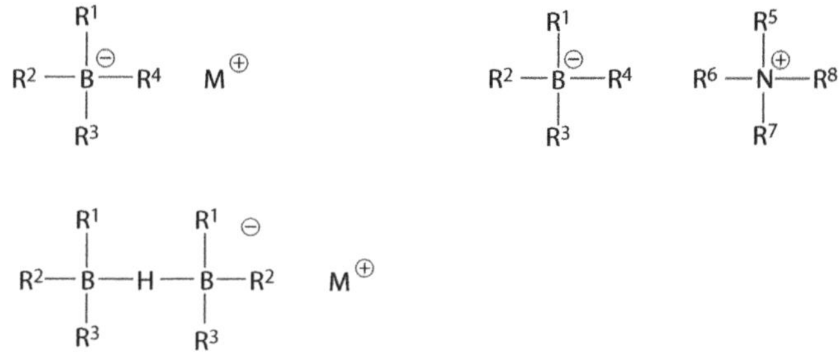

Figure 7.17 Examples of borate compounds used as SAA polymerization initiators. Here R^1 is C_{1-10} alkyl; R^2, R^3, R^4, R^5, R^6, R^7, R^8, which may be the same or different, are hydrogen atoms, C_{1-10}-alkyl or C_{3-10}-cycloalkyl, phenyl, or phenyl-substituted C_{1-10}-alkyl or C_{3-10}-cycloalkyl, M^+ is a metal ion [175, 176].

borate, tetramethylammonium phenyl triethyl borate [175], can be used as such initiators.

Another type of boron-containing initiators are borohydride initiators, for example, lithium triethylborohydride, sodium triethylborohydride, potassium triethylborohydride, and lithium 9-borabicyclo [3.3.1]-nonane (9-BBN) hydride [177]. The undoubted advantages of using such initiators are their availability, stability, and safety.

To illustrate shear strength achieved using lithium triethylborohydride as an initiator, Table 7.26 presents data for the joints with various substrates [177].

The other important types of boron-containing initiators are the intrablocked borate compounds [178], for example, of the following structures (Figure 7.18).

Table 7.26 Strength of adhesive joints bonded with SAAs containing lithium triethylborohydride used as initiator [177].

Substrates	**Shear strength, MPa**
Polyethylene/Polyethylene	8.0
Polypropylene/Polypropylene	7.9
Mild Steel/Polyethylene	8.8
Mild Steel/Polypropylene	6.9
Mild Steel/Polyolefin rubber	2.7

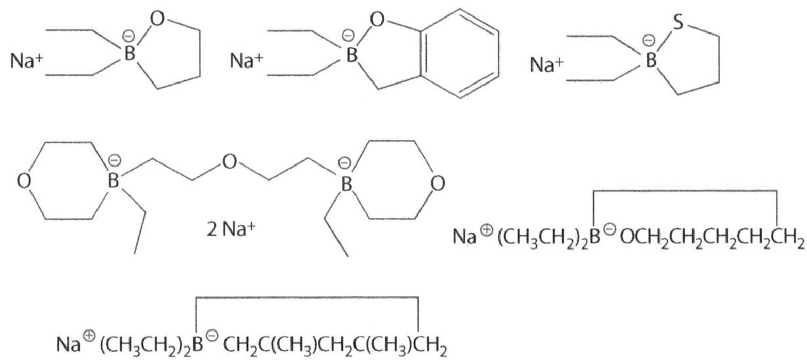

Figure 7.18 Examples of intrablocked borate compounds [178].

Comparison of the stability of SAAs containing TAB and borate complexes showed that in the case of a covalent bond of a boron atom with a nitrogen atom, borate complexes promote greater stability to SAAs. Moreover, initiators with intrablocked borate compounds provide SAAs with a prolonged open time [178].

Some additional examples of boron-containing initiators which allow obtaining stable adhesives with a controlled cure rate and increased strength the readers can find in [179–181].

7.4.4 Additives Modifying the Curing Stage

Application of modifying additives in SAAs, capable of affecting the initiation stage, improves their characteristics. Thus, the addition of peroxides and hydroperoxides (benzoyl peroxide, tert-butylperbenzoate, cumene hydroperoxide, etc.) increases not only the curing rate but also increases the adhesive strength by 1.2-1.5 times when bonding PP (Figure 7.19) [182].

Various quinones (p-benzoquinone, anthraquinone, ortho-benzoquinone) [183], when used as an additive in SAA, in contrast to conventional radical reactions, act as curing accelerators and not as inhibitors. In this case, the accelerating effect of quinones during the polymerization of MMA in the presence of trialkylborane crucially depends on the structure of the quinone and its concentration. In such reactions, trialkylborane acts as a chain mediator, and in this case, the mechanism of "living" controlled polymerization is realized [184, 185]. The corresponding scheme for the reaction is shown in Figure 7.20.

According to [186], the introduction of p-benzoquinone, naphthoquinone, and anthraquinone in an amount of 0.1-0.2 wt. % leads to an increase in the adhesion strength of PP joints. The dependence of type of failure on the content of p-benzoquinone is demonstrated in Figure 7.21.

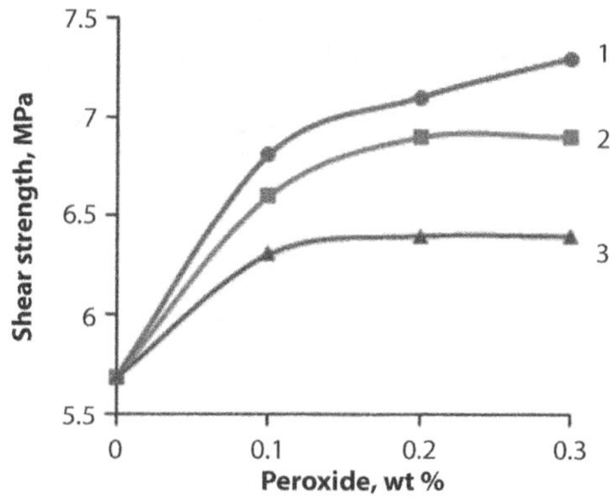

Figure 7.19 Effect of organic peroxide on adhesion strength in PP joints: 1 – benzoyl peroxide; 2 – lauryl peroxide; 3 – tert-butyl hydroperoxide [182].

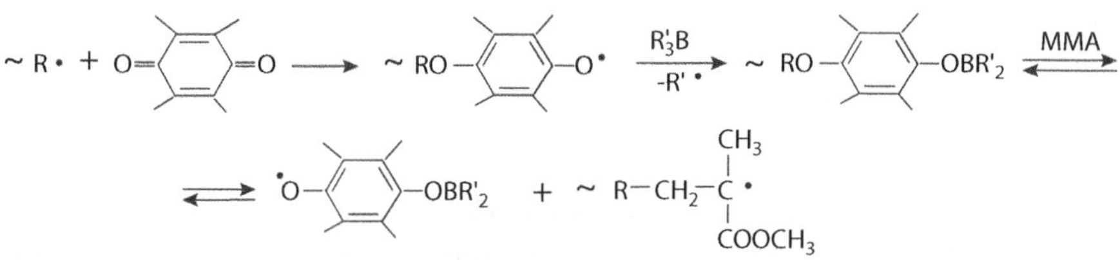

Figure 7.20 Scheme of the interaction of a growing polymer radical with p-benzoquinone [184].

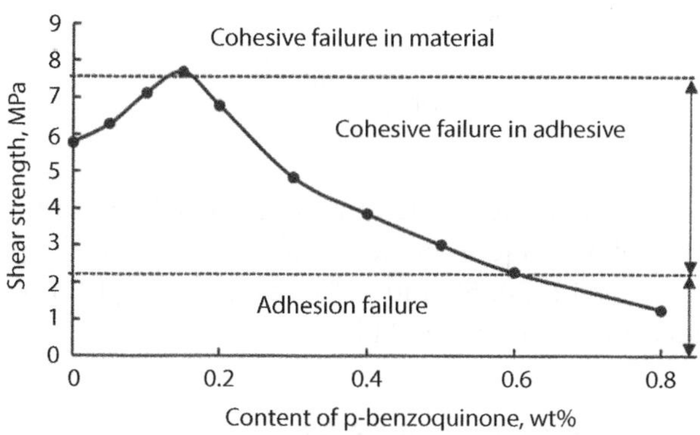

Figure 7.21 The impact of p-benzoquinone on the type of SAA failure upon bonding of polypropylene [186].

Table 7.27 Strength properties of adhesive joints made with SAA modified with small additions of p-benzoquinone and benzoyl peroxide [187].

Substrate	Hardener	Shear strength, MPa	
		initial composition*	modified composition**
PVC- PVC	$Et_3B{\cdot}NH_2(CH_2)_6NH_2$	3.1	6.6
	$Pr_3B{\cdot}\ NH_2(CH_2)_6NH_2$	3.5	7.9
	$Et_3B{\cdot}NH_2CH_2CH_2OMe$	2.8	6.8
PP-PP	$Et_3B{\cdot}NH_2(CH_2)_6NH_2$	1.9	5.5
	$Pr_3B{\cdot}\ NH_2(CH_2)_6NH_2$	1.6	4.6
	$Et_3B{\cdot}NH_2CH_2CH_2OMe$	2.7	1.8

*The initial composition contains methyl methacrylate, butyl acrylate, hydroquinone, methacrylic acid, poly(methyl methacrylate)

**The modified composition additionally contains 0.15 wt.% of p-benzoquinone and 0.1 wt.% of benzoyl peroxide

Besides, simultaneous use of p-benzoquinone and benzoyl peroxide in the SAA composition increases the strength properties when bonding PVC and PP in the presence of various TAB complexes (Table 7.27) [187].

A similar accelerating effect was also demonstrated by the substituted aromatic containing compounds with the following chemical structures:

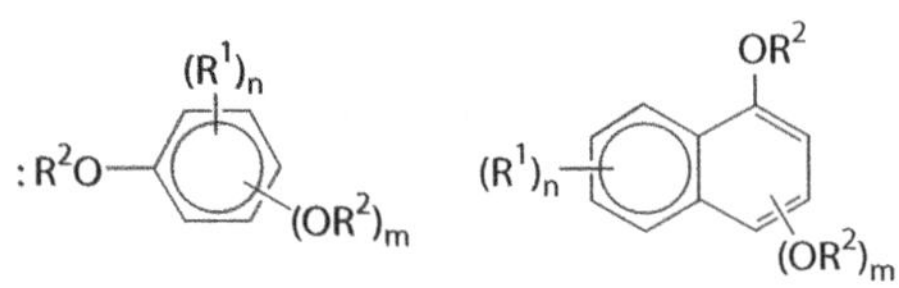

Another example of the accelerating effect is related to the application of methyl ether of hydroquinone, when after 1, 2, 3, 5 h of exposure results in strength gain of 12, 30, 58 and 70%, respectively [183], relative to the strength value, characteristic for 24 h.

7.4.5 Other Components of SAAs

The effect of crosslinking agent additives on the shear strength of SAAs used for PVC bonding studied in [182] for the group of compounds including triethylene glycol dimethacrylate (TEGMA), polyester dimethacrylate of diethylene glycol ester of phthalic acid (MDF-2) and oligocarbonate dimethacrylate (OCM-2), showed that the highest shear strength of 8.4 MPa was achieved by introducing MDF-2 in an amount of 1 wt.%

An alternative approach to increase the strength of adhesive joints made of PP and other low surface energy materials is to add in the SAA 5-10 wt.% of bis-imide compounds, for example, of the following structures [188]:

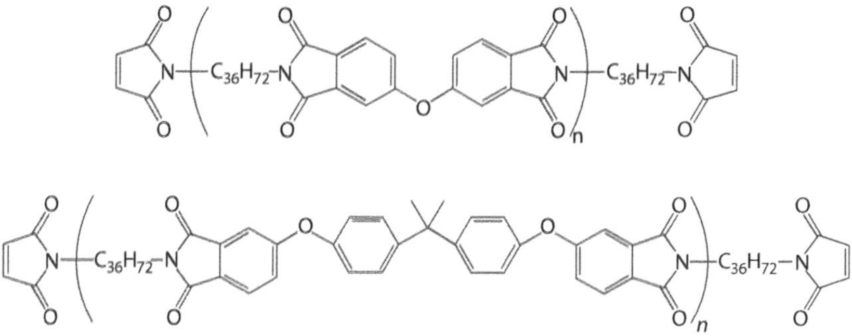

Including additional vinyl compounds, such as ethyl vinyl ether, n-butyl vinyl ether, t-butyl vinyl ether, cyclohexyl vinyl ether, 2-ethylhexyl vinyl ether, octadecyl vinyl ether, 1,4-butanediol divinyl ether, diethylene glycol divinyl ether, triethylene glycol divinyl ether, hydroxybutyl vinyl ether, or vinyl compounds of the type shown in Figure 7.22, into SAA formulations allows obtaining modified SAAs, which are stable and able to withstand accelerated thermal aging at 35 °C for more than 8 weeks [189].

The influence of various acrylic and vinyl aromatic urethane- and urea-containing oligomers on the adhesion property of the SAA used for

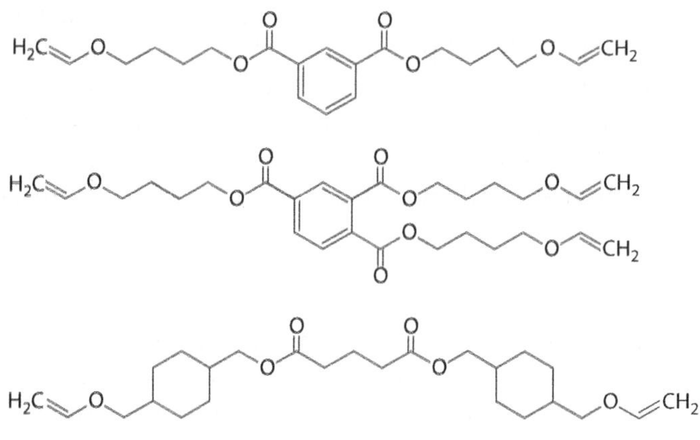

Figure 7.22 Examples of vinyl modifiers for SAAs [189].

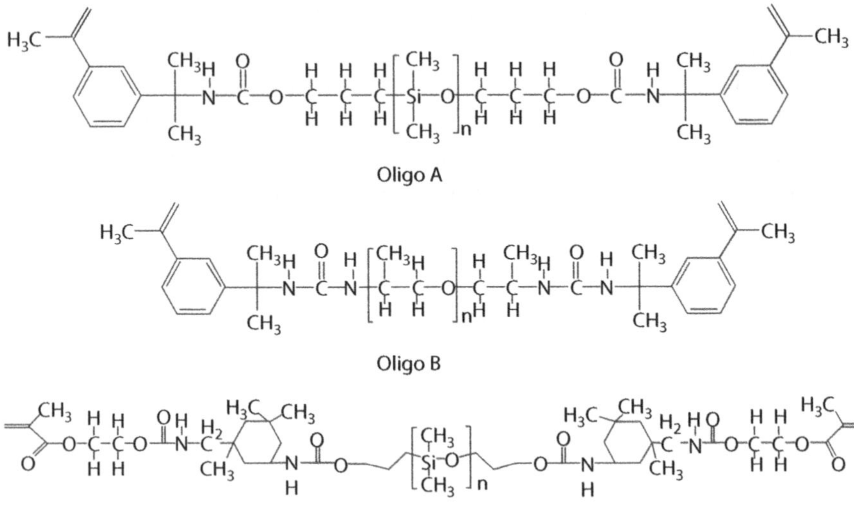

Figure 7.23 Chemical structures of oligomers used in SAAs for poly(ethylene terephthalate) bonding [190].

poly(ethylene terephthalate) (PET) bonding, was studied in [190]. It was shown that the introduction of oligomers such as oligosiloxane urethane with terminal α-methylstyrene groups (Figure 7.23 (Oligo A) in Table 7.28), oligooxypropylene urethane with terminal α-methylstyrene groups (Figure 7.23 (Oligo B) in Table 7.28), oligosiloxane urethane acrylate (Figure 7.23 (Oligo C) in Table 7.28) into an acrylate composition based on tetrahydrofurfuryl methacrylate and containing a tributylborane complex with hexamethylene diamine, leads to an increase in strength characteristics in comparison to compositions including industrial oligomers epoxy acrylate (D in Table 7.28) and polyether acrylate (E in Table 7.28).

It should be emphasized that the properties of the SAAs obtained in this work for PET bonding were significantly higher than those of the well-known commercially produced DP8010NS and Loctite 3030 adhesives, which showed shear strength of 5.24 and 1.52 MPa, respectively [190].

7.4.6 Hybrid SAAs

The chemistry of two-component SAAs makes it possible to obtain hybrid adhesives in which two different curing mechanisms can be realized simultaneously with the formation of semi- and interpenetrating networks (IPNs).

Table 7.28 Strength characteristics of SAAs with different oligomers used for poly(ethylene terephthalate) bonding [190].

Property	Modifier					
	-	**Oligo A**	**Oligo B**	**Oligo C**	**D**	**E**
Shear strength, MPa/ Failure mode	7.20/ cohesive	11.0/ cohesive	9.48/ cohesive	7.63/ cohesive	6.10 / adhesional	7.34/ cohesive
T-peel, N/mm/ Failure mode	4.91/ cohesive	5.34/ cohesive	4.94/ cohesive	4.16/ substrate	0.68/ adhesional	2.76/ cohesive

Hybrid SAAs can be obtained using various approaches. The first of these approaches is based on the release of the diamine during curing and its interaction, for example, with a diisocyanate which is introduced in the other part of the SAA. The second approach uses the introduction into one of the parts of reactive oligomers of various built-in blocks, which are capable of copolymerizing with (meth) acrylate upon curing by a radical mechanism, and additionally modifying the polymer structure.

Polyurea-poly(meth)acrylate interpenetrating polymer networks formed in a cured SAA obtained by the reaction of polyisocyanates with diamines had low shrinkage, high moisture and heat resistance, and were able to withstand continuous exposure to hot water without destroying the adhesive joints [191]. This was achieved by using a trifunctional isocyanate to unblock the tributylborane-methoxypropylamine or tributylborane-diaminopropane complex. In the same work, it was proposed to use styrene-butadiene-styrene block copolymer, hydrophobic polybutadienes with a terminal amine, and methacrylate group as thickeners. The addition of amine-terminated polybutadiene to the formulation also improves the properties of the resulting IPNs. Comparative data for the properties of 3 commercially available SAAs and of the composition proposed in [191] used for bonding polypropylene pipes for hot and cold water supply are presented in Table 7.29.

One more approach for the design of hybrid SAA was illustrated in [192]. In this study, the enhancement of adhesive and elastic properties of obtained adhesive was shown when modified with oligourethane (meth) acrylate. Dynamic mechanical analysis and transmission electron microscopy results demonstrated that polyurethane was evenly dispersed in the polyacrylate phase. Such a distribution of the components in the adhesive promotes cohesive strength increase of the polymer, while for bonding of the polypropylene such a polymer demonstrated an increase in the adhesive strength to some extent [192].

SAAs in which a diisocyanate (isophorone diisocyanate) acted as a deblocking agent and simultaneously reacted with the introduced polyetheramine (Jeffamine D-400) or polyether triol (CAPA-3050) to form polyurea (polyurethane) in a polyacrylate matrix, were obtained in [193]. IPN adhesives had high adhesion to materials with low surface energy over a wide range of polyurea (polyurethane) content. Including oligourethane (meth)acrylate (HEA-PU) into adhesives led to an increase in shear strength, in particular for polypropylene bonding (Figure 7.24).

Hybrid poly(acrylate/siloxane) SAAs for bonding polyolefins with increased elasticity and heat resistance (up to +150 °C) were investigated in [194]. The formation of poly(dimethylsiloxane) (PDMS) in situ in an acrylic monomer due to the interaction of hydroxyterminated PDMS with

Table 7.29 Comparative properties of industrial SAAs and SAA prepared in accordance with the patent [191].

Property	Proposed SAA [191]	DP 8005 (3M)	Loctite 3035 (Henkel)	PPX5 (Sci Grip)
A : B ratio	1:10	10:1	1:1	10:1
Peak exothermic temperature, °C	28.6	29.9	28.6	28.2
Lap shear strength at room temperature (PP), MPa	10	9.1	8.2	2.9
Lap shear strength at 80 °C (PP), MPa	7.0	1.8	0.9	0.6
Time to failure of plastic pipe joint during hydrostatic tests*	12 h	0	0	0

*Test conditions: The PP pipe joint was conditioned for 1 hour in a water bath at the test temperature (90° C) and then a constant hydrostatic pressure of 1.6 MPa was applied until the leakage was observed.

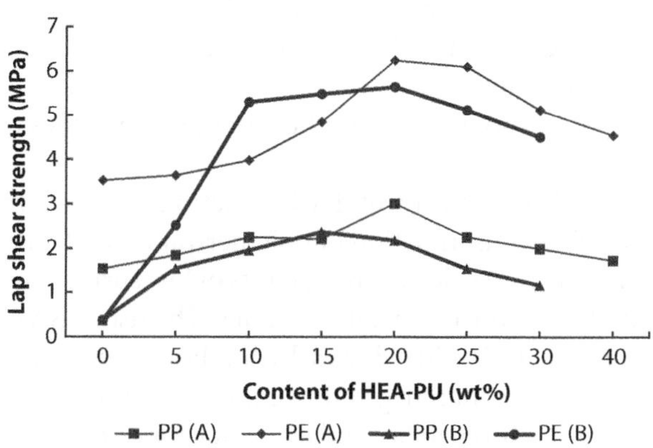

Figure 7.24 The effect of oligourethane (meth)acrylate (HEA-PU) on the strength property when bonding PP and PE by IPN adhesives: A- polyacrylate-polyurea: B- polyacrylate-polyurethane [193].

tetramethylorthosilicate leads to the formation of a two-phase structure upon curing. The introduction of acryloxypropyltrimethoxysilane (4-7 wt.%) promotes additional crosslinking and improves the compatibility of the acrylate and silicone phases. The best adhesion strength results when bonding PP and PET were obtained for the curing of SAA with the formation of a silicone phase due to the hydrosilylation reaction of divinyl-PDMS and hydride-terminated PDMS, catalyzed by the introduced platinum-containing compound. The introduction of reactive aziridine-containing compounds into SAA in the part containing the trialkylborane complex increased the strength of the adhesive bonds to Teflon, PE, and PP [194].

In order to increase the adhesive properties and the thermal stability of SAA, the fabrication of hybrid acrylate-epoxy adhesives was suggested in [195, 196]. In this case, it was not possible to use polyamine hardener (organoborane blocker) for curing the epoxy component, because the rate of radical polymerization initiated by borane is many times higher than the rate of interaction of an epoxy ring with an amine. The authors used Lewis acids (etherate BF_3, $ZnCl_2$, or $SnCl_4$) as catalysts for the iron-catalyzed reaction of opening the epoxy ring. The obtained hybrid SAA with the introduction of a mixture of epoxides (bis-phenol A diglycidyl ether and glycidyl methacrylate) had a fine two-phase dispersed structure and had increased adhesion to low surface energy substrates and significantly higher thermal stability (up to 170 °C).

In some cases, the hybrid adhesives demonstrate clear advantages. For example, the shear strength when bonding isotactic polypropylene with hybrid epoxyacrylate adhesive, curing simultaneously under the action of tributylborane-3-dimethylaminopropylamine and BF_3 • etherate and containing the crosslinking agent glycidyl methacrylate, reaches 17-18 MPa. This strength significantly exceeds the values obtained using separate adhesives [196]. Such hybrid SAA also had enhanced chemical stability in aggressive environments [195].

The effect of some epoxy resins on the strength properties of isotactic PP joints is given in Table 7.30 [196].

Thus, the radical mechanism of TAB oxidation in active media (monomers) allows simultaneous initiation of polymerization and chemical grafting of methacrylates, which results in strong joints of low surface energy materials both with each other and with metal substrates. The use of SAAs for bonding low surface energy materials in vehicles, hydraulics, pipelines, medical equipment, in the manufacture of refrigeration equipment, furniture, sports equipment, etc. confirms their reliability and efficiency for different substrates. This conclusion is illustrated by the shear strength values and the nature of joint's failure for various substrates bonded by 3M Scotch-Weld™ DP 8010NS adhesive, which contains organoboron initiation system (Table 7.31) [197].

Table 7.30 Shear strength of SAAs containing epoxy compounds for isotactic PP joints [195].

Epoxy name *	Shear strength, MPa
Bisphenol A diglycidyl ether	3.3
Trimethylolpropane triglycidyl ether	>3.6
Tris-(2,3-epoxypropyl) isocyanurate	>5.5
Tetraphenylethane glycidyl ether	>4.3
Bis-phenol A diglycidyl ether + 10% glycidyl methacrylate	>4.1

*Oxirane ring opening catalyst $BF_3O(CH_3)_2$

Table 7.31 Shear strength of adhesive joints made using various substrates [197].

Substrate	Shear strength, MPa	Type of failure*
HDPE	7.6	SF
LDPE	2.8	SF
UHMWPE	5.2	SF
PP	7.9	SF
ABS	8.6	SF
Polycarbonate	5.1	AF
Plexiglass (PMMA)	8.2	SF
PTFE	2.5	AF
PVC	12.1	SF
Polystyrene	3.9	SF
Fiber Reinforced Plastic (Epoxy)	19.5	CF
Al	12.3	CF
Cold-rolled steel	12.9	CF
Stainless steel	13.7	CF
Copper	10.3	CF
Galvanized steel	5.8	MM
Glass	4.6	SF

*SF – substrate failure, AF – adhesional failure, CF – cohesive failure in adhesive, MM – mixed mode (AF and CF)

7.5 Comparison of the Properties of SAAs and Other Reactive Adhesives

In this section, we will briefly compare the properties of modern SAAs with the properties of some other reactive adhesives, both used in industry and actively created and studied in laboratories. The comparison of properties of SAAs with corresponding properties of some other reactive adhesives, widely used for joining of metal substrates is given in Table 7.32.

Data presented in Table 7.32, allow concluding that the adhesion properties of SAAs are in many ways close or superior to those of epoxy adhesives and urethanes. In general, SAAs have a higher cure rate, impact resistance, and adhesion to plastics. Table 7.33 shows a comparison of strength parameters for joints of polymer and composite substrates based on the above adhesives [8]. A significant advantage of SAA is its ability to be used in thick gaps.

Another example of the advantageous properties of an SAA in comparison with epoxy adhesives is related to the faster strength gaining, in particular during the initial curing period. Thus, according to [198], after 5 and 10 minutes of curing, SAA can gain more than 70% and 80% of the total strength, respectively, while the epoxy adhesive reaches only 5% and 60% of the final strength in the same time.

As for the chemical resistance, SAAs are chemically quite resistant in many industrial environments, such as water, engine oil, diesel fuel, antifreeze, isopropyl alcohol and practically do not differ notably in these properties from epoxy and urethane adhesives [199].

It should be mentioned here that a distinctive feature, extremely attractive for the practical application of SAAs, is the ability to bond metals without any chemical or mechanical surface treatment prior to bonding. When comparing the shear strength of aluminum joints with different surface preparations before bonding (cleaning with isopropyl alcohol, cleaning with acetone, abrasion, and chemical etching) when using toughened epoxy adhesives and SAAs, the latter provide greater strength except for such a pretreatment as chemical etching, where SAAs are inferior to epoxy adhesive [199].

At negative temperatures, the modulus of elasticity of epoxy and acrylate adhesives cured in a block was nearly the same for both adhesives, while the yield strength was noticeably higher for the SAA. On the contrary, in the temperature range above zero, with an increase in temperature, the decrease in the strength characteristics of acrylate adhesive occurred faster than that of epoxy (Table 7.34) [200].

Table 7.32 Performance and processing features of reactive acrylic, epoxide, and urethane structural adhesives [8].

Performance features	Reactive acrylics	Epoxies	Urethanes
Benefits	Ease of processing, good impact resistance	Wide formulation range, high strength	Excellent flexibility, toughness
Temperature range, °C	− 40 ÷ + 204	− 40 ÷ + 204	− 30 ÷ +121
Fluid resistance	Very good	Excellent	Good
Adhesion to metals	Excellent	Excellent	Good
Adhesion to plastics	Excellent	Fair	Very good
Adhesion to glass	Very good	Excellent	Good
Adhesion to rubber	Poor	Fair	Good
Tensile strength	High	High	Medium
Elongation/flexibility	Medium	Low	High
Cost	High	Medium	Low
Limitations	Odor/flammable	Mixing required	Sensitive to moisture

(Continued)

Table 7.32 Performance and processing features of reactive acrylic, epoxide, and urethane structural adhesives [8]. (*Continued*)

Performance features	**Reactive acrylics**	**Epoxies**	**Urethanes**
Processing Features			
Cure temperature	Room temperature/heat	Room temperature/heat	Room temperature/heat
Work time (min)*	1– 60	5– 180	4– 120
Handling time	2– 60 min	2– 12 h	0.5– 24 h
Max gap fill (mm)	12.7	3	3

*Time available to re-position the substrates without effect on the final bond strength

Table 7.33 The influence of the adhesive type on the adhesion strength for bonding polymer and composite materials [8].

Substrate	Acrylic	Epoxy	Polyurethane
ABS-plastic	5.8 MPa (SF)*	2.6 MPa (AF)	4.4 MPa (AF)
Polycarbonate	8.0 MPa (SF)	2.0 MPa (AF)	7.4 MPa (SF)
Poly(methyl methacrylate)	8.8 MPa (SF)	2.4 MPa (AF)	6.7 MPa (SF)
Fiberglass-polymer composite	11.2 MPa (SF)	6.2 MPa (AF)	2.5 MPa (AF)
Gelcoat	5.4 MPa (SF)	5.3 MPa (SF)	5.5 MPa (SF)
Xenoy material – a mixture of poly(butylene terephthalate) and polycarbonate	8.4 MPa (CF)	3.5 MPa (AF)	7.8MPa (AF)

*The failure behavior is indicated in parentheses: SF – substrate failure, AF – adhesion failure, CF – cohesive failure in adhesive.

Table 7.34 Comparison of strength properties of cured adhesives depending on the test temperature [200].

Adhesive	Temperature, °C	Elastic modulus, MPa	Yield strength of the adhesive in tension, MPa
Two-component cold-cure toughened acrylic 3M Scotch-Weld™ DP 810	−20 +40	1053 303	51.11 28.96
Two-component cold-cure epoxy adhesive 3M Scotch-Weld™ DP 490	−20 +40	1023 812	38.47 32.74

Comparison of epoxy adhesive (Loctite 9432NA) obtained by thermal curing and acrylate adhesive of cold curing (3M DP810) for bonding aluminum substrates was given in [201]. Significant influences of the overlap length, adherend thickness, and adherend plastic deformation, as well as the impact of adhesive hardness and adhesive thickness were demonstrated for the joints bonded with both epoxy and acrylic adhesives. It was shown that epoxy adhesive performed better for short overlaps, while acrylic adhesive demonstrated higher fracture loads at larger overlaps. Such behavior is seemingly related to the ability of SAA for more even distribution of stresses along the adhesive contact line. In addition, SAA is better than epoxy for thicker adhesive layers [201].

Some of the recently developed SAAs like Plexus® demonstrate high fatigue resistance. Data presented in [202] on the resistance of different adhesive joints to cyclic loading allowed concluding notably lower ultimate stress in the Plexus®-containing joints than, say, in urethane or toughened epoxy and a higher number of cycles performed before the failure.

Thus, under certain conditions of bonding and loading of adhesive joints, SAAs have advantages over other structural adhesives.

To date, three main types of SAAs have been developed and fabricated: a) based on MMA with improved physical and mechanical properties, such as shear and peel strengths, impact resistance and resistance to cyclic loads and vibration, storage stability; b) based on other (meth)acrylic monomers with less odor and increased heat resistance, chemical resistance at elevated temperatures and high elongation; c) targeted for bonding a variety of metal and polymer materials, including low surface energy ones.

7.6 Summary and Outlook

This chapter aimed to highlight the advantages of structural reactive acrylic adhesives and to provide the spectrum of SAA formulations to meet specific needs and opportunities. In this context, we tried to show that SAA bonded joints offer advantages in comparison to other types of adhesives and joining methods. For example, distinct advantages of acrylic adhesives over metal joining methods such as welding, casting, brazing, riveting should be mentioned here. There are advantages in durability, fatigue strength, stress distribution, aesthetics, corrosion resistance and process simplicity/cost. As a substitute for riveting/welding, acrylic adhesives have been successfully used in bonding applications in the automotive, and truck/trailer industries. It is difficult to overestimate the benefits of adhesives for applications where mechanical attachment may be impossible or nearly impossible, such as plastic and composite bonding for recreational/marine vehicles, signs, and facades.

The advantages of acrylic adhesives over epoxy adhesives and urethanes were discussed in detail especially for joining low surface energy materials, and indicate the possibility and need for further intensive development of this field of adhesive technology.

Further progress in the field of structural acrylic adhesives, in the first place, is related to the development of synthetic chemistry and nanotechnologies. Purpose-oriented new chemical compounds used as matrices, modifiers, tougheners, initiators, etc, on the one hand, and nanofillers on the other hand, will provide the possibility to finely manipulate the physicochemical, thermal and mechanical properties of the adhesive layer, its interaction with the substrates and its durability in different, including harsh, environments.

In addition, the emergence of new materials, for example, superhydrophobic materials, also requires the design of new adhesives for the wide application of these new materials in traditional and new technologies. Intensive research in recent years has already led to the development of more environmentally-friendly reactive acrylics with improved properties. Such results indicate the beginning of the period of development of third-generation structural acrylic adhesives [131, 202–204] and allows us to be very optimistic about further achievements in this field of adhesive technology.

References

1. C.W. Boeder, Anaerobic and structural acrylic adhesives, in *Structural Adhesives. Topics in Applied Chemistry*. S.R. Hartshorn (Ed), pp. 217-247, Springer, Boston, MA (1986).
2.. P.C. Briggs, Jr. and L.C. Muschiatti, Novel adhesive compositions, US Patent 3890407 A, assigned to E.I. du Pont de Nemours and Co (1975).
3. R.F. Righettini, Structural acrylics, in: *Adhesion Science and Engineering. Surfaces, Chemistry and Applications*, Vol. 2, A.V. Pocius, D.A. Dillard and M. Chaudhury (Eds.), pp. 823-845, Elsevier (2002).
4. D.J. Damico, Reactive acrylic adhesives, in: *Handbook of Adhesive Technology*, A. Pizzi and K. L. Mittal (Eds.), 2nd ed., pp. 747-759, CRC Press, Boca Raton, FL (2003).
5. P.C. Briggs and G.L. Jialanella, Advances in structural adhesives, in: *Advances in Structural Adhesive Bonding*, D. Dillard (Ed.), pp. 132-150, Woodhead Publishing, Oxford, UK (2010).
6. E.M. Petrie, Thermosetting acrylic structural adhesives. SpecialChem (Mar. 20, 2013), https://adhesives.specialchem.com/tech-library/article/thermosetting-acrylic-structural-adhesives, accessed on 2021.01.07.
7. A.P. Sineokov, Z.S. Khamidulova and D.A. Aronovich, Structural acrylic adhesives. *Int. Polym. Sci. Technol.* 34(3), 57-61 (2007).

8. E. Pitia and J. Hill, Reactive acrylic adhesives, in: *Handbook of Adhesive Technology*, 3rd ed., A. Pizzi and K. L. Mittal (Eds.), pp. 349-365, CRC Press, Boca Raton, FL (2018).
9. N.A. Ustyuzhantseva, D.A. Aronovich, S.V. Balashov and G.A. Zhukova, Influence of the components of structural acrylic adhesive on the adhesive properties when gluing various materials. *Klei. Germetiky. Tehnologii* (Rus) 6, 6-12 (2016).
10. H. Suto, K. Yoda, J. Watanabe and L. Yang, Development of the low-odor and non-flammable second-generation acrylic adhesive and its applications. J. Adhesion Soc. Jap. 48, 127-136 (2012).
11. F. Fotoohi, A. Salimi, H. Bouhendi and K. Kabiri, Investigation of the mechanical and thermal properties of reactive AAEMcoMMA adhesive. *Polym. Bull.* 77, 5767-5782 (2020).
12. R.F. Righettini and J.A. Hatcher, Acrylic structural adhesive having improved T-peel strength, US Patent 7019075 B2, assigned to Lord Corp (2006).
13. N. Bianchi, F. Garbassi, R. Pucciariello and A. Apicella, Compositional influence on toughness of structural acrylic adhesives. *J. Mater. Sci.* 26, 434-440 (1991).
14. S.B. Osae, S.L. Brown and R.G. Phillipps, Methacrylate-based adhesive compositions, US Patent 20120302695 A1, assigned to Scott Bader Co Ltd. (2012).
15. X. Wang, D. Doe, P. Savory and G. Hovan, Acrylate / methacrylate adhesives initiated by chlorosulfonated polymer, US Patent 20060205850 A1, assigned to Illinois Tool Works Inc. (2006).
16. K.K. Hennemann and D.M. Lenz, Structural methacrylate/epoxy based adhesives for aluminum joints. *Int. J. Adhesion Adhesives* 89, 11-18 (2019).
17. P.S. Achary, D. Joseph and R. Ramaswamy, Study on a vinyl ester/methyl methacrylate based reactive acrylic adhesive toughened by hydroxyl terminated polybutadiene. *J. Adhesion* 34, 121-136 (1991).
18. A.D. Murzin, Development and study of the properties of polydieneurethanes obtained by structuring oligodienediols on double bonds and hydroxyl groups. PhD Thesis. Volgograd State Technical University. Volgograd, Russia (2016).
19. M.A. Vaniev, Composite materials obtained from monomer-polymer solutions under conditions of redox and photoinitiation, Dissertation for obtaining the degree of Dr. of Science in Engineering, Volgograd State Technical University, Volgograd, Russia (2014).
20. M. Li, K. Han, H. Xu and J. Xiao, Study on two-component acrylate structural adhesive. *AIP Conference Proceedings* 1995, 020023 (2018).
21. U. Khromiak, V. Levytskyi, K. Stepova and A. Tarnawsky, Synthesis and properties of adhesive polymer-methylmethacrylate materials. *Int. J. Polym. Sci.* 2018, 4905304 (2018).
22. M. Du, Y. Li, X. Lin and H. Zhai, Thermosetting acrylic structural adhesive and preparation method, CN Patent 101781536 B (2010).

23. R. Ito, T. Nishimura, A. Tamura and N. Takahashi, Radical-curable adhesive composition and adhesive, US Patent 20190169468 A1, assigned to Showa Denko KK (2019).
24. W. Wen, Sh. Ghazi and J. Zhang, Methacrylate structural glue, EP Patent 1506265 B1, assigned to Sika Technology AG (2005).
25. P. G. Barker and D. L. Kotzev, Multiphase acrylic adhesives, US Patent 20110139819 A1, assigned to Huntsman International LLC (2011).
26. R. De Cooman, W. Meyer and R. Flury, Elastic meth(acrylic) adhesive compositions, US Patent 20050004304 A1, assigned to Sika Schweiz AG (2005).
27. P. Egli and F. Cirillo, Elastic (meth)acrylate composition, WO Patent 2008151849 A1, assigned to Sika Technology AG (2008).
28. C. Bret, O. Fradin, M. Roux, E. Vaique, N. Valloir and R. Inoubli, Additive composition suitable for use in a structural adhesive, EP Patent 3328957 B1, assigned to Arkema France SA (2018).
29. M.A. Kropp and Z.J. Thompson, Structural acrylic adhesive, WO Patent 2013126377 A1, assigned to 3M Innovative Properties Company (2013).
30. A. Khongal, N. Chitnavis, D. Gosiewski, S. Deshpande and P. Carbutt, Methacrylate structural adhesive formulation and process for the use thereof, US Patent 10301514 B2, assigned to Illinois Tool Works Inc (2019).
31. C. Liu, J. Lin and J. Fan, Adhesive having rapid fixing structure, and preparation method thereof, CN Patent 103305130 B (2013).
32. S. Osae and P. Briggs, Adhesive compositions for bonding metals, US Patent 20070155879 A1, assigned to IPS Corp (2007).
33. K.J. Abbey, Acrylic adhesion promoters, US Patent 8859098 B2, assigned to Lord Corp (2014).
34. N. Chitnavis, M. Kulkarni, S. Deshpande, D.K. Doe, P. Carbutt and G. Schulz, Galvanized metal bonding adhesive formulation and process for the use thereof, US Patent 9771501 B2, assigned to Illinois Tool Works Inc (2017).
35. X. Wang, D.K. Doe, P.M. Savory and K.A. Lambert, Acrylic adhesives for metal bonding applications, US Patent 7776963 B2, assigned to Illinois Tool Works Inc (2010).
36. N. Sweeney, B. Deegan, L. Hurlburt, A.D. Messana and M. Galligan, Adhesive compositions, US Patent 20180320025 A1, assigned to Henkel IP and Holding GmbH (2018).
37. J. Huang and E.N. Wyman, Structural adhesives containing maleimide terminated polyimides, WO Patent 2010088648 A2, assigned to Lord Corp (2010).
38. B. Vazquez, C. Elvira, J. San Roman and B. Levenfeld, Reactivity of a polymerizable amine activator in the free radical copolymerization with methyl methacrylate and surface properties of copolymers. *Polymer* 38, 4365-4372 (1997)
39. D.S. Achilias and I.D. Sideridou, Kinetics of the benzoyl peroxide/amine initiated free-radical polymerization of dental dimethacrylate monomers: Experimental studies and mathematical modeling for TEGDMA and Bis-EMA. *Macromolecules* 37, 4254-4265 (2004).

40. I.D. Sideridou, D.S. Achilias and O. Karava, Reactivity of benzoyl peroxide/amine system as an initiator for the free radical polymerization of dental and orthopaedic dimethacrylate monomers: Effect of the amine and monomer chemical structure. *Macromolecules* 39, 2072-2080 (2006).
41. A.V. Zaikina, E.I. Yarmukhamedova, Yu.I. Puzin, and Yu.B. Monakov, Study of methyl methacrylate polymerization initiated with N,N-dimethyl-N-benzylamine-benzoyl peroxide system. *Izvestiya Vysshikh Uchebnykh Zavedenii, Seriya Khimiya i Khimicheskaya Tekhnologiya* (Russian) 53 (3), 86-89 (2010).
42. P. Garra, F. Morlet-Savary, C. Dietlin, J.-P. Fouassier and J. Lalevé, Charge-transfer complexes as new inhibitors/photoinitiators for on-demand amine/peroxide redox polymerization. *ACS Omega* 3, 6827–6832 (2018).
43. M. Worzakowska, The influence of tertiary aromatic amines on the BPO initiated cure of unsaturated epoxy polyesters with styrene studied by non-isothermal DSC. *J. Therm. Anal. Calorim.* 105, 987-994 (2011).
44. A. Zoller, D. Gigmes and Y. Guillaneuf, Simulation of radical polymerization of methylmethacrylate at room temperature using a tertiary amine/BPO initiating system. *Polym. Chem.* 6, 5719-5727 (2015).
45. V.P. Rybalko, E.I. Pisarenko, P.B. D'yachenko, S.A. Timerkhanov, A.A. Palamarchuk and A.S. Korchmarek, The influence of the structure of tertiary amines on low-temperature curing of a highly filled acrylic composite initiated with benzoyl peroxide. *Polym. Sci. Ser. D* 12, 231-235 (2019).
46. D.V. Gusev, Yu.R. Ebich, M.Ya. Kuzmenko, O.Yu. Poloz and O.M. Kuzmenko, Alkoxyaminosilanes as accelerators of hardening of anaerobic compositions. *Voprosy Khimii i Khimicheskoi Tekhnologii* (Ukr) 6, 51-59 (2008).
47. D.V. Gusev, Yu.R. Ebich, M.Ya. Kuzmenko, O.Yu. Poloz and S.M. Kuzmenko, Aminoalkoxytitanates as accelerators of benzene peroxide decomposition in anaerobic compositions. *Voprosy Khimii i Khimicheskoi Tekhnologii* (Ukr) 2, 72-77 (2009).
48. K. Kim, N.R. Singstock, K. Childress, J. Sinha, A.M. Salazar, S.N. Whitfield, A.M. Holder, J.W. Stansbury and Ch.B. Musgrave, Rational design of efficient amine reductant initiators for amine-peroxide redox polymerization. *J. Amer. Chem. Soc.* 141, 6279-6291 (2019).
49. G.O. Wilson, J.W. Henderson, M.M. Caruso, B.J. Blaiszik, P.J. McIntire, N.R. Sottos, S.R. White and J.S. Moore, Evaluation of peroxide initiators for radical polymerization-based self-healing applications. *J. Polym. Sci. A. Polym. Chem.* 48, 2698-2708 (2010).
50. E.P. Pedraza, C.I. Cano, J. VanDalen, R.B. Pipes and J.P. Youngblood, Reduction in fixture time of a two-component structural acrylic adhesive. *Int. J. Adhesion Adhesives* 28, 283-290 (2008).
51. P. Garra, C. Dietlin, F. Morlet-Savary, F. Dumur, D. Gigmes, J.-P. Fouassier and J. Lalevée, Redox two-component initiated free radical and cationic polymerizations: concepts, reactions and applications. *Prog.Polym. Sci.* 94, 34-56 (2019).

52. C.B. Musgrave III, K. Kim, N.R. Singstock, A.M. Salazar, J.W. Stansbury and C.B. Musgrave, Computational and experimental evaluation of peroxide oxidants for amine–peroxide redox polymerization. *Macromolecules* 53, 9736-9746 (2020).
53. G.R. Guzaliya D.M. Zemski and O.V. Karmanova, Polymerization of vinyl monomers in the presence of the initiating catalytic complexes "oxopropylidene tertiary aromatic amines – benzoyl peroxide. *Proceedings of the Voronezh State University of Engineering Technologies.* 79, (2017). http://doi.org/10.20914/2310-1202-2017-2-165-169
54. D.S. Achilias and I.D. Sideridou, Study of the effect of two BPO/amine initiation systems on the free-radical polymerization of MMA used in dental resins and bone cements. *J. Macromol. Sci., Part A: Pure Appl. Chem.* 39, 1435-1450 (2002).
55. J.E. Hill and K.J. Abbey, Amine co-accelerator for acrylic adhesives, US Patent 9896607 B2, assigned to Lord Corp (2018).
56. Q.L. Hibben, Low temperature curable adhesive compositions, US Patent 8334346 B2, assigned to Simpson Strong Tie Co Inc (2012).
57. A. D. Messana and L. Hurlburt, Two part curable composition, US Patent 10689504 B2, assigned to Henkel IP and Holding GmbH (2020)
58. P. Liu, Q. Zhu and K. J. Welch, Two part curable compositions, WO Patent 2021127128 A1, assigned to Henkel IP and Holding GmbH (2021).
59. J.-P Huang and T. Perry, Redox initiation system for acrylic adhesives, WO Patent 2021150819 A1, assigned to Lord Corporation (2021).
60. C. Elvira, B. Levenfeld, B. Vazquez and J. San Roman, Amine activators for the "cool" peroxide initiated polymerization of acrylic monomers. *J. Polym. Sci. A. Polym. Chem.* 34, 2783-2789 (1996).
61. J. Murray, C.F. Schuft and C.-M. Cheng, Adhesive compositions, US Patent 20150225626 A1, assigned to Henkel IP and Holding GmbH (2015).
62. V. D. McGinnis, Catalytic systems for two-park acrylic adhesive formulations, US Patent 4446246A, assigned to Three Bond Co Ltd (1984).
63. F. Tran and S.N. Kaul, Peroxide-free methacrylate structural adhesives, WO Patent 2012134735 A1, assigned to Huntsman Advanced Materials Americas LLC (2012).
64. D.P. Birkett, A.F. Jacobine, A.D. Messana, J.D. Schall, D. Mullen, M. Wyer and L. Hurlburt, Accelerators for two part curable compositions, US Patent 20140004354 A1, assigned to Henkel Ireland Ltd and Henkel Corp (2014).
65. Y. Hisha and K. Yoda, Curable composition, bonding method and assembly, US Patent 8067500 B2, assigned to Denka Co. Ltd (2011).
66. J.T. Fields, A. Garton and J.P. Bell, A reactive acrylic adhesive for bonding polyolefins. *J. Adhesion Sci. Technol.* 9, 627-645 (1995).
67. X. Xie and A. Schneider, Heat and moisture resistant acrylic adhesive composition, US Patent 9676922 B2, assigned to IPS Corp Weld-On Division (2017).

68. A. Kramer and S. Maier, (Meth)acrylate composition with reduced oxygen inhibition, EP Patent 1893701 B1, assigned to Sika Technology AG (2008).
69. A. Kramer and M. Haufe, (Meth)acrylic composition with reduced surface tackiness, EP Patent 2492329 A1, assigned to Sika Technology AG (2012).
70. I.C. Quarmby, Adhesive formulations, EP Patent 1190011 A2, assigned to Lord Corporation (2001).
71. R.F. Righettini and T.H. Dawdy, Free radical polymerizable compositions including para-halogenated aniline derivatives, US Patent 5932638 A, assigned to Lord Corp. (1999).
72. J. Huang, R.F. Righettini and F.G. Dennis, Adhesive formulations, US Patent 6225408 B1, assigned to Lord Corp (2001).
73. D. Farrell, C.B. McArdle and M. Doherty, Surface-promoted cure of one-part radically curable compositions, US Patent 20090288771 A1, assigned to Henkel IP and Holding GmbH (2009).
74. A. Curet, Composition for structural adhesive, US Patent 8480844 B2, assigned to Jacret, (2015).
75. V. Satheeshkumar and R. G. Narayanan, In-plane plane strain formability of adhesive-bonded steel sheets: Influence of adhesive properties. *Int. J. Adv. Manuf. Technol.* 76, 993–1009 (2015).
76. A. Frapsauce and C. Baron, Adhesive composition with retarding additive, WO Patent 2020126818A1, assigned to Huntsman Advanced Materials Licensing Switzerland GmbH (2020).
77. Z.J. Thompson and T. Kasahara, Adhesive precursor composition, two-part adhesive kit, and method of making an adhesive composition, WO Patent 2015200007 A1, assigned to 3M Innovative Properties Company (2015).
78. V.L. Horner and K.H. Knapp, Autoxydationsstudien an N,N-dialkylierten Anilinderivaten. *Makromolekulare Chemie* 93, 69-108 (1966).
79. T. Huver, H. Fischer and W. Emmerling, New multiple-constituent mixtures capable of free-radical polymerization and their use, WO Patent 1990002142 A1, assigned to Henkel Kommanditgesellshaft auf Aktien (1990).
80. B.J. Kneafsey and R.B. Barnes, Auto-oxidation systems for air-activatable polymerisable compositions, US Patent 6552140 B1, assigned to Loctite (R&D) Ltd, Dublin (2003).
81. T. Huver, H. Fischer, W. Klauck and G. Bolte, One-component reactive adhesive, US Patent 5744543 A, assigned to Henkel KGaA (1998).
82. B. Kneafsey, Air-activatable polymerisable compositions containing onium salts, EP Patent 0611813 A1, assigned to Henkel Loctite Ireland Ltd (1998).
83. E. Lyons and J.J. Leahy, An ageing study of air activated acrylic adhesives using dynamic mechanical thermal analysis, in *Ageing Studies and Lifetime Extension of Materials*. L.G. Mallinson (Ed.), pp. 55-61 Springer, Boston, MA (2001).
84. S. Kenig, H. Dodiuk, G. Otorgust and S. Gomid, Nanocomposite polymer adhesives: A critical review. *Rev. Adhesion Adhesives* 7, 93-168 (2019).

85. A. Garbuz, P. Bilym and D. Zubenko, Features of change of structure and adhesive properties for acrylic adhesives under the influence of the filler. *EUREKA: Physics and Engineering* 3, 13-16 (2016).
86. S. Zolotov, O. Pustovoitova, P. Firsov, H. Kais and S. Kamchatna, Stress-strained state of steel-adhesive bonding on the acrylic adhesives. *IOP Conf. Series: Mater. Sci. Eng.* 708, 012117 (2019)
87. E.M. Petrie, New Toughening Agents (Micro & Nano) for Structural Adhesives. SpecialChem. https://adhesives.specialchem.com/tech-library/article/toughening-agents-for-structural-adhesives (2020).
88. A. Tutunchi, R. Kamali and A. Kianvash, Steel-epoxy composite joints bonded with nano-TiO_2 reinforced structural acrylic adhesive. *J. Adhesion* 91, 663-676 (2015).
89. A. Tutunchi, R. Kamali and A. Kianvash, Effect of Al_2O_3 nanoparticles on the steel-glass/epoxy composite joint bonded by a two-component structural acrylic adhesive. *Soft Materials* 14, 1-8 (2016).
90. M. Chavooshian, R. Kamali, A. Tutunchi and A. Kianvash, Effect of silicon carbide nanoparticles on the adhesion strength of steel–epoxy composite joints bonded with acrylic adhesives. *J. Adhesion Sci. Technol.* 31, 345-357 (2017).
91. A. Tutunchi, K. Osouli-Bostanabad, M. Eskandarzade, R. Kamali and M. Chavoshian, Steel-epoxy composite joints bonded with nano reinforced structural acrylic adhesive. In: Proc. 5th Int. Biennial Conf. on Ultrafine Grained and Nanostructured Materials, UFGNSM15, Tehran (2015). https://www.researchgate.net/publication/284273322.
92. P.S. Shin, J.-H. Kim, J.-Y. Choi, D.-J Kwon, S.-I Lee and J.-M. Park, Comparison of mechanical and interfacial properties on chemical structures of acrylic and epoxy adhesives. *Composites Res.* 29, 79-84 (2016).
93. P.C.P. Watts, P.K. Fearon, W.K. Hsu, N.C. Billingham, H.W. Kroto and D.R.M. Walton, Carbon nanotubes as polymer antioxidants. *J. Mater. Chem.* 3, 491-495 (2003).
94. W.K. Oh, S.Y. Park and G.S. Choi, Adhesive composition, EP Patent 3412743 B1, assigned to LG Chem Ltd. (2019).
95. L. Hurlburt, N. Sweeney and Ph. T. Klemarczyk, Adhesive compositions containing reactive silsesquioxanes, US Patent 20190136102 A1, assigned to Henkel IP and Holding GmbH (2019).
96. C.F. Schuft, D.E. Frost and D.N. Mason, Acrylic adhesives, US Patent 8231758 B2, assigned to Henkel IP and Holding GmbH (2012).
97. J.C. Del Real, M. Cano de Santayana, J. Abenojar, M. Pantoja and M.A. Martinez, Influence of silanisation parameters with γ-methacryloxypropyltrimethoxysilane on durability of aluminium/acrylic adhesive joints. *J. Adhesion Sci. Technol.* 22, 1461-1475 (2008).
98. J.M. Arenas, J.J. Narbón and C. Alía, Optimum adhesive thickness in structural adhesives joints using statistical techniques based on Weibull distribution. *Int. J. Adhesion Adhesives* 30, 160-165 (2010).

99. E.R. Fotsing, F. Miron, Y. Eury, A. Ross and E. Ruiz, Bonding analysis of carbon/epoxy composites with viscoelastic acrylic adhesive. *Composites B.* 43, 2087-2093 (2012).
100. A. Purabgola, S. Rastogi, G. Sharma and B. Kandasubramanian, Surface preparation for structural adhesive joints, in: *Structural Adhesive Joints: Design, Analysis, and Testing*, K.L. Mittal and S.K. Panigrahi (Eds.), pp. 3-34, Wiley Scrivener, Beverly, MA (2020).
101. R. Chamochin, M. Cano de Santayana, J. Abenojar, M. Pantoja, Y. Ballesteros and J.C. Del Real, The effect of surface treatment on the behavior of toughened acrylic adhesive/GRP(epoxy) composite joints. *J. Adhesion Sci. Technol.* 24, 1903–1916 (2010).
102. M. Morra, V. Di Ruocco, E. Occhiello and F. Garbassi, Surface analysis of polymeric joints. *J. Adhesion* 41, 139-146 (1993).
103. V. Di Ruocco, M. Morra, E. Occhiello and F. Garbassi, On the effect of water on the durability of PP and PC/acrylic adhesive joints. *Angew. Makromol. Chem.* 194, 189-200 (1992).
104. K. Osouli-Bostanabad, A. Tutunchi and M. Eskandarzade, The influence of pre-bond surface treatment over the reliability of steel epoxy/glass composites bonded joints. *Int. J. Adhesion Adhesives* 75, 145-154 (2017).
105. H.B. Kim, K. Naito and H. Oguma, Fracture toughness of adherends bonded with twopart acrylicbased adhesive: Double cantilever beam tests under static loading. *Appl. Adhesion Sci.* 4, 1-11 (2016).
106. E. Paz, J.J. Narbón, J. Abenojar, M. Cledera and J.C. Del Real, Influence of acrylic adhesive viscosity and surface roughness on the properties of adhesive joint. *J. Adhesion* 92, 877-891 (2016).
107. S. Kawasaki, G. Nakajima, K. Haraga and C. Sato, Functionally graded adhesive joints bonded by honeymoon adhesion using two types of second generation acrylic adhesives of two components. *J. Adhesion* 92, 517-534 (2016).
108. M. Nakanouchi, C. Sato, Yu. Sekiguchi, K. Haraga and H. Uno, Development of application method for fabricating functionally graded adhesive joints by two-component acrylic adhesives with different elastic moduli. *J. Adhesion* 95, 529-542 (2019).
109. Yu. Sekiguchi and C. Sato, Effect of bond-line thickness on fatigue crack growth of structural acrylic adhesive joints. Materials, 14, 1723 (2021).
110. J.C. Del Real, M. Cano De Santayana, J. Abenojar and M.A. Martinez, Adhesive bonding of aluminium with structural acrylic adhesives: durability in wet environments. *J. Adhesion Sci. Technol.* 20, 1801–1818 (2006).
111. V. Di Ruocco, M. Morra, E. Occhiello and F. Garbassi, On the failure of water-immersed steel-acrylic adhesive joints. *J. Mater. Sci. Lett.* 10, 649-652 (1991).
112. K. Machalická, M. Vokác, M. Kostelecká and M. Eliášová, Structural behavior of double-lap shear adhesive joints with metal substrates under humid conditions. *Int. J. Mech. Mater. Design.* 15, 61-76 (2019).

113. Y. Hou and Q. Lu, A method to evaluate water resistance of acrylic adhesives. *J. Adhesion Sci. Technol.* 23, 1647–1656 (2009).
114. M. Kozłowski, A. Bula and J. Hulimka, Determination of mechanical properties of methyl methacrylate adhesive (MMA). *Architect. Civil Eng. Environ. (ACEE)* 11, 87-96 (2018).
115. A. Bula, S. Jurczyk, B. Chmielnicki, J. Hulimka and M. Kozłowski, Multi-frequency dynamic-mechanical thermal analysis of methyl methacrylate adhesive (MMA). *Int. J. Polym. Analys. Characteriz.* 24, 169-177 (2019).
116. A. Bula, M. Kozłowski and B. Chmielnicki, Analysis of methyl methacrylate adhesive (MMA) relaxation with non-linear stress–strain dependence. *Int. J. Adhesion Adhesives* 94, 40-46 (2019).
117. M. Kałuża and J. Hulimka, Methacrylate adhesives to create CFRP laminate-steel joints – preliminary static and fatigue tests. *Procedia Eng.* 172, 489-496 (2017).
118. M. Kałuża, J. Hulimka, J. Kubica and M. Tekieli, The methacrylate adhesive to double-lap shear joints made of high-strength steel - Experimental study. *Materials* 12, 120 (2019).
119. M. Kałuża, J. Hulimka and J. Kubica, Effectiveness of adhesive CFRP/steel joints - double-lap static tests, in: Proc. Int. Symp. Brittle Matrix Composites 11, Warsaw, 479-488 (2015). https://www.researchgate.net/publication/315370229_Pre
120. I. Amatya, Structural performance evaluation of dynamic message signs with adhesive and welded joints, Thesis for the Master of Science, South Dakota State University (2020). https://openprairie.sdstate.edu/cgi/viewcontent.cgi?article=5147&context=etd
121. K. Kamiyama, M. Mikuni and T. Matsumoto, Fracture propagation analysis on two component type acrylic adhesive joints. *Int. J. Adhesion Adhesives* 83, 76-86 (2018).
122. V.A. Silvestru, M. Drass, O. Englhardt and J. Schneider, Performance of a structural acrylic adhesive for linear glass-metal connections under shear and tensile loading. *Int. J. Adhesion Adhesives* 85, 322-326 (2018).
123. S. Samer and P. Hartmut, Epoxy- and acrylic-bonded joints for strengthening lightweight steel girders. *Stahlbau* 84, 577-583 (2015).
124. Q. Zhao, K. Jin and M.J. Buehler, Molecular modeling and mechanics of acrylic adhesives on a graphene substrate with roughness. *BioNanoScience* 6, 177-184 (2016).
125. R. Avendaño, R.J.C. Carbas, E.A.S. Marques, L.F.M. da Silva and A.A. Fernandes, Effect of temperature and strain rate on single lap joints with dissimilar lightweight adherends bonded with an acrylic adhesive. *Composite Structures* 152, 34-44 (2016).
126. M.D. Banea and L.F.M da Silva, Adhesively bonded joints in composite materials: An overview. Proc. Inst. Mech. Eng., Part L: *J. Mater. Design Applications* 223, 1-18 (2009).

127. I. Makoto, O. Masaki and T. Noriaki, Determination of traction-separation laws on an acrylic adhesive under shear and tensile loading. *J. Adhesion Sci. Technol.* 33, 646-659 (2019).
128. A. Tutunchi, R. Kamali and A. Kianvash, Adhesive strength of steel-epoxy composite joints bonded with structural acrylic adhesives filled with silica nanoparticles. *J. Adhesion Sci. Technol.* 29, 195-206 (2015).
129. M.Q. dos Reis, R.J.C. Carbas, E.A.S. Marques and L.F.M. da Silva, Numerical modelling of multi-material graded joints under shear loading. *Proc. Inst. Mech. Eng., Part E: J. Process Mech. Eng.* 234, 436-445 (2020).
130. 3M™ Scotch-Weld™ Structural Acrylic Adhesives. Next generation structural acrylic adhesives. Enhance productivity and performance. (2016) 3m-brochure--scotch-weld-next-generation-structural-acrylic-adhesives.pdf
131. E. Lertora, C. Gambaro, C. Mandolfino and M. Pedemonte, Environmental effects on methacrylate adhesive. *Welding International* 28, 372-379 (2014).
132. J. Moller, R. Hunter, J. Molina, A. Vizán, J. Peréz and L.F.M. da Silva, Influence of the temperature on the fracture energy of a methacrylate adhesive for mining applications. *Appl. Adhesion Sci.* 31, 1-13 (2015).
133. M. Angelidi, A.P. Vassilopoulos and Th. Keller, Ductility, recovery and strain rate dependency of an acrylic structural adhesive. *Construct. Build. Mater.* 140, 184-193 (2017).
134. Yu. Sekiguchi and C. Sato, Experimental investigation of the effects of adhesive thickness on the fracture behavior of structural acrylic adhesive joints under various loading rates Int. *J. Adhesion Adhesives* 105, 102782 (2021).
135. J. Ni, J. Min, H. Wan, J. Lin, S. Wang and Q. Wan, Effect of adhesive type on mechanical properties of galvanized steel/SMC adhesive-bonded joints. *Int. J. Adhesion Adhesives* 97, 102482 (2020).
136. Yu. Sekiguchi, M. Nakanouchi, K. Haraga, I. Takasaki and C. Sato, Experimental investigation on strength of stepwise tailored single lap adhesive joint using second-generation acrylic adhesive via shear and low-cycle shear tests. *Int. J. Adhesion Adhesives* 95, 102438 (2019)
137. M.M. Alves, L.M. Abreu, A.B. Pereira and A.B. De Morais, Determination of bilinear elastic-plastic models for an acrylic adhesive using the adhesively bonded 3-point bending specimen. *J. Adhesion.* 97, 553-568 (2021)
138. K.L. Mittal, The role of the interface in adhesion phenomena. *Polym. Eng. Sci.*, 17, 467- 473 (1977).
139. Y. Taira and Y. Imai, Review of methyl methacrylate (MMA)/tributylborane (TBB)-initiated resin adhesive to dentin. *Dental. Mater. J.* 33, 291–304 (2014).
140. D. Ahn, K.A. Wier, T.P. Mitchel and P.A. Olney, Applications of fast, facile, radiation-free radical polymerization techniques enabled by room temperature alkylborane chemistry. *ACS Appl. Mater. Interfaces* 7, 23902-23911 (2015).
141. C. Lv, Y. Du and X. Pan, Alkylboranes in conventional and controlled radical polymerization. *J. Polym. Sci.* 58, 14-19 (2020).

142. C. Lv, C. He and X. Pan, Oxygen-initiated and regulated controlled radical polymerization under ambient conditions. Angewandte Chemie, 130, 9574-9877 (2018).
143. M. Yu. Zaremski, V.V. Odintsova, A.V. Plutalova, M.E. Gurskii and Yu.N. Bubnov, Reactions of initiation and reinitiation in polymerization mediated by organoborane–oxygen systems. *Polym. Sci. Ser. B* 60, 162-171 (2018).
144. G.L. Jialanella, Advances in bonding plastics, in: *Advances in Structural Adhesive Bonding*, D. Dillard (Ed.). pp. 237-264. Woodhead Publishing, Oxford, UK (2010).
145. V.A. Dodonov, Yu.V. Zharov, Yu.N. Krasnov, V.V. Chesnokov and S.A. Sergeev, Composition for gluing thermoplasts, SU Patent 1609117 A1 (1996) (Application SU4654249/05A, 1989).
146. Yu.V. Zharov and Yu.N. Krasnov, Polymerizable compositions made with polymerization initiator systems based on organoborane amine complexes, U.S. Patent 5539070 A, assigned to 3M Innovative Properties Co (1996).
147. M.F. Sonnenschein, S.P. Webb, P.E. Kastl, D.J. Arriola, B.L. Wendt, D.R. Harrington and N.G. Rondan, Mechanism of trialkylborane promoted adhesion to low surface energy plastics. *Macromolecules* 37, 7974-7978 (2004).
148. M.F. Sonnenschein, S.P. Webb, O.D. Redwine, B.L. Wendt and N.G. Rondan, Physical and chemical probes of the bond strength between trialkylboranes and amines and their utility as stabilized free radical polymerization catalysts. *Macromolecules* 39, 2507-2513 (2006).
149. M. Li, Z. Zheng, S. Liu, W. Wei and X. Wang, Mechanism of adhesion of low surface energy materials treated with trialkylborane. *Int. J. Adhesion Adhesives* 31, 36-42 (2011).
150. S. Liu, Z. Zheng, M. Li and X. Wang, Study of the radical chemistry promoted by tributylborane. *Research on Chemical Intermediates* 38, 1893-1907 (2012).
151. V.A. Dodonov and T.I. Starostina, Radical bonding thermoplastics and materials with low surface energy using acrylate compositions. *Polym. Sci. Ser. D* 11, 60-66 (2018).
152. M.Yu. Zaremski, E.S. Garina, M.E. Gurskii and Yu.N. Bubnov, Organoboranes–atmospheric oxygen systems as unconventional initiators of radical polymerization. *Polym. Sci. Ser. B* 55, 304-326 (2013).
153. A.C. Moran, N.V. Lebedeva, S. Richert and M.D.E. Forbes, Spin trapping studies of amine-stabilized trialkylborane free radical initiator systems: kinetics and mechanism. *Appl. Magnetic Resonance* 41, 295-303 (2011).
154. Shujuan Liu, Zhen Zheng, Minrui Li and Xinling Wang, Effect of oxidation progress of tributylborane on the grafting of polyolefins. *J. Appl. Polym. Sci.* 125, 3335-3344 (2012).
155. Yu.V. Zharov, An acrylate compound for adhesion of inert thermoplastics with metals. *Polym. Sci. Ser. D* 8, 31-36 (2015).

156. M.F. Sonnenschein, S.P. Webb and N.G. Rondan, Amine organoborane complex polymerization initiators and polymerizable compositions, US Patent 6706831B2, assigned to Dow Global Technologies LLC (2004).
157. G.L. Jialanella, S. Feng, P. Nickias and T. Ristoski, The evaluation of alkyl borane compounds and imidazole borate complexes as initiators for acrylic adhesives bonding low energy substrates. *J. Adhesion* 90, 269-278 (2014).
158. G.L. Jialanella and T. Ristoski, Stabilized organoborane polymerization initiators and polymerizable compositions, US Patent 7247596 B2, assigned to Dow Global Technologies LLC (2007).
159. M. Sonnenschein, S. Webb, B. Wendt and D.R. Harrington, Amine organoborane complex initiated polymerizable compositions containing siloxane polymerizable components, US Patent 20040220363 A1, assigned to Dow Global Technologies LLC (2004).
160. V.A. Dodonov, M.V. Gulenova, K.V. Kiryanov, S.N. Ilyanov and R.A. Verkhovykh, Thermokinetics of curing acrylate adhesive composition under the action of trialkylborane-amine complexes with oxygen. *Klei. Germetiky. Tehnologii (Rus)* 11, 17-21 (2014).
161. S.P. Webb and M.F. Sonnenschein, Organoborane amine complex polymerization initiators and polymerizable compositions, WO Patent 2003038006 A1, assigned to Dow Global Technologies Inc (2003).
162. G.L. Jialanella, S.H. Feng, P.N. Nickias and T. Ristoski, Blocking agents for organoborane compounds, US Patent 20100105910 A1, assigned to Dow Global Technologies LLC (2010).
163. D. Ahn, M. Mustafa and P. Olney, Organosilicon functional boron amine catalyst complexes and curable compositions made therefrom, US Patent 7928174 B2, assigned to Dow Silicones Corp (2011).
164. K. Sehanobish, D.J. Falla, Sh. Wu and M.F. Sonnenschein, Method for joining substrates and objects, WO Patent 2005017005 A1, assigned to Dow Global Technologies Inc (2005).
165. M. Mohamed, Borane catalyst complexes with amide functional polymers and curable compositions made therefrom, US Patent 20090123703 A1, assigned to Dow Silicones Corp (2009).
166. I. Hiroyuki, A. Motohiro and K. Shigeki, Adhesive composition, film-like adhesive, adhesive sheet, connection structure and method for producing connection structure, KR Patent 20130069471 A, assigned to Hitachi Chemical Co Ltd (2013).
167. J.E. Garbe and J.R. Baran, Jr., Triorganoborane-amino functionalized nanoparticles, compositions, and methods, US Patent 9908904 B2, assigned to 3M Innovative Properties Co (2018).
168. E.J. Deviny and V.C. Marhevka, Polymerizable system with a long work-life, US Patent 6762261 B2, assigned to 3M Innovative Properties Co (2004).

169. J.R. Baran, Jr. and J.E. Garbe, Acid-modified nanoparticles, dual part polymerizable compositions, and methods, US Patent 10059781 B2, assigned to 3M Innovative Properties Co (2018).
170. A.V. Fedorov, A.A. Ermoshkin and D.C. Neckers Remote polymerization of coatings with the release of borane from its amine complex. *J. Appl. Polym. Sci.* 107, 147–152 (2008).
171. O.R. Wilson and A.J.D. Magenau, Oxygen tolerant and room temperature RAFT through alkylborane initiation. *ACS Macro Letters*, 7, 370–375 (2018).
172. N. Sweeney, B.J. Kneafsey, G. Coughlan and J. Houlihan, Two-part polymerisable compositions, WO Patent 2012160452 A1, assigned to Henkel Ireland Ltd (2012).
173. H. Chiemi and Y. Imai, Studies on MMA-TBB resin I. Comparison of TBB and other initiators in the polymerization of PMMA/MMA resin. *Dental Mater. J.*, 21, 314–321 (2002).
174. Ch. Inami, H. Shimiz, Sh.Suzuki, N. Haraguchi and Sh. Itsuno, Study on the performance of methyl methacrylate polymerization: Comparison of partially oxidized tri-n-butylborane and benzoyl peroxide with aromatic tertiary amines. *Dental Mater. J.*, 38: 430–436 (2019).
175. B. Kneafsey and G. Coughlan, Quaternary boron compound polymerisation initiators, polymerisable compositions, and uses thereof for bonding low surface energy substrates, US Patent 20040068067 A1, assigned to Henkel IP and Holding GmbH, Henkel Loctite Ireland Ltd (2004).
176. B.J. Kneafsey and G. Coughlan, Metal alkyl borohydride polymerisation initiators, polymerisable compositions, and uses thereof, US Patent 20040097673 A1, assigned to Henkel IP and Holding GmbH (2004).
177. B. Kneafsey and G. Coughlan, Polymerisation initiators, polymerisable compositions, and uses thereof, EP Patent 1201722 A1, assigned to Henkel Loctite Ireland Ltd (2002).
178. J.L. Kendall, R.F. Righettini and K.J. Abbey, Internally blocked organoborate initiators and adhesives therefrom, US Patent 6630555 B2, assigned to Lord Corp (2003).
179. Sh. Feng, G. Jialanella, P. Nickias and T. Ristoski, Amido-organoborate initiator systems, US Patent 20070083051 A1, assigned to Dow Global Technologies LLC (2007).
180. Acrylate based adhesive composition, KR Patent 20180115557 A (2018).
181. G. Michaud, B. Colin, F. Simon, S. Fouquay, V. Liautard and M. Pucheault, Two-component adhesive composition comprising a borane-amine complex, WO Patent 2021009447 A1, assigned to Bostik Sa, Centre National De La Recherche Scientifique, Universite de Bordeaux, Institut Polytechnique De Bordeaux (2021).
182. V.A. Dodonov, R.A. Verkhovykh, S.S. Lomakin and A.V. Gushchin, The nature of the effect of small additives of organic peroxides, oligoester acrylates and organometallic diacrylates on the rate of curing and adhesive capacity of the acrylate composition. *Klei. Germetiky. Tehnologii (Rus)* 2, 19-23 (2014).

183. G. Jialanella and D. Johnson, Accelerated organoborane amine complex initiated polymerizable compositions, US Patent 20050137370 A1, assigned to Dow Global Technologies LLC (2005).
184. V.A. Dodonov, Yu.L. Kuznetsova, M.A. Lopatin and A.A. Skatova, Reactions of poly(methyl methacrylate) radicals with some p-quinones in the presence of tri-n -butylboron in methyl methacrylate polymerization. *Russ. Chem. Bull.* 53, 2209-2214 (2004).
185. V.A. Dodonov and D.F. Grishin, The features of polymerization of certain vinyl monomers, radical initiators containing triisobutyl, in the presence of hydroquinone and benzoquinone. *Visokomolekulyarnye Soedineniya B (Rus.)* 35, 137-141 (1993).
186. V.A. Dodonov, S.S. Lomakin and M.V.Gulenova, Influence of small additions of *p*-quinones on the curing rate and adhesive strength of the acrylate composition. *Klei. Germetiky. Tehnologii (Rus)* 11, 29-33 (2013).
187. R.A. Verkhovykh, Modification of reactive (meth)acrylate adhesive compositions based on trialkylborane complexes with hexamethylenediamine, PhD thesis. Nizhny Novgorod University, Nizhny Novgorod, Russia (2015).
188. X. Xie, Adhesive composition for bonding low surface energy polyolefin substrates, WO Patent 2013151680 A1, assigned to IPS Corp (2013).
189. J. Houlihan, B. Kneafsey and N. Sweeney, Curable acrylate or methacrylate compositions, US Patent 9732257 B2, assigned to Henkel IP and Holding GmbH (2017).
190. J. Zhang, J.-W. Wang and R.-K. Ge, Two-component acrylic structural adhesive initiated by tributylborane for untreated PET adhesion. *J. Appl. Polym. Sci.* 135, 46612 (2018).
191. X. Xie and A. Schneider, Polyurea-poly(meth)acrylate interpenetrating polymer network adhesive compositions and methods of using the same, US Patent 20190211229 A1, assigned to IPS Corp (2019).
192. M. Li, Z. Zheng, S. Liu, W. Wei and X. Wang, Effect of prepolyurethane on the performances of poly(acrylates-co-urethane) copolymer. *J. Appl. Polym. Sci.* 118, 3203-3210 (2010).
193. M. Li, Z. Zheng, S. Liu, Y. Su, W. Wei and X. Wang, Polyurethane (urea)/polyacrylates interpenetrating polymer network (IPN) adhesives for low surface energy materials. *Polym. Advanc. Technol.* 23, 1077-1083 (2012).
194. M.F. Sonnenschein, S.P. Webb and B.L. Wendt, Poly(acrylate/siloxane) hybrid adhesives for polymers with low surface energy. *Int. J. Adhesion Adhesives* 28, 126-134 (2008).
195. M.F. Sonnenschein, S.P. Webb, B.L. Wendt and D.D. Harrington, Organoborane amine complex polymerization initiators and polymerizable compositions, US Patent 6762260 B2, assigned to Dow Global Technologies LLC (2004).

196. M.F. Sonnenschein, S.P Webb, R.C. Cieslinski and B.L. Wendt, Poly(acrylate/epoxy) hybrid adhesives for low-surface-energy plastic adhesion. *J. Polym. Sci. Part A: Polym. Chem.* 45, 989-998 (2007).
197. D. Shaw and M. William, Simpler adhesive bonding for low surface energy polyolefin plastics, *Design World* (May 1, 2013), https://www.designworldonline.com/simpler-adhesive-bonding-for-low-surface-energy-polyolefin-plastics/, accessed on 2021/01/07.
198. K. Takahashi, M. Inoue and N. Arita Two-part acrylic elastic epoxy that cures at room temperature // Three Bond TechnicalNews81_EN.pdf (2013) https://pdf.directindustry.com/pdf/three-bond/two-part-acrylic-elastic-epoxy-that-cures-room-temperature/16498-876521.html
199. Acrylic structural adhesives. Features and recent advancements. (2015), https://multimedia.3m.com/mws/media/1054052O/acrylic-adhesives-recent-advancements-white-paper.pdf
200. S. Sahellie, Study on the temperature effect on lap shear adhesive joints in lightweight steel construction, EngD thesis. Brandenburg University of Technology, Cottbus-Senftenberg, Germany (2015).
201. M. Eskandarian, N. Rousseau and F. Hamel, On the influence of joint geometry and adhesive bulk properties on quasistatic performance of lap-shear joints, in: Proc. 34th Annual Meeting of the Adhesion Society (2011). https://www.researchgate.net/publication/50277651
202. Plexus guide to bonding plastics, composites and metals. https://www.curbellplastics.com/Research-Solutions/Technical-Resources/Technical-Resources/Plexus-Adhesives-Guide-to-Bonding
203. J.E. Hill, Next-generation acrylic (LORD Corp.), *Adhesives & Sealants Industry Magazine* (2017), https://www.lord.com/sites/default/files/RE8016_Next-GenerationAcrylicAdhesivesEnhanceAssemblyOperations.pdf
204. J.E. Hill, New developments in next-generation acrylic adhesive technology. (2017), https://www.hisco.com/UserFiles/Documents/LandingPages/Lord/July2017/AcrylicAdhesiveTech.pdf

8

Application of Structural Adhesives in Composite Connections

M. D. Banea* **and H.F.M. de Queiroz**

Federal Centre of Technological Education in Rio de Janeiro (CEFET/RJ), Brazil

Abstract
The use of structural adhesives is increasing significantly with higher composite materials usage in aerospace, transport and marine industries. This chapter presents an overview of recent advances in adhesive bonding of composites. The main factors affecting the performance of composite (limited in this chapter to fibre-reinforced polymers) adhesive joints are presented. Critical aspects, including surface preparation, design and durability are briefly discussed. All these parameters must be taken into account during the design of composite bonded connections. Finally, it ends with recent developments and future trends.

Keywords: Adhesive joints, composite materials, surface preparation, design, durability

8.1 Introduction

Composite materials are increasingly being used in a wide range of applications in aerospace, transport and marine industries, mainly due to their superior strength- and stiffness-to-weight ratios in comparison to metals. The implementation of new materials requires development of new and more suitable joining processes. Adhesive bonding is the preferred joining technique for composite materials mainly because of high fatigue resistance and high strength-to-weight ratio of adhesively bonded joints [1]. Another important advantage of using adhesive bonding is the possibility to join

**Corresponding author*: mdbanea@gmail.com

K.L. Mittal and S.K. Panigrahi (eds.) Structural Adhesives: Properties, Characterization and Applications, (375–396) © 2023 Scrivener Publishing LLC

dissimilar materials, such as carbon fibre reinforced plastics (CFRPs) or glass fibre reinforced plastics (GFRPs) to metals in many applications [2].

Adhesive bonding has found applications in various areas from high technology industries such as aeronautics, aerospace, electronics, and automotive to traditional industries such as construction, sports and packaging. Typical applications of structural adhesives to bond composite materials include: structural adhesives for bonding face skins to honeycomb cores in sandwich panels, commercial aircraft fuselages, wind turbine blades, bonding and/or repair of composite pipes, sports equipment, medical prosthetics, etc.

Adhesives used in composite connections include: epoxies (having high strength and temperature resistance), cyanoacrylates (fast bonding capability to plastics and rubber but poor resistance to moisture and temperature), acrylics (versatile adhesives with capabilities of fast curing and tolerate dirtier and less prepared surfaces), polyurethanes (good flexibility at low temperatures and resistant to fatigue, impact resistance, and durability) and high-temperature adhesives (phenolics, polyimides and bismaleimides) [3]. The adhesive selection process is quite difficult as there is no universal adhesive that will fulfil every need and the selection of the proper adhesive is often complicated by the wide variety of available options [1, 3, 4]. Adhesive selection includes many factors such as: type and nature of substrates to be bonded, cure and adhesive application method and the expected environments and stresses that the joint will encounter in service. Also the cost of the adhesive may sometimes be an important factor in adhesive selection in a particular production situation [1]. Datasheets from adhesive manufacturers provide basic information about what type of materials can be bonded. Some adhesive suppliers provide software tools to help the engineer or designer to select the adhesive, although for critical components expert assistance is still needed.

The fundamentals and practices for adhesive bonding of composites are described in a large number of books and review articles [1, 5–10]. This chapter gives an overview of recent advances in adhesive bonding of composites. The main parameters affecting the strength of composite adhesive joints are presented. Critical aspects, including surface preparation, joint design and durability are briefly discussed. Finally, recent developments and future trends are presented.

8.2 Factors Affecting the Performance of Composite Adhesive Joints

It is well known that different materials and joint geometries give rise to distinct stress states in the adhesive layer, and the performance of adhesive joints depends on several factors: the surface preparation, joint geometry and geometrical parameters (adhesive thickness, overlap length, etc.), material parameters (adhesive and adherend properties) and environmental conditions [11–14]. All these parameters must be taken into account during the design of composite bonded connections.

8.2.1 Effect of Surface Preparation

The surfaces play an important role in the bonding process and are, perhaps, the most important process step governing the quality of an adhesively bonded joint [1, 15]. Therefore, it is essential to carry out the necessary surface preparation in order to ensure good joint strength and maintaining long-term structural integrity of bonded joints. The surface treatment increases the bond strength by altering the substrate surface by increasing surface free energy, increasing surface roughness or changing the surface chemistry [16–22].

Different surface treatments are available and depend on the material used as adherend [23]. For example, the surface preparation of metal adherends includes traditional methods such as grit blasting, mechanical abrasion, and acid etching. In contrast, for plastic adherends, sometimes achieving adhesion is quite challenging as plastic surfaces are very smooth, show poor wetting and have low surface free energy. Typical composite surface treatments include traditional mechanical abrasion/solvent cleaning techniques for thermoset composites, while thermoplastic composites require appropriate surface chemistry and surface topographical changes to ensure strong and durable bonds [23]. Peel ply (a release fabric placed over the layup and cured with the part) can be incorporated into the surface of thermoset composites during manufacture and removed prior to bonding (the peel ply must be free of contaminants and release agents), while for thermoplastic composites, corona discharge and plasma treatments can be effective methods to increase their surface free energies [24]. Moreover, in recent years, adhesive suppliers have launched modern adhesives that can work with oily or uncleaned material surfaces.

Several review articles are available in literature, which have summarised the past research efforts on the effects of different types of surface treatments on composite adhesive joints [19, 22, 25–27]. It is concluded that surface treatments prior to the application of adhesives are recommended in order to achieve maximum bond strength and long term durability. By increasing surface free energy, increasing surface roughness and altering surface chemistry, a more intimate bond can be formed, which allows for an increase in strength and durability. The particular surface treatment applied will depend on the requirements of the bond and the service conditions. The adherend type and geometry and production constraints, as well as cost considerations, may also be relevant factors when selecting a treatment process. In addition, legislative drivers strive to reduce the environmental impact and operator health and safety risks (restrictions have been introduced with respect to the use of certain chemical substances for surface treatments) which promote the need for more environmentally-friendly surface treatments.

8.2.2 Effect of Joint Configuration and Failure Mode

Common joint configurations used in composite connections include single-lap joints (SLJs), double-lap joints, scarf joints, and stepped-lap joints (see Figure 8.1). Many other configurations exist, with their specific benefits, such as peel joints, the joggle-lap joints and L-section joints among others. However, the SLJ is the most common joint configuration used and studied in the literature mainly due to its simplicity and efficiency [7]. On the other hand, one of the problems associated with this joint is the fact that the stress distribution (shear and peel) is concentrated at the ends of the overlap. This effect is further amplified when dissimilar (composite/metallic) materials are bonded, owing to the stress concentration asymmetry as a function of adherend stiffness (higher rotation under peel loading). This might lead to joint premature failures at the ends of the overlap. Therefore, it is important to ensure that the joint geometry is selected to avoid or minimise undesirable stresses that might cause premature failure of the adhesive joint.

It is well established in the literature that the joint failure mode is extremely important for the ultimate load carrying capacity of a bonded joint. When the failure mode is well reported, it can give insight into the crack propagation, stress concentrations and significant material properties that affect the joint performance [28]. Typical failure modes for fibre reinforced polymer (FRP) composite joints can be seen in Figure 8.2.

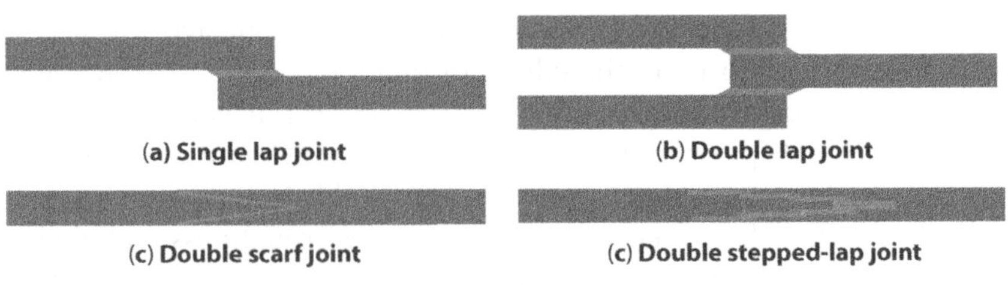

Figure 8.1 Typical joint configurations used in composite connections.

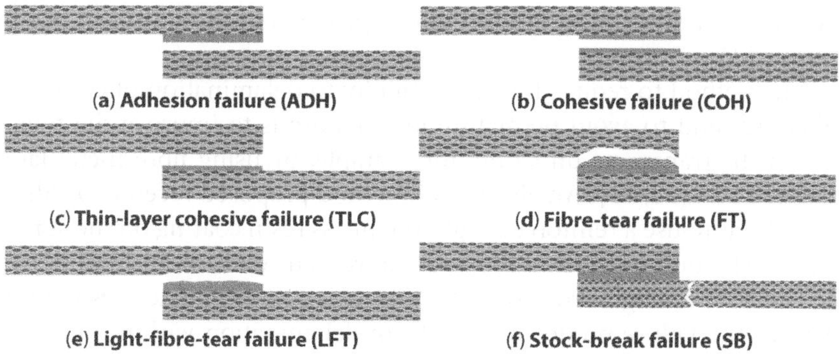

Figure 8.2 Typical failure modes in bonded joints of FRP composite adherends.

The type of failure is governed by the adherend and adhesive properties, joint geometry as well as the adherend/adhesive interface.

In general, cohesive failure within the adhesive or one of the adherends is the preferred type of failure as it means that the maximum strength of the materials in the joint has been reached. One of the issues in bonding composite adherends is the delamination failure (fibre-tear failure-FT). This is particularly true if the composite adherends are made of layers with different fibre orientations. It is well known that in the fibre direction, unidirectional composites can be very strong and stiff, whereas the transverse and shear properties are much lower. Thus, in order to avoid premature failure of the composite adherends, it is recommended to have the outer layers of the adherend with a direction parallel to the loading direction to avoid intralaminar failure of these layers (i.e., angle oriented fibres will tend to generate cracks that enter into the interlaminar space of the composite) [11]. Close to the interface failures such as, thin layer cohesive (TLC) and light fibre-tear (LFT) failures are also common failure modes in composite bonded connections [29, 30]. The TLC failure is characterized by crack propagation very close to the adherend/adhesive interface, where one side of the joint will present a thick adhesive layer while the other will show a light adhesive dusting with few delaminated fibres. On the other hand, the LFT failure type is characterized by the entirety of the adhesive layer on one side of the joint while the other presents some interface surface resin removal and the tearing of a few fibres. In general, TLC and LFT failure modes are caused by the peel stress loading (initiating at the overlap edges) in the thickness direction of the composite adherend, where the mechanical strength of the adherend is the weakest. It was stated that the maximum joint strength can be achieved when the cohesive failure of the adhesive and the delamination failure of the composite adherends occur simultaneously [31].

One method to reduce the susceptibility to delamination of composite adherends and to avoid premature joint failure is to improve the toughness in the transverse direction, for example, by using fibre-metal laminates (FMLs) [32, 33]. An alternative method proposed recently by Shang *et al.* [34] is to use a reinforced high toughness resin coating on the surface of a CFRP composite. This method showed an increase of 22% in average lap shear strength, compared to SLJs with non-toughened resin on the surface. The failure mode changed from delamination within the adherends as seen for the non-toughened to cohesive failure in the adhesive for the surface-toughened adherends. Another way to avoid delamination of composite adherends explored in the literature is to reinforce the adherends in the transverse direction. The transverse reinforcements in general increase the fracture toughness of the composite adherend (modes I and II) [35, 36]. However, the application of transversely reinforced composite adherends in adhesive joints is still limited. Bogdanovich *et al.* [37] studied co-cured and adhesively bonded joints of composite adherends reinforced with non-crimp 3D orthogonal woven glass fibre fabrics. Stitching the co-cured single-lap joints resulted in significant increase in joint strength. Moreover, the delamination failure was not observed within any of the 3D woven composite adherends owing to the effect of the z-binders, even in relatively low weight fraction (~2%).

The failure mode of adhesive joints with composite adherends is also influenced by the stacking sequence and fiber orientation of the different plies [38, 39]. For instance, Ozel *et al.* [40] stated that the lap shear strength of composite joints can vary up to 120% by just varying the composite layup. Kairouz and Matthews [41] tested SLJs with cross-ply adherends and also state that the surface layer orientation has a large influence on the failure mode and strength of these joints.

Temperature and moisture have higher influences on the failure mode of the bonded composite joints compared to the other parameters previously mentioned (e.g. adherend and adhesive properties, joint geometry, surface preparation) [7]. Several authors observed the changes in failure mode after ageing [42–44]. For example, Heshmati *et al.* [44] studied CFRP/steel and GFRP/steel double-lap joints, subjected to a number of environmental conditions (distilled water at 20 °C and 45 °C, immersion in salt water at 20 °C and 45 °C and exposure to 95% RH at 45 °C) for up to three years. They showed that the cohesive failure mode changed to a number of other failure modes upon ageing (see Figure 8.3). Therefore, understanding the

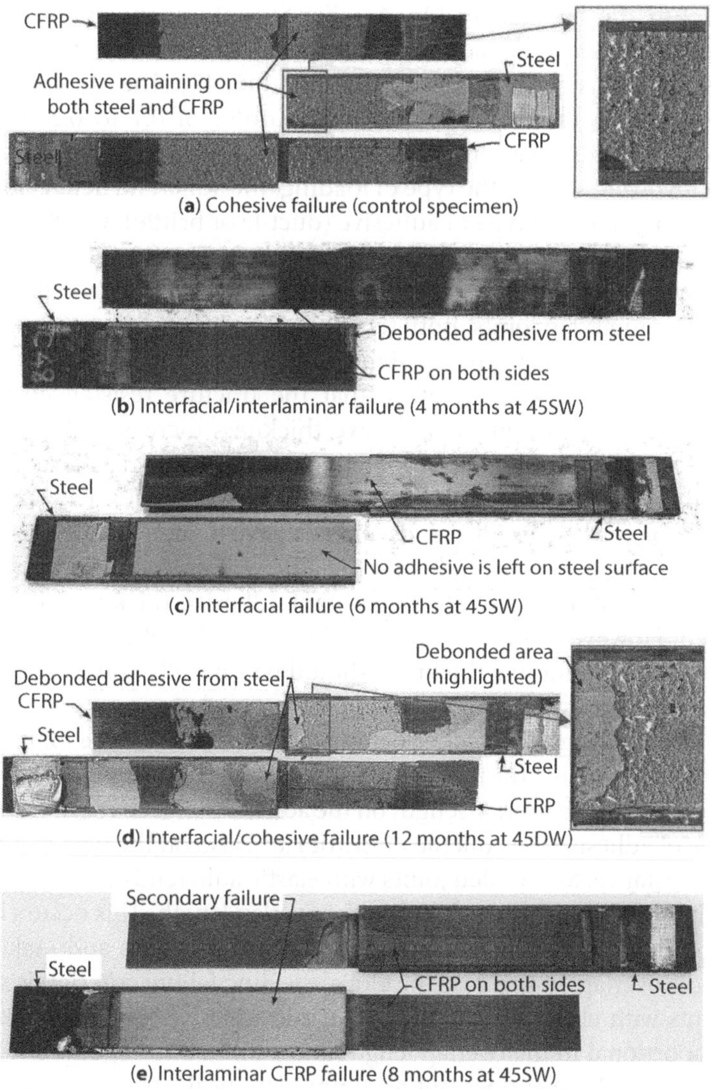

Figure 8.3 Examples of fracture surfaces and failure modes of CFRP/steel specimens. DW: distilled water, and SW: salt water. Reproduced with permission from [44].

failure mode of composite bonded joints is fundamental for the optimum design of a composite structure.

An important joint geometry parameter that affects the adhesively bonded joints performance is the adhesive thickness [45]. Several studies have examined the influence of adhesive thickness on joint strength and fracture mechanics tests [45–49]. For SLJs, it was shown that the joint strength decreases as the adhesive thickness increases, with the exception of SLJs bonded with elastomeric adhesives [46, 47]. The highest strength is obtained for adhesive thickness of the order of 0.1 to 0.5 mm [50]. Moreover, this observation is not generally applicable as there are other factors involved, such as the type of loading, the adherend behaviour (elastic or plastic), and the type of adhesive (ductile or brittle), which can modify the behaviour of joints as their thickness is varied [45]. Li *et al.* [51] studied the stress and strain distributions across the adhesive thickness in composite single-lap joints. Both tensile peel and shear stresses at the bond free-ends were shown to change significantly across the adhesive thickness.

It was shown in the literature that the fracture resistance increases up to a maximum when the adhesive thickness increases [45]. This was explained in terms of the development of the adhesive plastic zone and at higher thickness a plateau value is obtained. However, only a few results are available for composite adherends and adhesive thickness more than 2 mm [48, 52]. In the aerospace industry, the adhesive used to bond composites is generally a film, which allows constant adhesive thickness to be easily achieved. However, in many other applications, the adhesive is applied as a paste, and the adhesive thickness should be carefully controlled to avoid variations of adhesive thickness along the bonded area.

The overlap length is another parameter that can affect the joint strength [53]. While increasing the joint width increases the joint strength proportionally, the effect of the overlap length on the adhesive joint strength depends on the type of adhesive (i.e., ductile or brittle) and also on the type of adherend [4]. For instance, for bonded joints with elastic adherends and ductile adhesives, the strength is proportional to the overlap length. This occurs because ductile adhesives deform plastically as the load increases and make use of the whole overlap. In this case the global yielding failure criterion is suitable. For joints with elastic adherends and brittle adhesives, the joint strength is not proportional to the overlap length and a limited strength is attained. In this case the adhesive does not accommodate peak stresses at the ends of the overlap and failure is governed by these peaks (i.e., the adhesive is not capable to efficiently transfer the stresses across the entire overlap area). On the other hand, for composite adherends the overlap effect on the strength of the joints is mainly dictated by the composite transverse strength [54]. For joints with long overlaps, delamination is one of the main failure modes and performance limiting factors of adhesive joints with composite adherends [55]. Several researchers have conducted investigations to study the effect of overlap length on the behaviour of the adhesively bonded joints of composites [2, 54, 56–58]. For example, Neto *et al.* [54] studied the effect of different

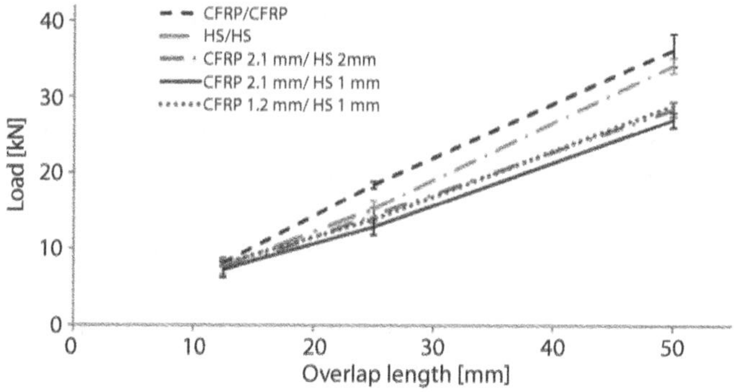

Figure 8.4 Average failure loads of SLJs (similar: hard steel and CFRP and dissimilar joints: CFRP/HS with different adherend thicknesses (1 and 2 for HS and 2.1 and 1.2 for CFRP) as a function of overlap length (12.5, 25, 50mm) and material (HS: hard steel; CFRP: carbon fibre reinforced polymer) [2].

overlap lengths on composite SLJs strength. They found that the failure load reached a plateau value when the overlap length was longer than 10 mm due to the delamination of the composite adherends. Seong *et al.* [56], studied the effects of overlap length, adherend thickness, and material type on the failure load and failure mode of composite-to-aluminium SLJs. They found that the bond strengths of the tested joints were lower than the metal-to-metal bonded joint strength and that the failure mode of all joints tested was delamination of the composite adherend. They also found a limiting value of the overlap length (30 mm) above which the joint strength was practically constant due to the limited ductility of the adhesive. It was concluded that high efficiency might not be obtained when the ratio of overlap length to the width of SLJs is much larger than 1.

Banea *et al.* [2] performed an experimental and numerical study on similar and multi-material (high strength steel, low strength steel and CFRP composite) SLJs. It was shown that for relatively short overlaps in SLJs bonded with modern tough structural adhesives, failure is dominated by adhesive global yielding and the influence of geometry and/or material on joint strength is not significant. Increasing the overlap length increases the load-bearing capacity of the joints (see Figure 8.4).

8.2.3 Effect of Mechanical Properties of Adhesive and Adherend Materials

The adhesive and adherend material properties influence the bonded joint performance and its effect should be taken into consideration in the design of composite adhesive joints [3]. It was shown in the literature that, in general, a strong and stiff adhesive will withstand higher stresses but will severely increase the stress concentrations at the joint edges, while a ductile adhesive will distribute more evenly the stresses along the bonded area, but will withstand lower stresses before failure as it is generally a less resistant adhesive [7]. Ductile adhesives significantly delay crack initiation and final failures of composite bonded joints, thus increasing the joints' strength [59, 60].

As has been previously discussed in this chapter (section 8.2.1), delamination is the main problem in joining composites and high adhesive toughness will directly influence the transverse stresses in the bonded adherends and this is directly linked to the initiation of intralaminar delamination failures. Thus, careful adhesive selection for composite connections is fundamental.

The properties of structural adhesives can vary greatly and an appropriate selection is essential for a proper joint design. Some typical mechanical properties values for different types of adhesives used for composite connections are presented in Table 8.1.

The adherend material also influences the adhesively bonded joint performance [4, 11]. It is well known that composite adherends have relatively low transverse strength and shear stiffness compared to their in-plane material properties. Therefore, it is important to understand the mechanisms of failure involved for each specific material, both adhesive and adherend, and how they are affected by these properties [11]. The effect of the adhesive and adherend materials on shear strength of SLJs was evaluated in several studies [2, 4, 11]. Most of the published work examines the influence of adherend stiffness on the joint strength and, in general, increasing the adherend stiffness has shown to improve the joint strength [7]. The joint strength can be improved by local geometry modification of adherends or adhesive [67]. There are several techniques available that aim at improving adherend/adhesive properties as well as local/global stress distributions (see Figure 8.5). In general, the goal of these techniques is to ultimately eliminate delamination failures and maximize efficiency through proper joint design. Local geometry optimization through taper and spew fillet shaping represents an effective way to reduce the high peak peel stresses at the bondline edges (i.e., smoother stress transfer at the material discontinuities). However, the overall joint architecture must be designed carefully so that both local and global modifications complement each other.

Through-the-thickness reinforcement is one of the most effective reinforcing methods that can be applied locally in critical areas in the composite structure such as Z-pin [68–72] and stitching [37, 73, 74] (see Figure 8.5c).

Table 8.1 Examples of typical adhesives used to bond composites with mechanical properties values.

Adhesive/reference(s)	Type	Tensile Modulus MPa	Tensile Strength MPa	Tensile Strain %
Betamate ™2096 [29, 30]	Epoxy	1600 ± 152	34.27 ± 1.52	8.04 ± 0.39
Betamate™2098 [61]	Epoxy	930	18	56
DP460 [40]	Epoxy	2077	-	-
FM73m [56, 62]	Epoxy	2800	43	-
Araldite AV138 [54]	Epoxy	4890 ± 810	39.45 ± 3.18	1.21 ± 0.10
SikaForce®7888 [2]	Polyurethane	2530 ± 16	31.12 ± 1.17	46 ± 7
SikaPower®4588 [4]	Epoxy-polyurethane	2000 ± 2	26.30 ± 0.54	5.1 ±0.03
BETAFORCE 9050L[48]	Polyurethane	300	18	60
EP21LV [60]	Epoxy	800	45	0.5
FM300 K [63]	Epoxy	3100	58	-
ITW Plexus MA310 [64]	Methacrylate	1860 ± 210	31.3 ± 1.01	3.4 ± 0.8
3M SA9820 [65]	Epoxy	1944.2	41.1	3.7
Loctite 9466 [66]	Epoxy	-	32	-

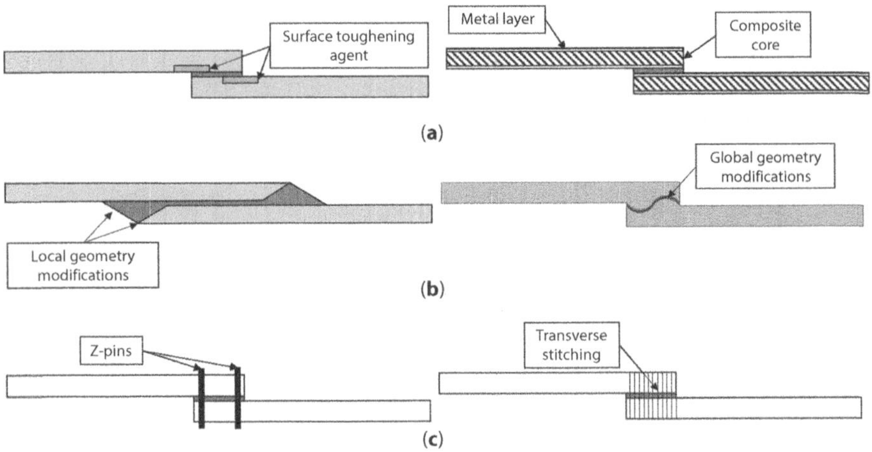

Figure 8.5 Schematic examples of the main modification techniques: (a) adherend surface modification, (b) adherend/adhesive geometry modification, and (c) transverse toughening.

Sarantinos *et al.* [75] have recently reviewed through-the-thickness 3D reinforcement methods and concluded that this technology was not explored in all loading or environmental conditions. Also, the effects of the various designs and manufacturing parameters need to be further investigated.

Nevertheless, most of the techniques to improve the joint strength investigated in the literature come with an increase in the complexity of the manufacturing process, which will increase the cost of the final products.

To summarise, the transverse properties of the composite adherend, the elastic properties of adhesive and adherend, adherend thickness, and loading mode ultimately govern the failure mode and bonded joint efficiency. If the stress in a composite adherend exceeds its transverse strength, delamination failure within the composite will initiate, and propagate interlaminarly as a function of its toughness. These stresses will increase with loading mode mixicity, high adherend axial modulus, low adherend thickness and high adhesive toughness [76]. Therefore, tailoring the properties of the adherend and adhesive (elastic properties in the in-plane and transverse directions as well as careful geometry selection) is crucial for maximizing composite bonded joint efficiency and decreasing or eliminating delamination failures.

8.2.4 Effect of the Environmental Conditions

Environmental considerations are a key factor in the design of bonded composite structures, particularly for load-bearing structures. For instance, the adhesives used in aircraft structures may be exposed to extreme temperatures, ranging from −50°C to 200°C (or higher). In tropical environments, high temperatures are accompanied by high humidity (almost 100% relative humidity under certain circumstances). Harsh environments are not limited just to aerospace applications. Automotive components will often be exposed to harsh environmental conditions (high temperatures and humidity) and chemical agents including salts and other chemicals used for melting ice, and air pollutants (ozone, sulphur dioxide and other smog constituents). Long-term performance and uncertainty relating to environmental durability still represent a critical barrier to the wide application of composite bonded joints in structural applications. Thus, environmental factors such as temperature and moisture should be well understood and studied as they directly influence the performance of the joints [77, 78].

It was shown in the literature that a prolonged exposure or even short term exposure to elevated temperatures will often produce irreversible chemical and physical changes within adhesives. As the temperature increases, the bond strength decreases [46, 53, 79]. Also, the moisture absorbed in a polymeric material can lead to a wide range of effects, both reversible and irreversible, including plasticization, swelling, and degradation. At temperatures below the glass transition temperature (T_g), polymer property reduction is reversible upon dehydration, whereas above (T_g), the properties are permanently degraded [77].

The presence of moisture in adhesive joints may not only weaken the physical and chemical properties of the adhesive itself but also the interface between the adhesive and the substrate. For composite bonded connections exposed to humid environments the mechanisms of degradation are quite different from adhesively bonded metal joints. Unlike metals, the work of adhesion for composite to epoxy joints remains positive in the presence of water [6], thus diminishes the likelihood of interfacial failure on ageing. In addition, the composite adherend will absorb water, which can affect the kinetics of water absorption into the adhesive. Temperature and moisture can also influence the mechanical properties of the composite matrix and the interface between fibre and matrix may be weakened in the presence of moisture [7].

Various studies were employed on the effects of various environments on some adhesive properties [77], but it is still necessary to address the performance of specific adherend-adhesive combinations and to analyse the environmental, fatigue, and fracture studies of bonded systems simultaneously. For example, it is known that moisture absorption results in varying degrees of plasticization, strength loss, and increased ductility of some epoxy adhesives. However, the effect of moisture on the fatigue and fracture properties of bonded joints employing these adhesives is still not fully understood. In addition, since adhesive joints are systems comprised of adherends, adhesives, and interphase regions, the performance of each of these components may strongly affect the performance of the joint. Thus, general knowledge of the behaviour of adhesives exposed to various environments must be supplemented by knowledge of the behaviour of specific bonded systems. Abdel-Monsef *et al.* [43] studied the influence of temperature and ageing on the properties of composite bonded joints. Double Cantilever Beam (DCB) and End Load Split (ELS) specimens (Mode I and mode II) under various temperatures (−55 °C, RT and 80 °C) and ageing conditions (wet-aged and non-aged) for four years were tested. It was found that the environmental factors affect the properties of the composite bonded joints.

The influence of environmental aspects has specific relevance for multi-material structures (e.g., composite/metal), where components with different reactions to the same environmental conditions might significantly alter the behaviour of the structure as a whole [6, 7, 44, 80]. This requires a renewed investigation of ageing effects, as well as a survey on the influence of environmental conditions like temperature, humidity, etc., to which these materials may be more sensitive.

8.3 Recent Developments and Trends

In recent days, one aspect that becomes more and more important is the fact that the materials can be recycled and repaired. In both cases, it is necessary to separate the joint into the bonded components. When it comes to recycling this separation is necessary so that the different materials can be reused on a qualitatively high level. Debonding-on-demand techniques have been developed for this purpose [81, 82]. For example adhesives modified with thermally expandable particles can be dismounted in a few seconds [61, 83]. Smart adhesives with self-healing properties are also studied as these can improve the durability of the structures and also are more economical when compared to repair of damaged structures [84].

Another trend is the use of sustainable composites which is expected to increase in the coming years. Cost and weight of vehicles can be partially reduced when natural fiber-reinforced composites (NFRCs) replace traditional glass fiber composites and aluminium in various components. Composites made of renewable materials have been used in interior and exterior body parts of cars, as well as trim parts in dashboards, door panels, seat cushions, back rests, and cabin linings [85, 86]. One example is the BMW 7 Series Sedan that uses natural fiber mats and a thermosetting acrylic copolymer for the lower door panel [87]. The continually growing demands for lightweight and fuel-efficient vehicles will further push the growth of NFRCs in the automotive market. Several researchers studied the replacement of GFRPs by natural plant fibers (e.g., jute, flax, hemp, sisal, ramie and curauá) reinforced composites [88]. Therefore, it is important to study their joining behaviour in order to utilize the full potential of natural fibre- reinforced composites [29, 30].

Finally, another area currently developing fast is the application of fibre reinforcements in additive manufacturing (AM, 3D printing) [89, 90]. 3-D printing is an emerging technology that will impact the future of composite structures as it allows the fabrication of complex geometries without the need for expensive tooling and moulds. The use of fibres as filament reinforcement presents many advantages. However, significant challenges remain to be solved such as: fabricating composite filaments, difficult quality control due to voids and porosities, and fibre orientation among others. Nevertheless, due to the limitations of 3D printers to produce large parts, the parts often have to be printed as several separate components and further joined together to obtain the final printed part. In addition, 3D printing can be used to produce only the most complex parts, which can be further combined with simple, non-printed parts from other materials to make the final product [91]. Thus, all these issues need to be addressed in future research. Nanotechnology is also a viable stand-alone technology or possibly integrated with 3-D printing and, as a result, exciting innovations in this area are sure to follow.

All these advancements open up new exciting opportunities for development in this field in the future.

8.4 Summary

The application of composite materials in industry is considered as one of the key technologies for significant weight reduction and higher specific strength and stiffness. Adhesive bonding is a viable technique for joining a wide range of materials including FRP composites. However, there are still some issues that need to be addressed, among which the following stand out: uncertainty about long-term behaviour, joint strength in severe environments, and recyclability issues. A lack of suitable material models and failure criteria has resulted in a tendency to "overdesign" composite structures. Safety considerations often require that adhesively bonded composite structures, particularly those employed in primary load-bearing applications, include mechanical fasteners (e.g. bolts) as an additional safety precaution. These practices result in heavier and more costly components. The development of reliable design and predictive methodologies can be expected to result in more efficient use of composite materials and adhesives.

References

1. M. D. Banea and L. F. M. da Silva, Adhesively bonded joints in composite materials: An overview, *Proc. Institution of Mechanical Engineers, Part L: J. Mater. Design and Applications* 223, 1-18 (2009).
2. M. D. Banea, M. Rosioara, R. J. C. Carbas, and L. F. M. da Silva, Multi-material adhesive joints for automotive industry, *Composites Part B.* 151, 71-77 (2018).
3. M. D. Banea, L. F. M. da Silva, and R. D. S. G. Campilho, Principles of adhesive bonding, in: *Joining of Polymer-Metal Hybrid Structures: Principles and Applications*, S.T. Amancio-Filho and L.-A. Blaga (Eds) pp. 3-27, John Wiley & Sons (2018).
4. M. D. Banea, L. F. M. da Silva, R. Carbas, and R. D. S. G. Campilho, Effect of material on the mechanical behaviour of adhesive joints for the automotive industry, *J. Adhesion ScIntl.. Technol.* 31, 663-676 (2017).
5. R. D. Adams, J. Comyn, and W. C. Wake, *Structural Adhesive Joints in Engineering.* 2nd ed. Chapman & Hall, London, UK (1997).
6. S. Budhe, M. D. Banea, and S. de Barros, Bonded repair of composite structures in aerospace application: a review on environmental issues, *Applied Adhesion Sci.*, 6, (2018).
7. S. Budhe, M. D. Banea, S. de Barros, and L. F. M. da Silva, An updated review of adhesively bonded joints in composite materials, *Intl.. J. Adhesion Adhesives* 72, 30-42 (2017).
8. M. A. Wahab (Ed), *Joining Composites with Adhesives: Theory and Applications*, DEStech Publications,. Lancaster, PA, USA (2015).
9. L. F. M. da Silva, A. Öchsner, and R. D. Adams (Eds), *Handbook of Adhesion Technology*, 2nd ed., Springer, Heidelberg (2018).
10. M. Davis and D. Bond, Principles and practices of adhesive bonded structural joints and repairs, *Intl.. J. Adhesion Adhesives* 19, 91-105 (1999).
11. M. D. Banea, Influence of adherend properties on the strength of adhesively bonded joints, *MRS Bulletin* 44, 625-629 (2019).
12. R. D. Adams, Strength predictions for lap joints, especially with composite adherends. a review, *J. Adhesion* 30, 219-242 (1989).
13. M. D. Banea, Dissimilar joining of PMCs to metals – adhesive bonding, in: *Encyclopedia of Materials: Composites*, D. Brabazon, (Ed)., pp. 324-333, Elsevier, Oxford (2021).

14. A. Purabgola, S. Rastogi, G. Sharma, and B. Kandasubramanian, Surface preparation for structural adhesive joints, in: *Structural Adhesive Joints: Design, Analysis and Testing*, K.L. Mittal and S. K. Panigrahi (Eds.), pp. 1-34, Wiley-Scrivener, Beverly, MA (2020).
15. A. Baldan, Adhesively-bonded joints in metallic alloys, polymers and composite materials: Mechanical and environmental durability performance, *J. Mater. Sci.*, 39, 4729-4797 (2004).
16. M. S. Islam, L. Tong, and P. J. Falzon, Influence of metal surface preparation on its surface profile, contact angle, surface energy and adhesion with glass fibre prepreg, *Intl. J. Adhesion Adhesives* 51, 32-41 (2014).
17. H. M. S. Iqbal, S. Bhowmik, and R. Benedictus, Surface modification of high performance polymers by atmospheric pressure plasma and failure mechanism of adhesive bonded joints, *Intl. J. Adhesion Adhesives* 30, 418-424 (2010).
18. L. F. M. da Silva, R. J. C. Carbas, G. W. Critchlow, M. A. V. Figueiredo, and K. Brown, Effect of material, geometry, surface treatment and environment on the shear strength of single lap joints, *Intl. J. Adhesion Adhesives* 29, 621-632 (2009).
19. M. Kanerva and O. Saarela, The peel ply surface treatment for adhesive bonding of composites: A review, *Intl. J. Adhesion Adhesives* 43, 60-69 (2013).
20. N. Encinas, B. R. Oakley, M. A. Belcher, K. Y. Blohowiak, R. G. Dillingham, J. Abenojar, and M. A. Martínez, Surface modification of aircraft used composites for adhesive bonding, *Intl,. J. Adhesion Adhesives* 50, 157-163 (2014).
21. T. Sinmazçelik, E. Avcu, M. Ö. Bora, and O. Çoban, A review: Fibre metal laminates, background, bonding types and applied test methods, *Materials & Design* 32, 3671-3685 (2011).
22. A. Baldan, Adhesively-bonded joints and repairs in metallic alloys, polymers and composite materials: Adhesives, adhesion theories and surface pretreatment, *J. Mater. SclIntl..* 39, 1-49 (2004).
23. P. Molitor, V. Barron, and T. Young, Surface treatment of titanium for adhesive bonding to polymer composites: a review, *Intl.. J. Adhesion Adhesives* 21, 129-136 (2001).
24. G. Scarselli, D. Quan, N. Murphy, B. Deegan, D. Dowling, and A. Ivankovic, Adhesion improvement of thermoplastics-based composites by atmospheric plasma and UV treatments, *Applied Composite Materials* 28, 71-89 (2021).
25. A. Yudhanto, M. Alfano, and G. Lubineau, Surface preparation strategies in secondary bonded thermoset-based composite materials: A review, *Composites Part A* 147, 106443 (2021).
26. S. Deng, L. Djukic, R. Paton, and L. Ye, Thermoplastic–epoxy interactions and their potential applications in joining composite structures – A review, *Composites Part A* 68, 121-132 (2015).
27. J. R. J. Wingfield, Treatment of composite surfaces for adhesive bonding, *Intl. J. Adhesion Adhesives* 13, 151-156 (1993).

28. Z. Benzait and L. Trabzon, A review of recent research on materials used in polymer–matrix composites for body armor application, *J. Composite Mater.* 52, 3241-3263 (2018).
29. H. F. M. de Queiroz, M. D. Banea, and D. K. K. Cavalcanti, Experimental analysis of adhesively bonded joints in synthetic- and natural fibre-reinforced polymer composites, *J. Composite .Mater.* 54, 1245-1255 (2020).
30. H. F. M. de Queiroz, M. D. Banea, and D. K. K. Cavalcanti, Adhesively bonded joints of jute, glass and hybrid jute/glass fibre-reinforced polymer composites for automotive industry, *Applied Adhesion Science* 9 (2021).
31. K.-S. Kim, Y.-M. Yi, G.-R. Cho, and C.-G. Kim, Failure prediction and strength improvement of uni-directional composite single lap bonded joints, *Composite Structures* 82, 513-520 (2008).
32. L. B. Vogelesang and A. Vlot, Development of fibre metal laminates for advanced aerospace structures, *J. Mater. Processing Technol* 103, 1-5 (2000).
33. S. Bano, S. Fida, and A. Israr, Design modification of lap joint of fiber metal laminates (CARALL), *Intl. J Mater Metall Eng* 11, 716-721 (2017).
34. X. Shang, E. A. S. Marques, J. J. M. Machado, R. J. C. Carbas, D. Jiang, and L. F. M. da Silva, A strategy to reduce delamination of adhesive joints with composite substrates, *Proc. Institution of Mechanical Engineers, Part L: J. Mater. Design and Applications* 233, 521-530 (2019).
35. R. B. Ladani, K. Pingkarawat, A. T. T. Nguyen, C. H. Wang, and A. P. Mouritz, Delamination toughening and healing performance of woven composites with hybrid z-fibre reinforcement, *Composites Part A* 110, 258-267 (2018).
36. C. Sun, P. Jia, C. Chen, A. Moradi, J. Zhou, M. Al Teneiji, W. J. Cantwell, and Z. W. Guan, The effect of carbon fibre stitching on the tensile behaviour of secondary bonded single- and double-lap composite joints, *Composite Structures* 265, 113774 (2021).
37. A. E. Bogdanovich, M. Dannemann, J. Döll, T. Leschik, J. N. Singletary, and W. A. Hufenbach, Experimental study of joining thick composites reinforced with non-crimp 3D orthogonal woven E-glass fabrics, *Composites Part A* 42, 896-905 (2011).
38. G. Meneghetti, M. Quaresimin, and M. Ricotta, Influence of the interface ply orientation on the fatigue behaviour of bonded joints in composite materials, *Intl. J. Fatigue* 32, 82-93 (2010).
39. R. Hazimeh, G. Challita, K. Khalil, and R. Othman, Experimental investigation of the influence of substrates' fibers orientations on the impact response of composite double-lap joints, *Composite Structures* 134, 82-89 (2015).
40. A. Ozel, B. Yazici, S. Akpinar, M. D. Aydin, and Ş. Temiz, A study on the strength of adhesively bonded joints with different adherends, *Composites Part B* 62, 167-174 (2014).
41. K. C. Kairouz and F. L. Matthews, Strength and failure modes of bonded single lap joints between cross-ply adherends, *Composites* 24, 475-484 (1993).
42. J. J. M. Machado, E. A. S. Marques, A. Q. Barbosa, and L. F. M. da Silva, Influence of hygrothermal aging on the quasi-static and impact behavior of

single lap joints using CFRP and aluminum substrates, *Mechanics Advanced Mater. and Structures* 28, 1377-1388 (2021).
43. S. Abdel-Monsef, J. Renart, L. Carreras, P. Maimí, and A. Turon, Environmental effects on the cohesive laws of the composite bonded joints, *Composites Part A* 155, 106798 (2022).
44. M. Heshmati, R. Haghani, and M. Al-Emrani, Durability of bonded FRP-to-steel joints: Effects of moisture, de-icing salt solution, temperature and FRP type, *Composites Part B:* 119, 153-167 (2017).
45. M. D. Banea, L. F. M. da Silva, and R. Campilho, The effect of adhesive thickness on the mechanical behavior of a structural polyurethane adhesive, *J. Adhesion* 91, 331-346 (2015).
46. M. D. Banea and L. F. M. da Silva, Static and fatigue behaviour of room temperature vulcanising silicone adhesives for high temperature aerospace applications, *Materialwissenschaft und Werkstofftechnik* 41, 325-335 (2010).
47. M. D. Banea and L. F. M. da Silva, Mechanical characterization of flexible adhesives, *J. Adhesion* 85, 261-285 (2009).
48. Y. Ogawa, K. Naito, K. Harada, and H. Oguma, Evaluation of crack growth behaviors under Mode I static loading for two-part polyurethane adhesives, *Intl. J. Adhesion Adhesives* 117, 103172 (2022).
49. R. D. S. G. Campilho, D. C. Moura, M. D. Banea, and L. F. M. da Silva, Adhesive thickness effects of a ductile adhesive by optical measurement techniques, *Intl. J. Adhesion Adhesives* 57, 125-132 (2015).
50. S. Maggiore, M. D. Banea, P. Stagnaro, and G. Luciano, A review of structural adhesive joints in hybrid joining processes, *Polymers* 13 (2021).
51. G. Li, P. Lee-Sullivan, and R. W. Thring, Nonlinear finite element analysis of stress and strain distributions across the adhesive thickness in composite single-lap joints, *Composite Structures* 46, 395-403 (1999).
52. R. Lopes Fernandes, S. Teixeira de Freitas, M. K. Budzik, J. A. Poulis, and R. Benedictus, From thin to extra-thick adhesive layer thicknesses: fracture of bonded joints under mode I loading conditions, *Engineering Fracture Mechanics* 218 (2019).
53. M. D. Banea and L. F. M. da Silva, The effect of temperature on the mechanical properties of adhesives for the automotive industry, *Proc. Institution of Mechanical Engineers, Part L: J. Mater. Design and Applications* 224, 51-62 (2010).
54. J. A. B. P. Neto, R. D. S. G. Campilho, and L. F. M. da Silva, Parametric study of adhesive joints with composites, *Intl. J. Adhesion Adhesives* 37, 96-101 (2012).
55. J. Li, Y. Yan, T. Zhang, and Z. Liang, Experimental study of adhesively bonded CFRP joints subjected to tensile loads, *Intl. J. Adhesion Adhesives* 57, 95-104 (2015).
56. M.-S. Seong, T.-H. Kim, K.-H. Nguyen, J.-H. Kweon, and J.-H. Choi, A parametric study on the failure of bonded single-lap joints of carbon composite and aluminum, *Composite Structures* 86, 135-145 (2008).

57. S. K. Mazumdar and P. K. Mallick, Static and fatigue behavior of adhesive joints in SMC-SMC composites, *Polym. Composites* 19, 139-146 (1998).
58. H. Luo, Y. Yan, T. Zhang, and Z. Liang, Progressive failure and experimental study of adhesively bonded composite single-lap joints subjected to axial tensile loads, *J. Adhesion SciIntl.. Technol.* 30, 894-914 (2016).
59. J. F. P. Owens and P. Lee-Sullivan, Stiffness behaviour due to fracture in adhesively bonded composite-to-aluminum joints *Intl.* Theoretical model, *Intl. J. Adhesion Adhesives* 20, 39-45 (2000).
60. J. F. P. Owens and P. Lee-Sullivan, Stiffness behaviour due to fracture in adhesively bonded composite-to-aluminum joints: I*INTL..* Experimental, *Intl. J. Adhesion Adhesives* 20, 47-58 (2000).
61. M. D. Banea, L. F. M. da Silva, R. J. C. Carbas, and S. de Barros, Debonding on command of multi-material adhesive joints, *J. Adhesion* 93, 756-770 (2017).
62. J.-H. Kweon, J.-W. Jung, T.-H. Kim, J.-H. Choi, and D.-H. Kim, Failure of carbon composite-to-aluminum joints with combined mechanical fastening and adhesive bonding, *Composite Structures* 75, 192-198 (2006).
63. M.-G. Song, J.-H. Kweon, J.-H. Choi, J.-H. Byun, M.-H. Song, S.-J. Shin, and T.-J. Lee, Effect of manufacturing methods on the shear strength of composite single-lap bonded joints, *Composite Structures* 92, 2194-2202 (2010).
64. R. A. Hunter-Alarcón, A. Vizán, J. Peréz, J. Leyrer, P. Hidalgo, B. Pavez, and L. F. M. da Silva, Effect of the natural aging process on the shear strength of FRP composite single lap joints, *Intl. J. Adhesion Adhesives* 86, 4-12 (2018).
65. N. Wolter, V. C. Beber, M. Brede, and K. Koschek, Adhesively- and hybrid-bonded joining of basalt and carbon fibre reinforced polybenzoxazine-based composites, *Composite Structures* 236, 111800 (2020).
66. O. Sayman, V. Arikan, A. Dogan, *INTL..* F. Soykok, and T. Dogan, Failure analysis of adhesively bonded composite joints under transverse impact and different temperatures, *Composites Part B* 54, 409-414 (2013).
67. X. Shang, E. A. S. Marques, J. J. M. Machado, R. J. C. Carbas, D. Jiang, and L. F. M. da Silva, Review on techniques to improve the strength of adhesive joints with composite adherends, *Composites Part B* 177, 107363 (2019).
68. M. S. Islam and L. Tong, Influence of pinning on static strength of co-cured metal-GFRP hybrid single lap joints, *Composites Part A* 84, 196-208 (2016).
69. P. Chang, A. P. Mouritz, and B. N. Cox, Properties and failure mechanisms of pinned composite lap joints in monotonic and cyclic tension, *Composites Sci. Technol.* 66, 2163-2176 (2006).
70. T. Koh, M. Isa, P. Chang, and A. Mouritz, Improving the structural properties and damage tolerance of bonded composite joints using z-pins, *J. Composite Mater.* 46, 3255-3265 (2012).
71. B. Beylergil, Y. Cunedioglu, and A. Aktas, Experimental and numerical analysis of single lap composite joints with inter-adherend fibers, *Composites Part B* 42, 1885-1896 (2011).

72. A. Arnautov, A. Nasibullins, V. Gribniak, *INTL.*. Blumbergs, and M. Hauka, Experimental characterization of the properties of double-lap needled and hybrid joints of carbon/epoxy composites, *Materials* 8, 7578-7586 (2015).
73. F. Aymerich, R. Onnis, and P. Priolo, Analysis of the fracture behaviour of a stitched single-lap joint, *Composites Part A* 36, 603-614 (2005).
74. J. R. Reeder and E. H. Glaessgen, Debonding of stitched composite joints under static and fatigue loading, *J. Reinforced Plastics Composites* 23, 249-263 (2004).
75. N. Sarantinos, S. Tsantzalis, S. Ucsnik, and V. Kostopoulos, Review of through-the-thickness reinforced composites in joints, *Composite Structures* 229, 111404 (2019).
76. B. R. K. Blackman, A. J. Kinloch, F. S. Rodriguez-Sanchez, and W. S. Teo, The fracture behaviour of adhesively-bonded composite joints: Effects of rate of test and mode of loading, .*Int. J. Solids Structures* 49, 1434-1452 (2012).
77. G. Viana, M. Costa, M. D. Banea, and L. F. M. da Silva, A review on the temperature and moisture degradation of adhesive joints, *Proc. Institution of Mechanical Engineers, Part L: J. Mater. Design and Applications* 231, 488-501 (2017).
78. H. S. Panda, R. Samant, K. L. Mittal and S. K. Panigrahi, Durability aspects of structural adhesive joints, in: *Structural Adhesive Joints: Design, Analysis and Testing,* K.L. Mittal and S. K. Panigrahi (Eds.), pp. 97-134, Wiley-Scrivener, Beverly, MA (2020).
79. M. D. Banea, L. F. M. da Silva, and R. D. S. G. Campilho, Effect of temperature on the shear strength of aluminium single lap bonded joints for high temperature applications, *J. Adhesion Sci. Intl. Technol.* 28, 1367-1381 (2014).
80. M. Heshmati, R. Haghani, and M. Al-Emrani, Environmental durability of adhesively bonded FRP/steel joints in civil engineering applications: State of the art, *Composites Part B* 81, 259-275 (2015).
81. M. D. Banea, Debonding on demand of adhesively bonded joints: A critical review, *Rev. Adhesion Adhesives* 7, 33-50 (2019).
82. M. D. Banea, Debonding of structural adhesive joints, in *Structural Adhesive Joints: Design, Analysis and Testing,* K.L. Mittal and S. K. Panigrahi (Eds.), pp. 135-158, Wiley-Scrivener, Beverly, MA (2020).
83. M. D. Banea, L. F. M. da Silva, and R. J. C. Carbas, Debonding on command of adhesive joints for the automotive industry, *Intl. J. Adhesion Adhesives* 59, 14-20 (2015).
84. M. D. Banea, L. F. M. da Silva, R. D. S. G. Campilho, and C. Sato, Smart adhesive joints: An overview of recent developments, *J. Adhesion* 90, 16-40 (2014).
85. L. Mohammed, M. N. M. Ansari, G. Pua, M. Jawaid, and M. S. Islam, A review on natural fiber reinforced polymer composite and its applications, *Intl. J. Polymer Science,* 243947 (2015).

86. J. Neto, H. Queiroz, R. Aguiar, R. Lima, D. Cavalcanti, and M. D. Banea, A review of recent advances in hybrid natural fiber reinforced polymer composites, *J. Renewable .Mater.* 10, 561-589 (2022).
87. R. Stewart, Automotive composites offer lighter solutions, *Reinforced Plastics* 54, 22-28 (2010).
88. P. Wambua, J. Ivens, and *INTL..* Verpoest, Natural fibres: Can they replace glass in fibre reinforced plastics, *Composites ScIntl.. Technol.* 63, 1259-1264 (2003).
89. X. Wang, M. Jiang, Z. Zhou, J. Gou, and D. Hui, 3D printing of polymer matrix composites: A review and prospective, *Composites Part B* 110, 442-458 (2017).
90. J. Saroia, Y. Wang, Q. Wei, M. Lei, X. Li, Y. Guo, and K. Zhang, A review on 3D printed matrix polymer composites: its potential and future challenges, *J. Advanced Manufacturing Technology* 106, 1695-1721 (2020).
91. D. K. K. Cavalcanti, M. D. Banea, and H. F. M. de Queiroz, Mechanical characterization of bonded joints made of additive manufactured adherends, Annals of "Dunarea de Jos" University of Galati, Fascicle XII, Welding Equipment and Technology 30, 27-33 (2019).

9

Naval Applications of Structural Adhesives

Bikash Chandra Chakraborty

Naval Materials Research Laboratory, Defense Research and Development Organization, Ambernath, India

Abstract

Adhesives for naval vessels are increasingly being used for reduction of weight arising from metal work and for the development of multilayer elements for both load-bearing and vibro-acoustic damping application. Some examples of adhesive application in naval vessels are gaskets of doors, hatches and domes, base frames of machines, joining Fiber Reinforced Plastics (FRP) elements of superstructures, aluminum-FRP bonded panels, acoustic enclosures, underwater acoustic tiles, pipe coatings, etc. Apart from the inherent strength and adhesive bond strength, the naval adhesives are required to be qualified for resistance to seawater, and in some special applications resistance to hydrocarbon oils and dilute mineral acids. Among all evaluations, most common is the diffusion phenomenon by water and electrolyte molecules. This chapter briefly describes different adhesives, especially based on toughened epoxy resin, their seawater ageing, and degradation of mechanical properties with some examples. Special application of adhesive-sealant combination is discussed in brief for both underwater acoustic lining and acid-resistant rubber lining of submarines. Two common diffusion models such as Fickian and dual-Fickian are discussed here with examples of two adhesive plaques, one with neat epoxy and the other with 2.5% nanoclay. Water absorption and deterioration of flexural strengths of two FRP sandwich laminates bonded by two experimental adhesives of epoxy and epoxy nanocomposite are shown here as an example of seawater ageing effect on bonded joints.

***Keywords*:** Adhesive, naval, rubber, epoxy, vibrodamping, acoustic tile, diffusion, flexural strength, Fickian

Email: bikash051954@gmail.com

K.L. Mittal and S.K. Panigrahi (eds.) Structural Adhesives: Properties, Characterization and Applications, (397–444) © 2023 Scrivener Publishing LLC

List of Abbreviations

Abbreviation	Description	Unit
ABS	Acrylonitrile-Butadiene-Styrene	
Al-FRP	Aluminium-Fiber Reinforced Plastic	
CBS	*n*-Cyclohexyl-2-benzothiazolesulfenamide	
CFRP	Carbon Fiber Reinforced Plastic	
CR	Poly(chloroprene) Rubber	
CTPEGA	Carboxyl terminated poly(ethylene glycol) adipate	
DGEBA	Diglycidyl ether of bisphenol A	
DICY	Dicyandiamide	
DMA	Dynamic Mechanical Analysis	
DOPA	3,4-Dihydroxyphenylalanine	
DSC	Differential Scanning Calorimetry	
ENC	Epoxy-Nanoclay	
EPDM	Ethylene Propylene Diene Monomer	
FRP	Fiber Reinforced Plastic	
FTIR	Fourier Transform Infrared Spectroscopy	
GPa	Giga Pascal	10^9 N/m^2
HTPDMS	Hydroxyl terminated poly(dimethylsiloxane)	
Hz	Hertz (unit of frequency)	cycle/s
IPN	Interpenetrating Polymer Network	
MBTS	Mercaptobenzothiazyl disulfide	
MPa	Mega Pascal	10^6 N/m^2
NBR	Acrylonitrile-Butadiene Rubber	
NR	Natural Rubber	
PDMS	Poly(dimethylsiloxane)	
PEPA	Polyethylene polyamine	
PF	Phenol-Formaldehyde	
phr	Parts per hundred of rubber (by weight)	
PMMA	Poly(methyl methacrylate)	
PU	Polyurethane	

PVC	Poly(vinyl chloride)	
RCC	Reinforced Cement Concrete	
RF	Resorcinol-Formaldehyde	
RH	Relative Humidity	
RTV	Room Temperature Vulcanizing (silicone)	
SBR	Styrene-Butadiene Rubber	
SEM	Scanning Electron Microscopy	
SONAR	Sound Navigation and Ranging	
SRB	Sulfate Reducing Bacteria	
TBPF	*tert*-Butyl Phenol-Formaldehyde	
TGMDA	Tetraglycidylmethylene dianiline	
TAST	Thick Adherend Shear Test	
TMTD	Tetramethylthiuram disulfide	
UTS	Ultimate Tensile Strength	MPa
UV	Ultraviolet (ray)	

List of Symbols with Units

Symbol	Description	Unit
b	Thickness of substrate in simplified dual-Fickian expression	m
C	Concentration	moles/liter
D, D_1, D_2	Diffusivity	m^2/s
D_0	Maximum diffusivity	m^2/s
E_A	Activation energy of diffusion	kJ/mole
G	Ratio of % water absorbed at any time to the maximum % water absorbed (by mass)	dimensionless
h	Thickness of substrate in Fickian expression	m
j	A number	dimensionless
l	Half of the substrate thickness in dual-Fickian expression	m
M_t	Water absorbed at time t (by mass)	%

M_∞	Maximum water absorbed (by mass)	%
R	Universal gas constant	kJ/mol.K
t	time	s
T	Temperature	K
W, W_0	W=Weight of substrate after water absorption W_0= Weight of substrate before water absorption	kg
x, y, z	Cartesian axes	
μ	Dynamic viscosity	Pa.s
π	3.1428	dimensionless

9.1 Introduction

Structural adhesives are an important component of marine structures. There are innumerable places onboard where joints are used. In many places, welding, riveting and bolting are not the appropriate solution. In addition, the modern lightweight marine vessels use FRP composites and FRP-aluminium panels bonded with structural adhesives in place of steel. Thus, adhesives and sealants which are basically polymer-based products are increasingly being used in marine industry. However, the adhesives and sealants for marine application must have sufficient and satisfactory ageing against seawater, oils, lubricants, alkali, acid, etc., apart from atmospheric degradation by ultraviolet rays and ozone. In the earliest days of seafaring, natural sealants such as pine resins, natural rosins from trees were applied liberally, requiring regular re-treatments. Later, man-made derivatives such as pitch or tar-like tackifying adhesives from petroleum, coal tar, or some plants, proved to be more resilient. Present-day adhesives are the products of synthetic polymers and polymer nanocomposites, developed through extensive research on their suitability in various service conditions in ships, submarines and off-shore structures.

Many special types of adhesives and sealants are used in naval ships and submarines. Most applications involve joining aluminium, steel and FRP items onboard ships, and most investigations focus on strong, tough thermoset based adhesive bonding, such as epoxy, vinyl ester, phenolic, etc. Various types of laminations, joints, linings, deck covers, protective layers on battery pit of submarines, pipe coatings for vibration damping, outer lining for acoustic stealth measures, vibroacoustic lining inside SONAR compartments, and adhesive joining of FRP structural components are some of the important applications in naval vessels. Some of these are applicable to commercial ships as well. With the development of off-shore structures, mainly for oil rigs and underwater pipelines, more sophisticated and long-lasting adhesives are being used for bonding concrete and FRP. The strengths of adhesive bonding between metals, FRP and polymeric laminates and coatings are similar for naval application as for other engineering systems, with a few additional requirements such as fatigue due to constant vibration of surfaces when the internal machinery like the engine, pumps, motors etc. operate, and in case of outer surfaces, ageing due to marine atmosphere is to be considered. A special case is adhesively-bonded acoustic lining on outer hull of submarines. The adhesive, if not a tough material, would suffer brittle failure in repeated diving of the submarine. The hydrostatic pressure swing (normally 0.01 to 1 MPa) due to depth change acts as a low frequency dynamic force, and hence may cause brittle failure due to low frequency fatigue of the adhesive.

Seawater is an electrolyte (mainly 3.5% NaCl) and is alkaline with an average pH of 8.3. Therefore, it is very corrosive to metals, and the alkalinity accelerates hydrolysis of esters, amides etc. groups if present in adhesives. Diffusion of seawater through the adhesive layer is highly detrimental and careful studies are required to select/develop such adhesives which are less prone to deterioration in saline atmosphere. In special cases like the engine room, the adhesive bond may be affected by hydrocarbon (engine oil) ageing, and in battery pit of submarines by dilute sulfuric acid. Apart from these, the conventional studies on structural adhesives in naval application are definitely the thermal ageing and physical ageing.

9.2 Types of Marine Adhesives

Adhesives used for naval applications are:

A. Flexible Rubber-based Adhesives
   i) Rubber to Rubber: for multilayer lamination
   ii) Metal to Rubber: adhesives for bushes, seals, gaskets, pipe coating, acoustic and vibroacoustic coating, special stealth rubber coating on superstructures.
   iii) Adhesives for special rubber linings of hatches, doors and windows
B. Engineering Thermoset-based Adhesives
   i) Metal to acoustic tile lining
   ii) Metal to vibroacoustic multilayer tiles
   iii) Metal to FRP bonding

iv) FRP to FRP bonding in superstructures
v) Acid-resistant adhesive for battery pit protective lining
vi) Hydrocarbon oil/fuel-resistant adhesive in engine room

9.2.1 Essential Characteristics

The design of adhesives primarily depends on the materials of the surfaces to be joined. For example, when a rubber is to be joined to a metal surface, there is large difference in strains due to thermal or mechanical stresses as the rubber has elastic modulus nearly 1-10 MPa compared to steel 200GPa and aluminium 70GPa. The adhesive must have enough flexibility with good cohesive and adhesion strengths to accommodate shear stress due to disproportionate strain in metal and rubber. The adhesive is generally tough like a typical thermoplastic elastomer with acceptable flexibility for good bond strength with the metal and the elastomer adherends. For rubber-to-rubber bonding, the adhesive must be quite flexible, matching those of the two joining rubber surfaces. On the other hand, for metal-to-metal and FRP to FRP bonding, there can be a high modulus and tough adhesive. All adhesives are essentially polar when metal, FRP and polar rubbers (Nitrile, Neoprene, Polyurethanes, Hypalon) are to be joined. The toughness of all types of adhesives is essential to have a long fatigue life in a naval vessel. For fatigue, the most deteriorating frequency band of hull vibration is quite wide, generally 10 Hz to 2000Hz depending on the propeller, engine, pumps and other rotating machinery. A general study of hull vibration (partly damped) shows vibration intensity of about 125 decibels (dB) in a frequency range of 10-50 Hz in a typical warship, while it is preferred that underwater radiated noise level due to hull vibration must not be more than 75 dB. This necessitates rugged and highly efficient damping treatment of hull internal surfaces, base frames, hydraulics and decks. In all these applications, tough and compatible adhesives and sealants are required. Therefore, the fatigue life of adhesives for internal hull, deck, piping etc. is important. For an underwater hull lining or a small underwater unmanned vehicle, the low frequency dynamic force of sea waves, generally 0.08-0.1 Hz, is important for low frequency fatigue. It might be quite severe in reduction of service life of a bonded joint.

In special cases such as underwater hull lining or multilayer polymeric coating, there can be a combination of two or more adhesives or can be layer-wise different adhesives for different materials. In metal-to-rubber tile bonding, two adhesives are preferred because the metal side adhesive is generally a toughened epoxy composition, and the rubber side is a softer one, particularly rubber-epoxy blended formulation. The pair of such adhesives acts as a tie-coat for metal and rubber surfaces.

For naval application, the selection or development of an adhesive requires a time-dependent study of service life, since frequent replacement or repair in naval vessels are not possible due to operational requirement of the vessel. A schedule of long refit of about 6 years may be considered for minimum service life of an adhesive bond, except in very critical areas like engine room and battery pit (submarine), where the adhesives must have much longer life, beyond 10 years.

9.2.2 Flexible Adhesives

Flexible adhesives are the simplest kind of adhesives used for rubber-to-metal bonding in rubber lining on mild steel/aluminum structures and in rubber-to-rubber bonding in multilayer rubber lining on pipes, floors, vibroacoustic tiles, etc. Generally, a rubber such as styrene-butadiene rubber (SBR), natural rubber (NR), nitrile rubber (NBR) or neoprene rubber (CR) is dissolved in toluene/xylene/methyl ethyl ketone (MEK)/methyl isobutyl ketone (MIBK) or a non-toxic solvent such as ethyl acetate to make a homogeneous viscous liquid along with zinc/magnesium oxide and a resol type phenyl-formaldehyde resin [1–3]. Under ambient temperature, the rubber gets cross-linked due to zinc/magnesium oxide while the resol provides sufficient tackiness to improve bonding. In the adhesive formulation, other typical rubber compounding ingredients such as stearic acid, antioxidants such as 2,6-di-*t*-butyl-4-methylphenol and accelerators such as alkyl thiourea are used. Other ingredients could be according to the requirement of the color, consistency and mechanical properties of the adhesive.

Krishnan *et al.* [1] developed three types of rubber adhesives based on nitrile rubber/phenol-formaldehyde (PF) resin/resorcinol-formaldehyde (RF) resin/para *t*-butyl phenol-formaldehyde (TBPF) resin and compared their adhesive bond peel strengths for joining rubber to mild steel. All three types showed higher strength after prolonged curing for 7 days. The nitrile-RF resin adhesive showed maximum peel strength at some particular composition, while lap shear strength of nitrile-TBPF adhesive was maximum. Poh and Ong [2] experimented with a blend of SBR rubber and natural rubber in different proportions and a PF resin as tackifier for development of a pressure-sensitive adhesive for bonding paper to poly(ethylene terephthalate) sheets and determined T-peel, 90° and 180° peel strengths. The tack was maximum for particular proportions of SBR and PF resin based on total rubber mass. However, all peel strengths (T, 90° and 180°)

were found to be maximum (30-40 N/m) for a different composition than the former. Reactive resol is an alkylated phenol-formaldehyde resin and is generally cured with magnesium oxide. Such cross-linked resol provides high heat resistance and bond strength [3]. Han *et al.* [4] developed a hot vulcanized rubber-to-metal adhesive using poly(vinyl butyral) modified resol and nitrile rubber in methyl isobutyl ketone (MIBK) solvent to obtain about 3.5 kN/m peel strength for an optimum composition. Addition of a silane coupling agent further improved the peel strength. A butyl phenol-formaldehyde (resol) is used for polychloroprene (Neoprene) based adhesive for metal to rubber bonding [5]. Masao *et al.* [6] patented a special adhesive based on a copolymer of carboxylated chloroprene and chloroprene with rosin acid metal salt for bonding metal with rubber, especially natural rubber.

These flexible adhesives based on rubber are not suitable for application where the bonded item is used under water, or may come in contact with acid, alkali or hydrocarbon. The adhesives thus used are made from polar polymers, and seawater actually acts as a plasticizer to such rubbers and weakens the inherent strength and bond strength as well. Rubbers are characterised by large molecules with amorphous structure, low cross-link density and hence high free volume at ambient conditions where diffusion of any gas or liquid is higher than in highly cross-linked thermoset adhesives. The segmental motion in a rubber molecule due to thermal excitation at ambient conditions results in high free volume and greater passage to the small diffusing molecules such as water, electrolyte, oil, etc. The solubility of the diffusing species inside the adhesive matrix also depends on the polarity of the adhesive. However, neoprene is much more resistant to hydrocarbon oil, acid, alkali and seawater compared to nitrile rubber. Therefore, neoprene-based rubber lining and adhesives are more preferred in marine application as a versatile rubber adhesive for rubber to metal bonding in superstructures. However, in many systems inside the ship/submarine, nitrile-phenolic resol combination with zinc/magnesium oxide as curing agent has shown good peel strength of 3-4 kN/m. Job [7] studied the effects of temperature of application and the time of curing of the adhesive on the peel strength of neoprene, nitrile and natural rubber adhesives for rubber-to-rubber joining. The adhesive was in the form of a dough, reinforced by carbon black and contained various tackifying substances like wood rosin, pine tar, phenolic resol, etc. The curing system was a combination of zinc oxide and magnesium oxide, with accelerators NA22 (ethylene thiourea) and thiocarbamide. The peel strength of neoprene-to-neoprene sheets with this adhesive cured at 40°C was about 5 kN/m, and when cured at 70°C, it was about 7 kN/m.

Flexible adhesives based on room temperature vulcanizing silicone (RTV Silicone) and polyurethane (PU) are much better in marine applications. The two main advantages of these adhesives are their superior thermal stability and longer life in atmospheric and underwater ageing. Silicone polymers can be di-alkyl siloxanes, such as poly(dimethyl siloxane), or alkyl aryl siloxanes, such as poly(methyl phenyl siloxane) or di-aryl siloxanes, such as poly(diphenyl siloxane). The aromatic moiety in the backbone imparts stiffness and higher resistance to seawater ingress. However, the tensile strength of a room temperature vulcanized poly(dimethylsiloxane) (PDMS) adhesive /sealant is in the range of 2-4.5 MPa and lap shear bond strength is 0.4-1 MPa. Addition of 30phr (parts per hundred resin) by weight of nanoparticles of silica modified PDMS in an RTV silicone such as PDMS crosslinked by alinilo-methyl-triethoxysilane in presence of a organo-titanium complex as catalyst improved the tensile strength by 1.75 times and elongation-at-break improved by about 2.5 times as reported by Wu *et al.* [8]. The lap shear strength of aluminium-to-aluminium bond with the developed silicone adhesive improved from 0.5 to 1.8 MPa using nanosilica.

A special class of cyclic silicone adhesives was developed by Indulekha *et al.* [9] wherein 2,4,6,8-tetramethyl-2,4,6,8-tetravinylcyclotetrasiloxane and 1,3,5,7-tetramethylcyclotetrasiloxane were used in combination and was self-cured by platinum complex catalyst-assisted hydrosilylation reaction. The molecules of these precursors are 8-membered rings of Si-O bond and are highly stable. The adhesive curing is done at room temperature to obtain tack-free, transparent and strong adhesive. The resulting adhesive bond was excellent with a very high thermal stability (nearly 600°C).

Polyurethanes (PUs) are versatile adhesives because of enormous variety of polyols and combined curing system of isocyanate and amine is possible. Polyether, polyester and caprolactam based polyols are most common, wherein the combinations of alcohols, acids and alkyl oxides/glycols are chosen to balance flexibility, crystallinity, cohesive and adhesion strengths, acid and oil resistance and tolerance to seawater degradation. The chain length and chemistry of the polyol decide the quantity of both urethane and urea moieties in the cross-linked PU structure. Lower contents of urethane and urea groups improve the seawater ageing life by lower hydrolysis rates. Polyether based PUs are better in underwater application as they are less susceptible to hydrolysis compared to polyester and polycaprolactam based PUs. Polyurethane adhesives are very widely used under the trade name "Chemlock" for rubber-to-metal bonding prior to rubber vulcanization. In marine application, special adhesives and sealants made of polyether triols, hydroxyl terminated diene oligomers, and a combination of aromatic diisocyanate and diamines are found to be extremely durable not only in underwater application, but also in sulfuric acid environment inside the battery pit of submarines. A large variation in the above composition is possible by varying the functionality of the polyols, and proportions of two or more diamines and diisocyanates. The system might be of two components, one with polyol-amine mix and the other is the diisocyanate in solvent. Reinforcing fillers like carbon black, fumed silica etc. can be added to enhance strength, and also catalyst like dibutyl tin dilaurate or triethylamine are sometimes required for the secondary hydroxyl groups in polyols, such as poly(oxypropylene) triol, because the reaction of secondary hydroxyl group with isocyanate is quite slow. However, too fast curing could lead to lower bond strength. The normal ambient curing of the adhesive/sealant is for about 7 days to achieve a minimum peel strength of more than 3 kN/m. Fully cured adhesive strip would have tensile strength approximately 7.5 MPa, with % elongation at break about 350% [10]. The adhesive on full curing is a flexible elastomeric material and extremely good for metal-to-rubber bonding, gap filling and for sealing purpose. The seawater and acid resistances of such PU adhesives/sealants are superior to common rubber adhesives. The extents of urethane and urea groups are controlled by the polyol chain length and functionality, so that the PU is less prone to hydrolysis. Comparing the performance of PU adhesive with silicones, both have good underwater life, with silicones having lower mechanical strength but much higher service temperature (500-560°C) [11].

Two special elastomeric adhesives for naval ships and submarines are briefly described below.

9.2.2.1 Bonding Multilayer Rubber Tiles

Naval ships and submarines use vibroacoustic rubber tiles which have multiple layers of rubber sheets bonded together. Most of the rubber sheets might have special cavity design, which is required to isolate underwater sound and also to enhance structural damping of the hull plates inside or around a SONAR compartment. The SONAR system is housed in a seawater-filled compartment and hence the compartment requires seawater-resistant acoustic rubber lining. For the rubber-to-rubber bonding, an acceptable formulation could be neoprene-phenolic resol, because of excellent water resistance of neoprene. The strength of such adhesive bond is no less than 3 kN/m for rubber-to-rubber bond.

The effects of the adhesive layer and interface with successive rubber layers on vibration damping are neglected, because from a general knowledge on viscoelastic damping by rubber tiles, it is known that the thin layer of the adhesive does not alter the stiffness or damping to any significant extent. Moreover, the flexible nature of the cured adhesive does not create any impedance mismatch between the rubber tiles on either side, and hence the propagation of vibration energy through the successive layers is not affected. Neoprene of the adhesive layer has a low loss factor at room temperature than the viscoelastic sheets used in vibrodamping. Therefore, the transfer of vibration energy to successive layers will not be reduced much. This is a unique requirement for maximum utilization of damping by the rubber layers. Nitrile rubber-resol based flexible adhesives are also employed in cases where the rubber layers are made with nitrile-poly(vinyl chloride) (PVC) or nitrile-styrene butadiene rubber (SBR) blended compositions.

9.2.2.2 Bonding Silicone Rubber Gaskets

The polymeric materials intended for use onboard ship or submarine (except for underwater use) require stringent fire ratings and limits of toxic gas emission on burning. Most polymers emit various toxic gases such as hydrocarbons, acetonitrile, hydrocyanic acid vapor, hydrogen halides, carbon monoxide, etc. apart from carbon dioxide [12]. Silicone rubber is preferred for use as gaskets linings on hatches, doors, windows of ships and submarines due to high atmospheric (UV, ozone, etc.) and thermal stability and much superior flame retardancy compared to other rubbers [13–16], although fluoropolymers like Viton and polychloroprene (Neoprene) are also good choice, except for hazard of toxic gas emission in case of fire. Previously, natural rubber was used as it is very cost-effective, has low creep, and can be used for quite long time when protected by antioxidant and antiozonant. However, many synthetic rubbers are now available with better ageing characteristics and longer service life compared to natural rubber. Carbon black reinforcement also enhances the resistance to atmospheric degradation. However, with more and more advancement in this area, silicones have replaced almost all such gaskets because of much superior quality and most importantly almost zero fire hazard in terms of fire retardancy and toxic gas emission. On combustion, the char yield (silica) is also very high, particularly with special polycyclic silicones. Bonding of silicone gasket to the steel hatch door is best achieved by a room temperature curable silicone adhesive, reinforced by nanosilica and carbon black. A fully cured adhesive bond could have sufficient peel strength of 3-4 kN/m and lap shear strength of more than 1.5 MPa when produced as a nanocomposite.

In order to improve both the cohesive and the adhesion strengths of pure silicone adhesives, a new type of epoxy modified silicone polymer is developed [17–20]. The inherent tensile strength of one particular variety of an epoxysilane cured film is about 5-8 MPa with about 10-25% elongation at break, and the pull-off adhesion strength (ASTM D4541-17) for steel-to-rubber is about 4-7 MPa.

It is well established that the adhesion is directly related to the surface energies of polymeric adhesive and the adherend. Mittal discussed various surface energetic criteria against the joint strength [21]. According to his analysis, the adhesion strength almost linearly increases with surface energy. Rath *et al.* developed a PDMS modified epoxy with urethane linkage for a low surface energy coating, which can adhere to a metal surface or epoxy primer surface [22]. The hydroxyl terminated poly(dimethylsiloxane) (HTPDMS) oligomer is capped with a diisocyanate and subsequently the hydroxyl group of the epoxy reacts with the NH-group of the urethane linkage, thus lightly cross-linking the entire chain, resulting in a highly polar and strong material. The cured material has a dynamic modulus of 2-6 MPa depending on silane content and high damping factor (1.0-1.5) at the ambient temperature. Thus, the material is quite suitable as an adhesive for metal to silicone rubber linings and gaskets.

Similarly, a highly branched urethane modified silane polymer has been synthesized by Pandey *et al.* [23]. The diisocyanate capped HTPDMS is reacted with triethanolamine to get highly branched urethane modified silane, which showed rubbery nature with glass transition temperature (T_g) below -50°C, having a tensile strength of 15-20 MPa, and elongation at break about 40-45%. Although the viscoelastic loss factor at T_g was significantly high, the polymer was highly resilient (low loss) at ambient temperature.

The advantages of these types of silane-modified adhesives are that they are not only tough and strong in adhesion, but also superior in flame retardancy, resistant to water diffusion and thermal/UV/ozone stability. However, these adhesives cannot be used in acidic environment, such as in the battery pit of submarines.

9.2.3 Thermoset-Based Marine Adhesives

Engineering thermosets are, in general, highly cross-linked structures synthesized from oligomers chemically cross-linked by a small molecule reactant as cross-linker. This results in densely cross-linked product with high Young's modulus, around 2-5 GPa, similar to engineering thermoplastics. However, strength is also accompanied by brittleness in such thermosets, which is a drawback for fatigue life and possibility of catastrophic failure as well. Toughness is an essential property for the thermoset-based adhesives in marine application, since a ship or submarine is constantly under vibration due to large rotational machines. Thermoset-based adhesives are used onboard ship for joining metal-to-metal, metal-to-rubber tile, pipe coatings, FRP-to-FRP, etc. In recent advancements, stealth superstructures are being made with FRP to reduce radar signatures. In all naval and undersea applications including commercial lightweight boats and off-shore constructions, the use of glass and carbon fiber based FRPs is rapidly increasing [24–28]. In structural construction, FRP-to-FRP joining is critical and use of a strong, tough adhesive is required which can have almost matching elastic modulus and strength of the matrix polymer or the FRP. One way to achieve similar mechanical strength as the FRP is to prepare a glass fiber reinforced adhesive as a dough molding compound (DMC) for joining any configuration using a long and thick bondline. Many applications are on the superstructure of a ship. There are two main applications: to repair cracks in metallic panels of superstructures and to join lightweight structures made of composite or aluminium to the steel hull. The polymeric adhesives being electrically insulating, they isolate dissimilar metals in joining to prevent bimetallic corrosion, such as with steel to aluminium. Since carbon fiber FRPs (CFRPs) are electrically conducting, the glass fiber reinforced adhesives are preferred to provide isolation of metal from CFRP in joints and hence prevent corrosion of the metal.

Thermoset-based adhesives are well documented in literature, including their behaviour to seawater exposure. A few references are cited in this chapter [29–38]. Toughened epoxy resin is most versatile material and needs no styrene or other monomer to crosslink, such as for polyester and vinyl ester resins, and does not have to be cured at high temperature like phenolic resin. Examples of the behavior of toughened epoxy-based structural adhesives in shock, vibration and impact were discussed in detail elsewhere [39].

However, metal-to-rubber and metal-to-composite joints by thermoset-based strong adhesives require special consideration in terms of load transfer and compatibility to the two different surfaces. As an example, a common rubber needs different adhesive systems when used inside the battery compartment than on outer hull. In many applications, a single adhesive is not used, rather a combination of two adhesives is employed, that too on a primer applied metal surface. The reason for two adhesives is simply the different stiffness requirement on each surface, metal and rubber, or composite, or ceramic tiles. A special solvent containing epoxy putty has been designed using plasticizer and suitably pigmented with iron oxide, talc, etc. for adhering to the steel hull. The best performance is obtained when the hull surface is grit blasted to Sa 2½ standard* and primed with an epoxy-red oxide primer coating of about 50 microns [40]. The adhesive also serves as a leveling putty, which provides an even surface to the steel hull. Secondly, the adhesive bonds an acrylate cured unsaturated polyester resin which is grafted onto a rubber sheet. The grafting of polyester on the rubber provides a harder and highly polar (polyester) surface which is a modified rubber surface. The purpose of this modification is to bond the highly polar epoxy putty with this hard polar surface. The polyester is made of succinic or adipic or phthalic acid, maleic anhydride and ethylene glycol, while the crosslinker is a dimethacrylate. Typical bond strength in pull-off test for this system is more than 3.5 MPa.

Two types of vibrodamping sheets of 50/50 Nitrile-PVC blend (colored white and blue) were developed for damping of mild steel structure. These two sheets are intended to be used on steel hulls for constrained layer damping, especially in the engine base frame and deck inside the ship/submarine to reduce the hull vibration by at least 10 dB (90% in terms of vibration power) in the mid frequency range (100-2000Hz). A study was done to bond these two types of damping sheets to mild steel base using various types of adhesives. The steel plate was cleaned and blasted to Sa 2½ standard before application of the adhesive. The dimensions of the steel plate were 25 mm (w) x 2 mm (t) x 180 mm (L), and the rubber sheets were of the same length and width, but 6 mm thick. The length of the adhesive bondline was 45 mm. The test was a 180° peel test, and the result is expressed in force per unit width (kN/m). One of the adhesives was a commercial product by Huntsman CIBA and known as Araldite AW106 with hardener HY953U, which is a flexible but strong epoxy composition [41]. The adhesive has a lap shear strength of about 27 MPa for aluminum-to-aluminium joint, 2.5-7.0 MPa for engineering plastics (PVC, PMMA, ABS etc.) and 18-20 MPa for CFRP-CFRP composite joint. The second adhesive, developed by Ratna *et al.* [42] was a toughened epoxy which is a reaction

* Note: Sa 2½ is a Swedish standard of cleaning a metal surface. It specifies "very thorough blast cleaning" of metal surface by shot blasting or grit blasting before any coating or adhesive application.

product of LY556 epoxy resin and carboxyl terminated poly(ethylene glycol) adipate (CTPEGA) cured with HY953U.

This adhesive has a lap shear strength of 13.5 MPa for aluminium-to-aluminium joint and 2.5 kN/m 180° peel strength for nitrile rubber-to-steel joint. The 180° peel strength data for initial, oven aged at 70°C for 168 hr (7days) and after seawater immersion for 30 days is shown in Table 9.1. Initially and after oven ageing, both adhesives are almost similar in performance. The strength after seawater immersion was drastically reduced to 5-10% of original strength. This could be due to ingress through the edges, as the edges were not sealed by wax to prevent large water ingress. Therefore, it is quite evident that the adhesives are not for bonding rubber to metal when used under water, and modified epoxy is far less preferable, unless protected by an impermeable sealant on edges.

A common area of interest in modern warships and small lightweight boats is aluminum-to-FRP bonding, particularly in superstructures [33, 43, 44]. An experiment on modified epoxy-based adhesives was carried out to evaluate the adhesion and inherent strengths compared to AW106+HY953U adhesive. The commercial adhesive AW106+HY953U was used as such, and three different adhesives were made using a toughened epoxy wherein 20% toughening rubber was chemically reacted with AW106 and LY556 epoxy resins, and two hardeners N, N' *bis* (2-aminoethyl) ethane diamine (HY951) and HY953U (a blend of aminopropyl propane diamine and diamine of a fatty acid) were used. Ultimate tensile strength, % elongation at break and Al-FRP lap shear strength were determined. Table 9.2 shows the results. Although HY951, which is a short chain diamine, gave the highest ultimate tensile strength (UTS), but the bond strength was very poor compared to HY953U, which is a high molecular weight diamine.

From Table 9.2, it is seen that the overall properties including flexibility, cohesive and adhesion strengths and applicability are acceptable for a moderately viscous adhesive composition (CTPEGA modified LY556 with HY953U hardener). Being a toughened adhesive, the fatigue life under ship structural vibrations and dynamic force due to sea waves too will be better than for normal structural adhesives.

Bordes *et al.* [30] studied long term strength of epoxy-steel adhesive bonding in seawater in terms of water diffusion kinetics and double lap shear strength after seawater immersion. The adhesive was a commercial epoxy-based product. The authors opined that the simple data of degraded failure stress upon water ingress is not enough to analyse the performance of a structural bond. The extent of water penetration in the bondline is also important.

Table 9.1 Adhesion strength of rubber-to-steel sheet using various adhesives.

Conditions	Panel no.	Damping sheet -1 (White)		Panel no.	Damping sheet -2 (Blue)	
		AW 106 + HY 953U	CTPEGA modified LY556 + HY 953U		AW 106 + HY 953U	CTPEGA modified LY556 + HY 953U
180° Peel →		kN/m	kN/m		kN/m	kN/m
Initial Data	1	3.56	2.4	5	4.24	2.56
	2	4.0	2.69	6	2.48	2.8
	3	3.84	2.25	7	3.2	3.2
After air ageing 70°C, 168 h	1a*	1.8	3	5a	2.4	3
	2a	2.4	3.2	6a	2.4	2.8
	3a	2.0	3.2	7a	2.8	2.6
After seawater immersion 30 days	1w**	0.55	0.06	5w	0.28	0.23
	2w	0.43	0.05	6w	0.28	0.19
	3w	0.39	0.057	7w	0.27	0.21

'a' *indicates air oven aged sample.*
** 'w' *indicates seawater aged sample.*

Table 9.2 Tensile properties and adhesive bond strengths of aluminium-to-FRP joints.

Adhesives	Ultimate Tensile Strength (UTS), MPa	% Elong. at break	Lap Shear Strength, MPa	Comment
AW 106+HY953U	32	8–10	16	High viscosity
Modified AW106+HY 953U	24	8–10	14	Very high viscosity
Modified LY556 +HY 951	58	8–10	4	Low viscosity
Modified LY556 +HY 953U	19-22	20–25	10	Moderate viscosity
Modified LY556 +HY 953U. Cured for 2h at 70°C.	19-20	20–25	12	Moderate viscosity

Alia *et al.* [33] studied natural seawater ageing of two adhesives for 9 months, one based on polyurethane and the other based on vinyl ester resin. The water absorption in polyurethane adhesive attained a critical value and reduced drastically with time, while the water uptake in vinyl ester resin adhesive fluctuated with time over 180 days. The chemical change in polyurethane adhesive was not significant, but the vinyl ester showed many new groups as seen in FTIR spectra. The glass transition temperature of the vinyl ester was reduced after the exposure, but for polyurethane, the change was not significant. However, PU adhesive suffered more loss in tensile strength compared to the vinyl ester adhesive. Rudawska [38] studied the compression properties of a commercial marine-grade epoxy adhesive (Trade Name EPIDIAN 53) cured with triethylenetetramine hardener, containing non-reacting diluent, and a modified version which was loaded with $CaCO_3$ filler. Both varieties showed degradation in properties when immersed in 3.5% sodium chloride aqueous solution for maximum 3 months, as revealed by the microscopic examination of the morphology. There was marginal increase in compressive strength upon exposure to salt water, but the authors did not report the bond strength after the salt water ageing.

A static marine structure like a raft or rig needs a very stable adhesive bonding for a very long period, coterminous with major repair schedule of the structure. Cruz *et al.* [45] studied two commercial structural marine adhesives based on epoxy resins which are intended to be used for bonding vinyl ester-carbon fiber composite (CFRP) with reinforced cement concrete (RCC) marine structures. The bond strength (pull-off) with CFRP was 3-4 MPa before ageing. One adhesive showed degradation even in one year exposure to water and the epoxy chemically reacted with water to release small molecules, thereby showing a 16% reduction in water absorption in one adhesive, while the other one continued increase in water uptake beyond 2 years. However, the bond strengths for RCC-to-CFRP by these two adhesives after water immersion were not studied. Cabral-Fonseca *et al.* [46] studied the same CFRP-RCC bonding using three epoxy-based adhesives filled with calcium carbonate and silicate fillers. The adhesives had different glass transition regions. The mechanical properties were determined before and after immersion in pure water, salt water (3.5%) and alkaline water for 18 months at different temperatures The authors observed that the best adhesive absorbed only 1.5 % salt water after about 12000h and retained about 80% flexural strength, but in alkaline solution the strength drastically reduced to 37%, and the alkaline water uptake was quite high. However, the authors did not report the bond strength on ageing in any of the aqueous media used. An interesting study on restoration of marine equipment using epoxy-based adhesives was carried out by Sapronov *et al.* [47]. The epoxy was a low molecular weight diglycidyl ether of bisphenol A and the hardener was a polyethylene polyamine (PEPA), taken at various proportions. The maximum direct pull-off (tensile force) strength for steel-to-steel bonding and lap shear strength were reported to be very high. The performance also varied with the temperature and duration of curing.

9.3 Application on Naval Platform

Structural adhesives made from rubber-PF resin or modified epoxy-rubber and other conventional adhesives based on poly(methyl methacrylate), polyester resin, etc. are good for structures not intended for underwater use or in areas where acid or alkali solutions could be present. However, Rubber-PF adhesives get swollen in engine oils, hydrocarbon fuels and lubricants inside the marine vessel. Although the rubbers such as neoprene and nitrile are good in seawater application and in acidic or alkaline atmosphere, the other ingredients such as PF tackifiers, zinc/magnesium oxides are easily degraded in such environments. The modified epoxy-based structural adhesives are very common in marine industry, but limitation of such functionalized adhesives is very clearly seen in the seawater immersion results in Table 9.1. On the other hand, most versatile epoxy-based adhesives are likely to suffer significant loss of bond strength if the edges are exposed to the seawater in underwater bonded structures, as already discussed above. A very appropriate selection would be a multiple adhesive-sealant combination in case of underwater acoustic tiles which are used on the outer hull of a submarine and also inside the flooded SONAR compartments for sound and vibration isolation.

9.3.1 Vibrodamping Arrangements

A typical structural adhesive-bonded damping arrangement of a base frame of machines, supports, plate elements and sandwich type damping arrangement of a pipe in a ship are shown in Figure 9.1.

(a) (b) (c)

Figure 9.1 Adhesively-bonded structural elements with vibrodamping coatings: (a) Base frame, (b) Machinery support (c) Pipe coating (Cross section).

The thin damping sheet (< 2 mm thick) is bonded between two metallic structural components by a toughened structural adhesive in frames and pipes as shown in Figure 9.1. The structures are exposed to atmosphere and not intended for underwater application. The adhesive (yellow and red lines) can also be a high damping coating, thus serving both the purpose of joining metallic elements and vibration damping, such as a damping type epoxy-based adhesive developed by Ratna *et al.* [48].

The rubber-PF adhesives are also used on hydraulic piping, where a vibrodamping coating is bonded to the metal or FRP pipe with the adhesive. The adhesive being rubber based and flexible can withstand fatigue under constant vibration. In order to enhance damping, a constraining layer made of thin metallic sheet is bonded over the damping coating by the same adhesive. A sketch of the pipe coating with adhesive is shown in Figure 9.1(c).

9.3.2 Underwater Application

Underwater application of adhesives and sealants is an important area. The use of water-resistant adhesives and sealants under water has been known for last four decades, but such products are not much reported in literature. Way back in 1967, Kerr *et al.* [49] carried out one experiment on the effects of some hostile environments on aluminium-to-aluminium lap shear bond strength of an epoxy-based adhesive. The authors used a diglycidyl ether of bisphenol A (DGEBA) epoxy resin with epoxidised oil as reactive diluent (to make flexible adhesive) and cured by *bis*-(4-aminophenyl)-methane. The environments were vacuum, dry oxygen, ethanol vapor and water vapor, all at 90°C for about 55 days. Interestingly, dry oxygen and ethanol did not affect the bond strength to any significant extent, thus confirming no chemical reaction or oxidative degradation of the adhesive. However, water environment at 90°C drastically reduced the bond strength to one-third of unaged strength. The reason could be swelling, plasticization and possibly some chemical reaction with water at this elevated temperature. A more important observation was that the liquid ingress and degradation of the inherent strength of the adhesive was far more severe in ethanol than water. The study showed that the determination of bond strength is absolutely essential to qualify an adhesive by ageing in the service environments. Similar observation was made by Zanni-Deffrages and Shanahan [50] in their research on effect of water diffusion on properties of a toughened epoxy adhesive based on DGEBA and tetraglycidylmethylene dianiline (TGMDA) with dicyandiamide (DICY) as curing system. The bulk adhesive specimen was tested for tensile and compressive properties and the bond strength for adhesively bonded stainless-steel specimens (SS-to-SS) was determined in shear mode by a special loading arrangement. The tensile strength of the bulk adhesive decreased only by 10%, but the adhesion strength (lap shear) decreased by 33% when exposed to 70°C at 100% RH (air saturated with water vapor) for 120 days. The authors derived a correlation between time-dependent water uptake and corresponding tensile strength and estimated the diffusivity in the adhesive material.

The above studies [49, 50] and the results in Table 9.1 show that the adhesive-adherend interface is most affected by the diffusing molecule, rather than the bulk of the adhesive material. The interface is never perfect in most cases and very minute defects might lead to rapid deterioration of the interface. The diffusing water molecule plasticizes the adhesive and forms hydrogen bonding with polar adhesive, thus reducing the interface bond strength. The interface can be improved by a mechanical process of creating roughness on the metallic adherend, and also by chemical etching of the elastomer surface for better inter-locking of the adhesive and adherend. Grit blasting of metal to Sa2½ standard is the best process for steel, while surface roughening by emery paper is suitable for aluminium panels. For rubber surfaces, adhesion can be improved by creating microdefects or oxidation of the surface by strong acids such as chromic acid [51], or strong oxidizing agents such as caustic soda-naphthenic acid, which is also used for etching of aluminium surface [52]. Gamma radiation induced surface oxidation is another method of adhesion improvement [53, 54]. Secondly, a unique cleaning agent N,N-Dichloro-p-chlorobenzene sulfonamide [55] is used for cleaning the joining surfaces from oil, grease, etc. during the process of rubber lining. Additionally, for a longer life of the bonded structure immersed in water, edge sealing by a highly water-resistant sealant should be done.

Keenan [56] discussed in brief the properties and applications of epoxy-based adhesives. The author discussed surface preparation of steel components for best bond strength of epoxy adhesives. It was shown that the unprepared surface led to only 20% of the bond strength compared to the prepared surface. Secondly, the pot-life is important, as it depends on the ambient temperature while mixing, because the reaction rate of the resin with the hardener depends on the temperature. In addition, the batch size for mixing can also change the pot-life, since in large volume, the core of the mix would have higher temperature during the exothermic reaction due to poor thermal conduction of adhesive materials. One such epoxy-based adhesive is commercially produced by Wessex Resins and Adhesives [57], which has a pot-life of 45 min for 2kg mix and 100 min for 500g mix. The adhesive provides lap shear strength of 16 MPa for steel-to-steel bonding. A novel toughened structural adhesive for naval ships was developed with a functionalized rubber and an epoxy resin by Patri *et al.* [58]. The curing system for the rubber part was zinc oxide or a combination of zinc oxide and magnesium oxide with accelerators. Hexamethylene diamine was used for curing the epoxy resin. Ferric oxide was used as a filler. Ethyl acetate was used as a solvent. The most important property of this adhesive was a pot-life of 8h for 500g batch at ambient temperature. The long pot-life enables use of substantial quantity of adhesive mix to be applied on a large area of the ship structure. The cured adhesive provides a 90° peel strength of more than 3.0 kN/m as reported, as the rubber (SBR) strips ruptured during the tests, rather than adhesion failure. Best performance of this adhesive was realised when the cleaning agent as in ref. [55] was used on both adherend surfaces prior to application of the adhesive. The same adhesive was tested for the blue and white damping sheets (refer to Table 9.1) to bond to mild steel, and the 180° peel strengths were 3.4 and 2.2 kN/m, respectively.

A unique research report from The Michigan Technological University [59] describes the possibility of biomimicking the natural glue produced by mussels that adhere to rocks, underwater hull of ships and docks. A specific amino acid found in mussel foot proteins, called DOPA (3,4-dihydroxyphenylalanine), which is related to dopamine, is the main component of natural polymeric adhesive for bonding to any underwater surface. However, materials mimicking this underwater adhesion are widely used for skin or bone adhesion, and might be used under water for sensor assembly.

The application of adhesive for underwater structures needs some more treatments and protection by a sealing compound which would prevent water ingress into the joint. A typical sketch of multilayer adhesive bonding and edge sealing is shown in Figure 9.2.

In a typical example of multilayer tile manufacturing and application on metal panels in the seawater flooded compartment of a ship/submarine, the alternate solid and perforated or metamaterial type rubber layers as shown in Figure 9.2(b) are bonded by a high strength structural adhesive as described in ref [58]. The tile is bonded to the metal panels by application of two adhesives: the first layer being the epoxy putty as described in ref [40] to about 250-350 μm cured film thickness onto the metal, and subsequently three layers of the structural adhesive [58] totaling about 250-300 μm dry thickness. Thus, at the metal surface, the adhesive is strong epoxy-based [40] and the rubber tile is bonded with the rubber-epoxy structural adhesive. The adhesively-bonded tiles and the metal-to-tile bond all require an edge sealing to prevent water penetration through the bondline edges. The sealing compound can be a polysulfide sealant or a polyurethane sealant. A typical seawater-resistant polyurethane sealant as described in ref [10] is applied on the edges covering the adhesive lines on all edge surfaces. The highly viscous liquid polyurethane mix can be best applied by a putty blade for uniformity of the sealing line. Typical sequential drying and inter-coat application periods are 8h for epoxy putty adhesive, 20 minutes for each layer of structural adhesive and 24h curing of the adhesive system before sealant application. The sealant-adhesive system is allowed to cure for minimum 3 days.

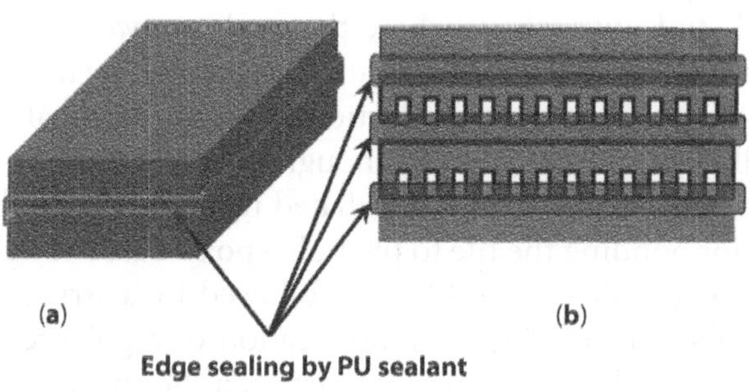

Figure 9.2 Adhesive and sealant application for vibrodamping coating of naval structures: (a) simple constrained layer damping coating; (b) multilayer vibroacoustic tiles.

Acoustic tile applied onto the outer hull of a submarine or part of underwater hull of a ship is generally quite thick, from about 40 mm to 60 mm rubber, either in single or two layers. Additionally, the tiles are designed as metamaterials, which are rubber tiles with inclusions of different materials like steel, aluminum, FRP or even air cavities of widely varying size and shapes, such as right circular cylinder, conical hole, cylindrical hole, thin circular button and similar air cavity, etc. These inclusions absorb mechanical energy (sound and vibration) by resonance at desired frequency bands [60–65]. Recently, Fu *et al.* [65] published a comprehensive review of types of underwater acoustic absorber tiles with different cavities, inclusions, fillers, etc. For such complicated construction and added mass, these tiles have significant self-weight and require quite durable and high bond strength for firmly adhering onto the hull for long duration. A 40 mm thick tile could be 45-60 kg in weight depending on cavity or solid inclusions. In case of double-hull submarines, multilayer thick tiles are also applied onto the inner surface of the outer hull, especially near the noisy or vibrating machinery of the submarine to isolate the vibration and to reduce radiated sound in the seawater. Apart from seawater resistance, the adhesives for these tiles are required to withstand fatigue due to low frequency dynamic force, as the submarine changes diving depth from surface to approximately 100m (hydrostatic pressure of 1MPa) under water during patrolling operation.

None of the available literature on the development of underwater acoustic lining deals with adhesive or other process of bonding the tiles onto the hull. The bonding with adhesive is very important due to possible high hydrostatic compression at great depths in the sea and degradation of the bondline exposed to the seawater. Commonly, the tiles are fitted with adhesive and stud-nut system, where the studs are welded to the hull at regular intervals and the rubber tiles are thus secured by the nut tightened to adequate torque. The excess projection of the stud is finally cut to flush with the hull. Such periodic studs, though of metal, do not significantly contribute to acoustic reflection or radiated noise. There can be two types of adhesives for bonding the tile to the hull: epoxy-based toughened adhesive on the metal hull surface [39,40], followed by a structural adhesive such as described in ref [58], with application of a polyurethane sealant similar to one in ref [10] for gap filling between the consecutive tiles, thus covering the bondline too. Since rubber is incompressible in bulk, it has a tendency to shear along the surface under hydrostatic compression (at high depth in sea). Therefore, the joining surface of the rubber (onto the metal) may be modified by a highly polar thermoset such as unsaturated

Figure 9.3 Underwater acoustic tile: (a) Cross section of a metamaterial tile; (b) Assembly of 4 tiles on a portion of hull.

polyester, or vinyl ester resin cured properly so as to improve the compatibility with the toughened epoxy layer [66, 67]. The epoxy layer, with a viscous consistency should be effective in fairing the hull roughness too, thereby facilitating rubber tile fixing. Alternately, water-resistant polyurethane (PU) adhesives can be used for fixing such acoustic tiles made from high damping polyurethane nanocomposites [68–70]. In such combination, the PU tile surface can be modified with another harder PU adhesive for better compatibility with the toughened PU nanocomposite adhesive.

A typical sketch in Figure 9.3 shows cross section of an acoustic absorbing tile as a metamaterial consisting of both cylindrical (button shape) air cavities and solid inclusions. The tiles are fixed onto a steel hull by two adhesive layers and an arrangement of 4 underwater acoustic tiles made of metamaterial, fixed onto a portion of a hull showing gap sealing by polyurethane sealant in between the tiles. The PU sealant is highly resilient to accommodate lateral strain (in shear) of the rubber tiles due to underwater hydrostatic pressure.

9.3.3 Acid-Resistant Rubber Bonding

Diesel electric submarines are most common in all international naval fleet today. The submarines use rechargeable batteries, mainly lead-acid batteries for propulsion and general power requirements, while the diesel engine charges the discharged batteries when the submarine comes up (snorkeling) to suck in air for combustion of the diesel. Spillage of acid occasionally inside the battery pits is common, particularly during filling and due to rolling-pitching of the submarine in higher sea state. The acid is a dilute sulfuric acid (24-25%) which is extremely corrosive to steel hull and appendages. Conventionally, acid-resistant coatings are applied onto the battery pit floors and supports to about 500 microns thickness. A comprehensive review of organic coatings for acid-resistant application is published by Møller *et al.* and have reported acid resistance of vinyl ester, or epoxy or polyurethanes [71]. Polyurethanes and epoxy paints are used in many submarines, but with a service life of only 3-4 years. The battery pit is a difficult place for frequent paint renewal and needs a long-life solution.

An interesting protection of the battery pit is by 2-3 mm thick rubber sheets, firmly bonded to the hull [72, 73]. Several rubbers, especially Butyl, SBR, Viton, Neoprene, EPDM, Hypalon (Chlorosulfonated polyethylene), silicone and fluorosilicone are known to be resistant to sulfuric acid of different concentrations even at higher temperature than ambient, particularly for 3 molar acid (29% acid) at about 70°C [74].

Most of these are non-polar (except Hypalon and Neoprene) and Viton is of low surface energy and are difficult to provide a strong adhesive bond with steel hull plates using structural adhesives which are essentially highly polar, such as epoxy resin. Neoprene stands out as the optimum choice in this regard. However, diffused dilute sulfuric acid in the edges of the rubber sheets might degrade the adhesive, unless adequately protected. Polyurethane sealants can be used for edge sealing, but needs suitable modification for better resistance to acid diffusion. To develop an acid-resistant adhesive which would be more effective in acidic media, nanocomposites of structural adhesives based on epoxy resin and polyurethane could be considered [75–77]. Rath and coworkers studied polyurethane-urea-clay nanocomposites in detail for solvent diffusivity, morphology and dynamic mechanical properties to show enhancement in strength and solvent resistance of the nanocomposites compared to the pristine polymer [78–80].

9.3.3.1 Example

In a study, a 2 mm thick neoprene rubber sheet vulcanized with conventional curing process was developed, wherein an organically modified nanoclay was used as reinforcement along with certain acid-resistant mineral fillers. A small quantity of a white filler was used to obtain white vulcanized neoprene sheets. The surfaces of the steel base and support frames are cleaned from any acid, contaminated water, minerals and corrosion products by washing with dilute aqueous alkali, followed by washing with high power water jet, subsequently dried with hot compressed air and blasted to Sa 2½ standard and immediately primed with epoxy paint for best adhesion bond.

An epoxy-based solvent-free coating was used as a primer [81] on the steel surface, followed by three layers of the structural adhesive [58] for bonding the neoprene sheet. The edge sealing was done by a modified polyurethane sealant [80], which is non-black and also nanoclay reinforced to obtain a white color, similar to the neoprene. The acid penetration was monitored regularly by sulfur mapping of the cross section of the adhesive layer and rubber samples and graded the performance accordingly. The theoretical calculation of lifetime under acid immersion of the total system (steel + epoxy primer + adhesive + neoprene sheet edge sealed) showed a service life of about 12 years at ambient temperature.

9.4 Diffusion of Water in Adhesive Matrix

For marine structural components with adhesively-bonded joints, both fatigue due to dynamic forces and degradation due to seawater and environmental ageing are important. However, seawater is the most damaging environment for bond strength of all adherend-adhesive combinations. Degradation of the adhesively-bonded joint due to seawater ingress is more severe than fatigue, because most adhesives are toughened materials and hence capable of attenuating vibrational energy. Nanocomposites of epoxy and other polymers are a possible solution to reduce water ingress and degradation of adhesives as studied by some researchers [82–90]. Burla [82] studied a cyclic sorption-desorption of water along with a prestress which gave information on the repeated sorption phenomenon for the Cloisite 10A nanocomposites of epoxy and vinyl ester resins.

Seawater ageing is most severe because of dissolved salts and alkalinity. In addition, seawater contains chlorides, bromides, iodides, sulfates and carbonates of sodium, magnesium, potassium, calcium and also traces of heavy metals like Fe, Mn, Cd, and Pb. Some of these directly deposit on the diffusion sites in the matrix and can affect the bond strength. Since adhesion failure can be a catastrophe in a naval structure, prediction of absorption of seawater due to diffusion in the adhesive is important.

There are, however, only a few very important mathematical models of basic diffusion phenomenon applied to thin laminates, ideally suited for adhesives. Some of the physical phenomena and governing predictive models are discussed here, which are also verified with some experimental data.

9.4.1 Fickian Diffusion

Fick's law of three-dimensional diffusion of a molecule in a solid matrix can be written as:

$$\frac{\partial C}{\partial t} = D\left(\frac{\partial^2 C}{\partial x^2} + \frac{\partial^2 C}{\partial y^2} + \frac{\partial^2 C}{\partial z^2}\right) \tag{9.1}$$

The above equation is valid only if there is no chemical reaction between the solid object and the diffusing molecule. In other words, the diffusion is purely a physical process. Further, the above partial differential equation is for a time-dependent diffusion rate, where C denotes concentration of diffusing species along the *x, y, z* directions at an instant *t*, while *D* is the diffusivity, a property of the pair of penetrant and matrix, and is constant for the pair at a constant temperature and pressure. The second-order dependence of instantaneous concentration in all directions indicates a fast rate of mass transfer initially, followed by a slow increase up to a constant maximum value. An asymptotic rise is typical of mass transfer phenomenon, where the rate of mass transfer depends on rate of change of instantaneous concentration gradient. Hence, with time, as the water ingress progresses, the concentration gradient at a point gradually reduces, hence the diffusion is prolonged. A straightforward solution to this complex equation is quite complicated. However, an experimental arrangement can be made to study diffusion of water in an adhesive matrix in one direction, taking a semi-infinite lateral section in a seemingly infinite medium, so that the concentration at the interface of the laminate and the water is considered as constant. Thus Eq. (9.1) can be modified for unidirectional diffusion at constant temperature and pressure as:

$$\frac{dC}{dt} = D\left(\frac{d^2 C}{dx^2}\right) \tag{9.2}$$

A typical sample of a thin, cured adhesive laminate is used to study diffusion in the direction of thickness as shown in Figure 9.4. Since the panel is exposed on both sides to water, the diffusion takes place from all exposed surfaces to the bulk along the thickness. The thickness for one-dimensional diffusion study must be a maximum of 2% of the lateral dimension, length or width, whichever is smaller. Hence, the diffusion at the edges is neglected and the area for diffusion is taken as twice the area of one face. If one face is blocked, one can take area of only the exposed face. However, a thickness more than 4 mm is considered a thick sample in case of a densely cross-linked thermosets such as phenolics, epoxy or vinyl ester adhesives.

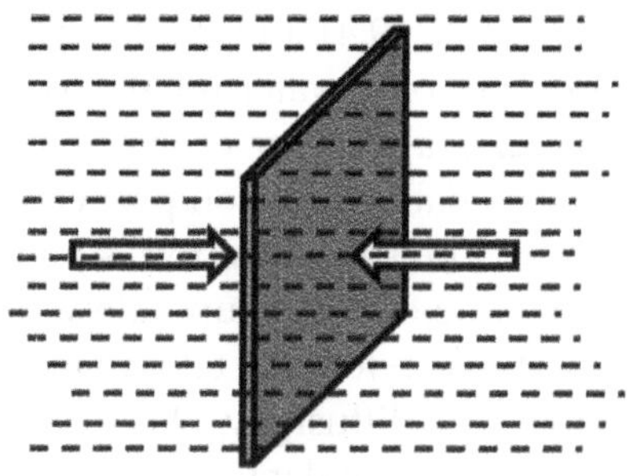

Figure 9.4 A sample of adhesive laminate immersed in water for one-dimensional diffusion study.

The solution to the unidirectional diffusion rate given in Eq. (9.1) can be derived as:

$$G = \frac{M_t}{M_\infty} = 1 - \sum_{j=0}^{\infty} \frac{8}{(2j+1)^2 \pi^2} \exp\left[-\frac{D(2j+1)^2 \pi^2 t}{h^2} \right] \tag{9.3}$$

where, G is the ratio of instantaneous water uptake (M_t) and maximum uptake at infinite time (M_∞), and h is the thickness of the laminate along the x-axis. The number j can be any number up to a very high value, and the accuracy is better with higher value of j. However, it is very tedious to calculate for j from low value (<1.00) to very high value (10^6). An approximate equation can be written instead, which could be conveniently used without much error:

$$G = 1 - \exp\left[-7.3 \left(\frac{Dt}{s^2} \right)^{0.75} \right] \tag{9.4}$$

Here, $s=h$, when two sides are exposed, and $s=2h$ when only one side is exposed and D is the diffusivity. The equation can be modified to the following expression to evaluate directly the time required to attain a certain water uptake (less than saturation):

$$t = \frac{s^2}{D} \left[-\frac{1}{7.3} \ln(1-G) \right]^{1/0.75} \tag{9.5}$$

All parameters in Eqs. (9.3), (9.4) and (9.5) can be obtained by carrying out an experiment on periodic water uptake for a sample laminate at a fixed temperature and thus a number of experimental set-ups at various fixed temperatures can enable diffusion property to be used for time-temperature superposition. The superposition is a method for estimation of effective service life or time of maximum allowable water uptake for a matrix-seawater system at any reference temperature, within the experimental temperature range. The diffusivity, D is determined by the following expression from time-mass gain data:

$$D = \frac{\pi}{16}\left(\frac{M_t / M_\infty}{\sqrt{t}/h}\right)^2 \tag{9.6}$$

M_t should be taken from initial data, when (M_t/M_∞) vs $t^{1/2}/h$ is linear. Thus, D is calculated from the initial slope. To determine diffusivity, one has to carry out the experiment on water ingress up to saturation to obtain the value of M_∞, which should be a constant quantity beyond a certain time. The above expression for diffusivity has two major drawbacks: (1) accuracy of linearity in the initial portion of the curve from $t=0$, and (2) experimental accuracy of the maximum water uptake (M_∞), because in many cases, the water simultaneously reacts with the polymeric adhesive and releases small molecules in the bulk water (reverse diffusion). Hence, the mass increase is not exclusively due to water uptake. The hydrolysis of the adhesive resin may cause more water ingress due to formation of highly polar groups (such as -COOH, -OH) generated in the resin structure. Additionally, in case of seawater, the salt deposits prevent the natural ingress of water. The models of diffusion are thus approximate, and cannot consider hydrolysis, reverse diffusion and salt deposition (in case of seawater).

The diffusion coefficients in solids at different temperatures are often found to be well predicted by an Arrhenius equation:

$$D = D_0 e^{-E_A/RT} \tag{9.7}$$

where D is the diffusion coefficient at temperature T (absolute scale), D_0 is the maximum diffusion coefficient, E_A is the activation energy of diffusion, and R is the universal gas constant.

An approximate dependence of the diffusion coefficient on temperature in liquids can often be found using Stokes–Einstein equation, which predicts that:

$$\frac{D_1}{D_2} = \frac{T_1}{T_2}\frac{\mu_2}{\mu_1} \tag{9.8}$$

where:

T_1 and T_2 denote temperatures 1 and 2, respectively, D is the diffusion coefficient (cm^2/s), T is the absolute temperature (K), μ is the dynamic viscosity of the solvent (Pa·s) at these two temperatures.

9.4.1.1 *Example*

A typical example of the diffusion data and calculation of diffusivity is shown here. One adhesive formulation was based on DGEBA epoxy (LY556) cured with a long chain poly(ether diamine) (Jeffamine 400) as developed by Ratna *et al.* [48], while the other was the same formulation, but with 2.5% Cloisite 10B. The samples were designated as ENC 0% and ENC 2.5%, respectively. The samples were properly cured according to a predetermined cure cycle. The glass transition temperature of the resin was 40°C as determined by DSC. The dimensions of the cured adhesive sample were 150 mm (length) x 100 mm (width) x 2 mm (thickness). All edges were sealed by 2 mm thick wax, which was enough to prevent ingress of water via the edges. The samples were immersed in natural seawater up to a mark in a glass container, and occasionally distilled water was added as needed. The entire experiment was conducted at room temperature of 30°C, RH varying marginally from 55-60%. The water was agitated slowly by a low-speed stirrer for uniformity in temperature and salt concentration.

All the samples were saturated after about 6700 h (280 days). There could be mass loss beyond saturation, hence the experiment was not extended further. Figure 9.5 shows the % water uptake with time ($\%M_t$) calculated as the ratio of weight difference to the original weight of the sample laminate:

$$M_t = \left(\frac{W - W_0}{W_0}\right) \times 100 \tag{9.9}$$

The experimental data was used to calculate the variables of Eq. (9.6) and the plot for initial 1500 h was taken. The $t^{1/2}$/h data was taken up to a limit where the R-square of the linear regression was more than 0.99. The slope of the linear plot was used to calculate diffusivity according to Eq. (9.6).

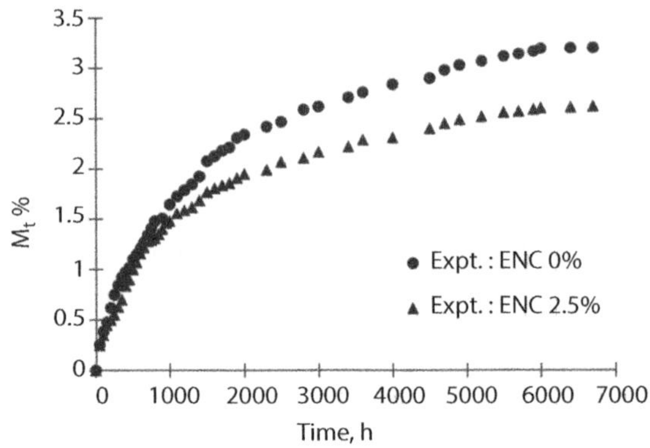

Figure 9.5 Water uptake by flexible epoxy resin based laminate and its clay nanocomposite: ENC0% - 0% Cloisite 20B; ENC2.5% - 2.5% Cloisite 10B.

The results of the experiment are listed in Table 9.3. The saturation water ingress for the epoxy laminate was 3.2% and that for the clay nanocomposite was 2.62%. Clay nanocomposites are known to improve barrier property of polymer matrices [87–90]. Compared to this result, a DGEBA epoxy cured with an aliphatic diamine showed about 2% water uptake at 40°C after 6000 h [57, 87] and for the 2% nanoclay loaded epoxy resin, the maximum water uptake was about 1.5% [87].

The experimental results are compared with the Fickian diffusion expression given in Eq. (9.4) and the plots of water uptake with time for both adhesive laminates are shown in Figure 9.6. There is good agreement between the experimental data and the theoretical prediction by Fickian diffusion for the epoxy without nanoclay, and some deviation is observed for the clay nanocomposite. This clearly shows that diffusion of water in such clay loaded epoxy resin matrix is not a simple unidirectional diffusion with a constant diffusivity.

In case of epoxy resin, the water ingress results in plasticization and increase in free volume of the matrix. This is evident from the increase in compressive strain after 90 days exposure in salt water (3.5% aqueous NaCl) from 6.4% to

Table 9.3 Seawater diffusivity in epoxy and nanocomposite based adhesive laminates.

Sample	Thickness, mm	% M_∞ *	D, cm^2/s
Epoxy (ENC 0%)	2.13	3.2	5.518×10^{-10}
Nanocomposite (ENC 2.5%)	2.09	2.62	6.532×10^{-10}

* *% M∞: it is the maximum percent of water uptake by weight of the sample.*

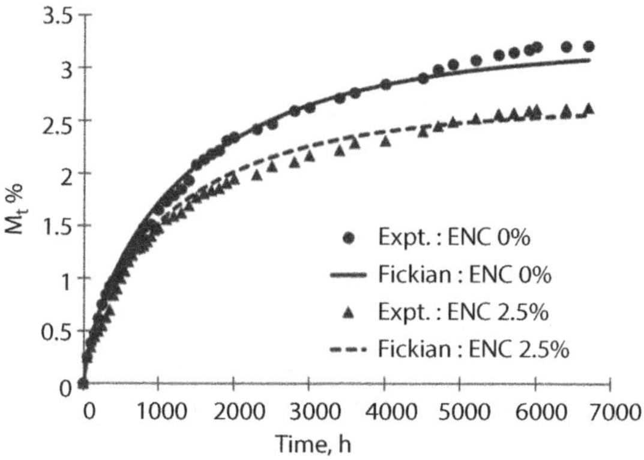

Figure 9.6 Comparison of experimental data with Fickian diffusion.

7.7%, in a detailed study on an epoxy adhesive [38]. Considering this change is due only to mass of the water in the matrix, the change in strain is quite significant compared to the mass change, which is only possible by a plasticizing effect and increase in free volume. The effect of plasticization is also to increase the segmental mobility of the cross-linked chains of epoxy resin, thereby reducing the glass transition temperature and decrease in relaxation time of the segments. The relaxation time of the epoxy sample with nanoclay actually increased from 3.4×10^{-4} s for neat epoxy to 5.3×10^{-4} s (as calculated from the complex shear modulus determined by DMA) as a result of stiffening by the intercalated/exfoliated clay in the epoxy. Along with providing a tortuous path for water diffusion, the clay also restricts the segmental mobility thus reducing the free volume, when used in low loading, as studied by Patil *et al.* for epoxy-Cloisite 10A nanocomposites using positron annihilation technique [91]. Therefore, the barrier property also changes with the mobility of the segments. On a longer exposure, the adhesive may undergo hydrolysis to produce small molecules like glycol. When the small molecules are produced due to hydrolysis, a quasi-equilibrium is established between the intake of water and diffusion of small molecules from the matrix to the bulk water medium. This phenomenon might be associated with decrease in mass gain as can be seen from many studies.

It is seen from Figure 9.6 that the theoretical Fickian model is not exactly in agreement with the experimental data. This is quite obvious due to errors in experimental weight measurement, since the samples, after removal from water, are wiped to remove water from all surfaces. This is never perfect, considering the fact that the water absorption as such is very low and hence the error is significant. The excessive drying effort, on the other hand, causes removal of absorbed water to some extent, causing some error.

9.4.2 Dual-Fickian Prediction

It was found that the nature of water diffusion in a clay containing epoxy resin could not be predicted by a single Fickian diffusion expression. It could be due to the two distinctly different diffusivities, one initial and the second one due to internal physicochemical changes in the adhesive film. It could be due to increase in volume, plasticization, formation of more polar groups for more hydrogen bonding with diffusing water, thereby more ingress and absorption simultaneously. The theory of dual-Fickian diffusion essentially assumes that there is no mass loss due to backflow of any small molecule produced due to hydrolysis, and the total mass uptake is a summation of two maximum mass uptakes due to two diffusivities.

The general expression is a simple summation of two single diffusion expressions:

$$M_t = \left(1 - \frac{8}{\pi^2}\sum_{n=0}^{\infty}\frac{1}{(2n+1)^2}e^{\frac{-D_1(2n+1)^2\pi^2 t}{4l^2}}\right)M_{1\infty} + \left(1 - \frac{8}{\pi^2}\sum_{n=0}^{\infty}\frac{1}{(2n+1)^2}e^{\frac{-D_2(2n+1)^2\pi^2 t}{4l^2}}\right)M_{2\infty} \quad (9.10)$$

The approximate simplified expression is [92, 93]:

$$M_t = M_1\ 1 - \exp\ -7.3\ \frac{D_1 t}{b^2}^{0.75} + M_2\ 1 - \exp\ -7.3\ \frac{D_2 t}{b^2}^{0.75} \quad (9.11)$$

Taking the data of ENC 0% and ENC 2.5%, two diffusivities for each material were calculated from two distinctly different slopes as seen in Figure 9.5. The diffusivities are listed in Table 9.4.

The dual-Fickian plots for both adhesives ENC 0% and ENC 2.5% are shown in Figure 9.7. It is observed from the nature of the dual-Fickian curves that the epoxy- nanocomposite is better predicted by this theory rather than neat epoxy, which was better fitted by simple Fickian model as seen in Figure 9.6. For the epoxy resin adhesive, the water uptake predicted by both single and dual-Fickian models was almost similar to experiment, but the long-term data was better fitted by the dual-Fickian model,

Table 9.4 The two diffusivities for dual-Fickian Diffusion.

Epoxy Adhesive	Diffusivity	Value, cm^2/s
ENC 0%	D_1	5.518×10^{-10}
	D_2	1.432×10^{-10}
ENC 2.5%	D_1	6.532×10^{-10}
	D_2	1.294×10^{-10}

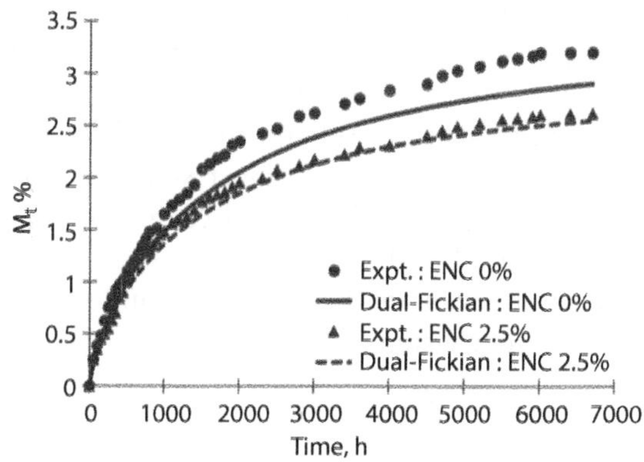

Figure 9.7 Dual-Fickian fits for experimental water diffusion in epoxy and its clay nanocomposite.

and hence, it is always better to use this model to know the ultimate water uptake in the long run. This is important, since the plasticization effect, which results in increase in compliance and drastic deterioration of bond strength largely depends on extent of water absorbed.

9.4.3 Effect on Flexural Strength

One of the most important aspects of deterioration of an adhesive due to water ingress is loss of mechanical strength in flexure. The strength of an adhesively-bonded joint is often determined in bending. For example, a beam or a plate as a hanging span between two rigid supports. The interface in the adhesively-bonded joint is the weakest part. Therefore, failure in the structure would initiate from the bending mode. Generally, butt joint, T-peel, 180° peel, single lap shear, double lap shear, thick adherend shear test (TAST), Arcan test, etc. are commonly used for evaluating joint strength. Detailed description of bond strength test methods is documented by Ilioni [37]. However, when laminates are used in structures as multilayers and sandwich construction as in many naval applications, flexural loads are experienced more by the structural elements, compared to other stresses. In addition, the deterioration of flexural strength is more than tensile or shear strength for bonded joints in water ageing [94, 95]. It is, therefore, important to determine flexural strength of a bonded beam, especially in three-point bending. In the ships and submarines, load bearing structural elements are adhesively-bonded panels, and are subjected to different loadings in static, steady state dynamic, and transient modes. Few critical examples are sandwich hull (with damping treatment) of underwater unmanned vehicles, multilayer metal-composite construction of superstructures, multilayer silent acoustic enclosures in ships, bonded joint in composite SONAR domes, base of equipment support with damping treatment, bulletproof shields of fast attack boats, and many such high-performance multilayer structures of naval vessels. Evaluation of synthetic seawater ingress and corresponding deterioration of flexural strength of the bonded sandwich laminates was carried out, where the purpose was to evaluate the life of the bonded joint under bending stress when used under water.

9.4.3.1 Example

A specially modified epoxy resin (toughened DGEBA) cured with triethylene tetramine was used to form a 4 mm thick laminate for sandwich beam construction, with 2 mm and 1.5 mm FRP sheets bonded by the epoxy-based adhesive (T_g about 60°C) of about 0.5 mm thickness, and the edges sealed by wax (melting point: 80°C), and an adhesive of the same epoxy resin loaded with 6% nanoclay (Nanomer 130E). The sketch of a sandwich laminate is shown in Figure 9.8.

Both sandwich beams were exposed to synthetic seawater at various isothermal conditions (20°C, 30°C, 40°C and 50°C) for some periods. For comparison, a single monolithic FRP laminate of the same epoxy resin and

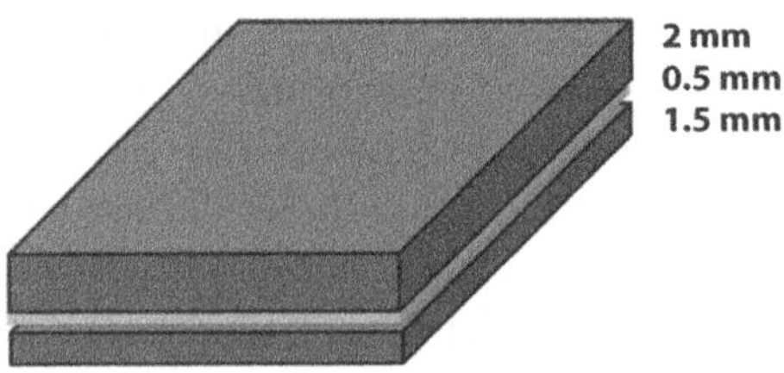

Figure 9.8 An experimental FRP sandwich laminate.

an E-glass fabric was made and studied for seawater degradation by measuring flexural strength at similar isothermal temperatures (20°C to 50°C). The edges were sealed to prevent water ingress through the edges. Figure 9.9 shows the flexural strength values on seawater ageing of single FRP laminate. The average value dropped by only 9% after 5 months. It was not significant compared to the bond strength decrease as seen in Table 9.1 and in literature too [30, 37].

The water ingress and flexural strength to failure were monitored for the sandwich laminates bonded by epoxy and nanocomposite adhesives. Due to sealed edges, the diffusion was assumed to be one-dimensional and through the surfaces of the laminates (both surfaces exposed) only along the thickness direction. Water absorption was measured up to 50 days only to determine diffusivity, but the flexural strength was measured for 5 months to observe the trend in degradation of the bond strength. The seawater diffusion was studied to establish both Fickian and dual-Fickian trends with good agreement with experimental data (R^2>0.97). The diffusivity of the panels did not change much due to change in the adhesive composition (neat and nanocomposite) since their surfaces were not directly exposed to the seawater. Rather, the trend was similar to the typical epoxy-glass fabric composites. The maximum seawater uptake was about 4.8% at 30°C and the diffusivity was in the range of 1.0×10^{-4} to 2.0×10^{-4} mm^2/h. Figures 9.10(a) and (b) show the result of neat epoxy laminate and nanocomposite, respectively for absorption of synthetic seawater for 50 days, when almost saturation was observed. There is drastic increase in seawater uptake beyond 20°C for both neat epoxy laminate and 6% nanocomposite.

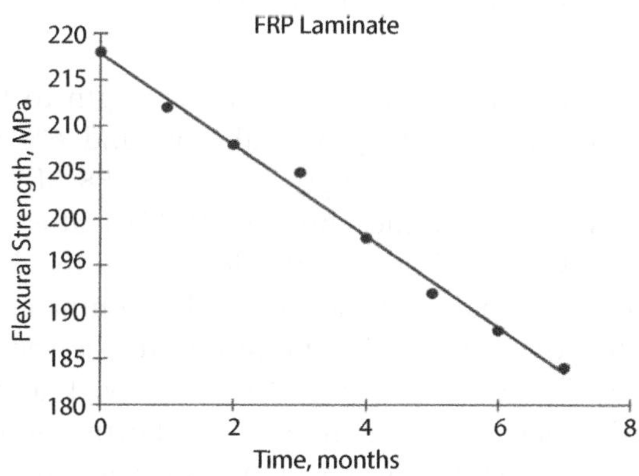

Figure 9.9 Flexural strength of a single FRP laminate under seawater ageing (average values) at temperatures ranging from 20°C to 50°C.

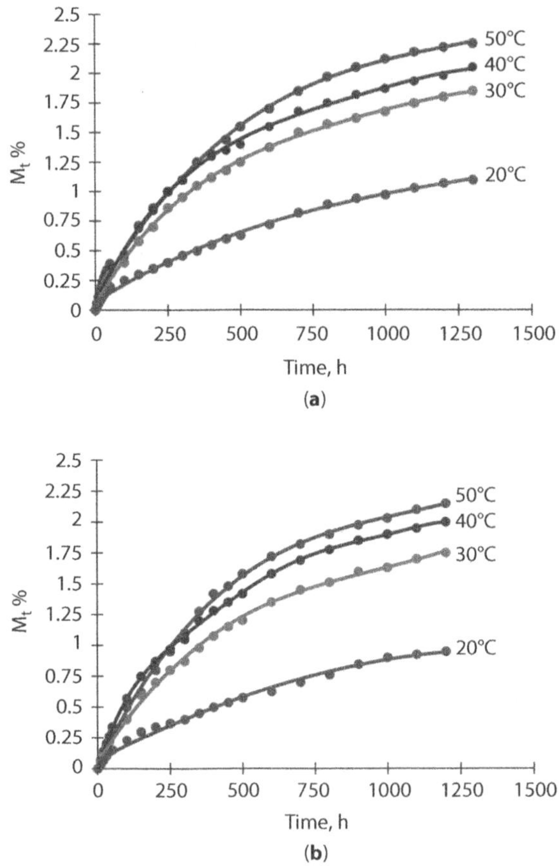

Figure 9.10 Seawater absorption (M_t %) with time for a limited period (50 days): (a) neat epoxy adhesively bonded system; (b) 6% nanoclay composite adhesively bonded system.

Figures 9.11(a) and (b) show the flexural strength to failure for the sandwich laminates with neat epoxy adhesive and 6% nanocomposite adhesive, respectively. The strength was about 88 MPa initially, and deteriorated to much lower value, about 52-60 MPa, after 5 months. The strength should have been higher, considering the much higher value (about 200-220 MPa initially) of the single FRP laminate, but in case of bonded laminates, the failure takes place due to the separation of the adhesive bond at the lower interface of the sandwich at a much lower stress, because that part is weakest in bending. Furthermore, on adsorption of even small amount of water at the interface, the plasticization by water leads to much loss in adhesion forces due to loss of hydrogen bonding and van der Waals forces.

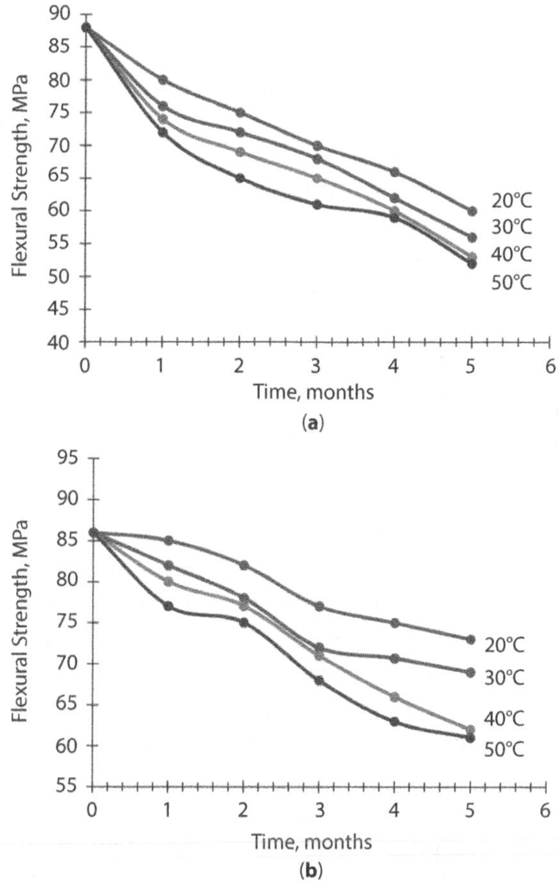

Figure 9.11 Flexural strength during seawater ageing for 150 days at various temperatures. (a) with neat epoxy adhesive, (b) with 6% nanoclay composite

Whereas for an FRP laminate, the effect is not so pronounced, since the laminate is a single homogeneous composite, without any dissimilar layer bonded to it. Hence overall strength is not reduced significantly even on saturation by seawater. However, if the study was extended beyond one year, there could be deterioration of the fiber resulting in weakening of the fiber-resin interface [96].

The applications of such sandwich construction of layered FRP and multilayer elements with multiple materials like aluminium-FRP are being studied and also being used in critical underwater objects, where the ultimate bending stress of the bonded element should be sufficient to withstand hoop stress due to the hydrostatic force along with the impact of waves and dynamic force in the sea.

9.5 Summary

Adhesives of different chemistries and formulations are used in naval ships and submarines for diverse purposes. Commonly used adhesives as in other industrial sectors are rubber adhesives and structural adhesives based on thermosetting resins, with a modification for marine-grade characteristics. Earlier common adhesives were mostly formulations of natural adhesives from plants and trees, and subsequently pitch and tar. With rapid advancement in polymers such as methacrylates, thermosets such as vinyl ester, unsaturated polyester, polyurethane, polyurea, phenol-formaldehyde, urea-formaldehyde, epoxy resin, etc., a vast choice is available now depending on the nature of the adherends. Nanotechnology has further improved and expanded the scope of new developments in adhesive field. Metal adherends are obviously most used materials for bonding, followed by FRP and engineering thermoplastics. Considering the strength and bond strength requirements, epoxy resin stands out as the best choice, provided some toughening is imparted by means of either physical mixing or chemical reaction or choosing long chain epoxy or oligomeric curing agent (diamines). However, marine qualification is needed.

One major application of special adhesives for naval vessels is bonding of acoustic and vibration damping coating on panels, hull structures, base frames and supports of machinery and hydraulic piping. These coatings are made of specially designed thick elastomeric sheets. Adhesives are required in all damping arrangements such as free layer damping, constrained layer damping and multilayer vibroacoustic tiles. Naval platforms use soft rubber-based adhesives, rubber-phenolic blended adhesives and rubber-epoxy combined formulations depending on acoustic and vibrodamping lining systems. Other special adhesives are used for SONAR domes; doors, hatches and windows of ships; superstructures, hull lining, and acoustic tiles on external hull, etc. However, edge sealing and sealing of the bondline are essential to prevent excessive water ingress. Therefore, in addition to adhesives, special sealants such as polyurethane, polysulfide, silicone, Viton, etc. are also employed over the adhesive joints. Another special adhesive-sealant combination using nanotechnology has a great potential for rubber lining inside the battery pit of submarines, where the lining system has to withstand dilute sulfuric acid for long period.

In most cases, the seawater ageing studies of adhesives are reported for 12-24 months and mostly in synthetic seawater or aqueous sodium chloride solution. Natural seawater contains more ingredients and microorganisms, some of which also cause deterioration of polymers, such as sulfate reducing bacteria (SRB), which generate hydrogen sulfide. It is required to undertake a comprehensive study on the effect of marine environment on bonded joints by exposing in two types of environments: one in a humid and hot chamber simulating ship superstructure, and the other ageing under seawater, using sea rafts and periodic evaluation of flexural bond strength for sandwich laminates. There may not be a direct correlation between water ingress and reduction of bond strength, but a large amount of data can possibly be used to make a performance matrix which might be a graphical representation of the three variables: time, water ingress, and adhesive bond strength, with temperature as parameter. The performance data for each adhesive-adherend combination can be generated. Slow cycle fatigue and mid-frequency fatigue are other important evaluation methods for bonded joints of metal to rubber and FRP to rubber, in both air and seawater media, since the hull lining and panels are subjected to very low frequency (0.08 to 0.1 Hz) dynamic force of sea waves and under high depth in the sea (hydrostatic pressure), and mid-frequency (10 Hz to 2000 Hz) vibration due to machinery.

The experiments on such scale should be aimed at designing bonded structural elements with a service life coterminous with major refitting programme of the naval ship.

References

1. P.S.G. Krishnan, N. Venkatesan and K. Adhinarayanan, Effect of phenolic resin structure on the bonding properties of nitrile rubber-based adhesives, J. Adhesion. Sci. Technol. 18, 1483-1495 (2004).
2. B.T. Poh and L.N. Ong, Adhesion properties of styrene-butadiene rubber (SBR)/Standard Malaysian Rubber (SMR L)-based adhesives in the presence of phenol formaldehyde resin, EXPRESS Polym. Lett. 1, 654–659 (2007).
3. Phenolic Resins for The Rubber Industry – An Overview, Product development information, Ram Charan Company, Vol. 2, Issue 8, pages 1-6 (2012), www.ramcharan.org, accessed on May 9, 2022.
4. K. Han, M. Li, H. Xu, and J. Xiao, Study on hot vulcanized adhesive for rubber and metal bonding. Proc. 2018 International Symposium on Mechanics, Structures and Materials Science (MSMS 2018), AIP Publishing (1995).
5. High methylol phenolic resin: Polyols & Polymers Pvt. Ltd., Vapi, Gujarat, India. https://www.polyolsandpolymers.net/contact. Accessed on May 9, 2022.
6. O. Masao, K. Gaito, and H. Shogo, Chloroprene rubber latex adhesive composition. European Patent No. EP 3 508 547 A1. Assigned to Denka Company Limited. Tokyo, Japan (2019).

7. L. Job, Studies on adhesives for bonding rubber to rubber and rubber to textiles. Ph.D. Thesis, Department of Polymer Science and Rubber Technology, Cochin University of Science & Technology, Kochi, India (1999).
8. L. Wu, X. Wang, L. Ning, J. Han, Z. Wan and M. Lu, Improvement of silicone rubber properties by addition of nano-SiO_2 particles, J Appl Biomater Funct Mater, 14 (Suppl 1), S11-S14 (2016).
9. K. Indulekha, R.S. Rajeev, K.N. Ninan and C. Gouri, Polycyclic silicone polymer as novel single source precursor for the facile synthesis of nanostructured SiC, Mater. Chem. Phys. 206, 64-70 (2018).
10. M. Patri, U.G. Suryavanshi and P.C. Deb, A process for the preparation of an improved polyurethane-based sealant. Indian Patent No. 51/2006. Assigned to The Chief Controller, Research and Development, Ministry of Defence, Govt. of India, New Delhi (2006).
11. Silicone Rubber – Technical Literature, AZO Materials, Manchester, UK. (2001) https://www.azom.com/properties.aspx?ArticleID=920. Accessed on May 10, 2022.
12. T.L. Junod, Gaseous emissions and toxic hazards associated with plastics in fire situations - a literature review, NASA Technical Note No. TN D-8338. Washington D.C. (1976).
13. Hatch Seals-Marine Gaskets-Hatch Cover Rubber Packing: Simolex Rubber Corporation. USA. https://simolex.com/product/hatch-seals-marine-gaskets-watertight-gaskets-hatch-cover-rubber-packing. Accessed on May 11, 2022.
14. Watertight Rubber Hatch Packing Seals: AQUASEAL Rubber Ltd. U.K. https://www.aquasealrubber.co.uk/watertight-rubber-hatch-packing-seals. Accessed on May 11, 2022.
15. Ship Rubber Door Seals: Ronfell Group. UK. https://ronfell.com/ship-rubber-door-seals. Accessed on May 11, 2022.
16. Hatch Seal and Marine Rubber Gaskets: Accurate Rubber Corporation. USA. https://www.accuraterubber.com/extruded-silicone-rubber-gasket-seals/hatch-seal-marine-rubber-gasket. Accessed on May 11, 2022.
17. S.S. Hou, Y.P. Chung, C.K. Chan and P.L. Kuo, Function and performance of silicone copolymer. Part IV. Curing behavior and characterization of epoxy–siloxane copolymers blended with diglycidyl ether of bisphenol-A, Polymer 41, 3263-3272 (2000).
18. M. Ochi and S. Shimaoka, Phase structure and toughness of silicone-modified epoxy resin with added silicone graft copolymer, Polymer 40, 1305-1312 (1999).
19. T.H. Ho and C.S. Wang, Modification of epoxy resins with polysiloxane thermoplastic polyurethane for electronic encapsulation: 1., Polymer 37, 2733-2742 (1996).
20. F. Zhao, Q. Sun, D.P. Fang and K.D. Yao, Preparation and properties of polydimethylsiloxane-modified epoxy resins, J. Appl. Polym. Sci. 76, 1683-1690 (2000).

21. K.L. Mittal, The role of the interface in adhesion phenomena, Polym. Eng. Sci. 17, 467-473 (1977).
22. S.K. Rath, J.G. Chavan, S. Sasane, A. Srivastava, M. Patri, A.B. Samui, B.C. Chakraborty and S.N. Sawant, Coatings of PDMS-modified epoxy via urethane linkage: Segmental correlation length, phase morphology, thermomechanical and surface behavior. Prog. Org. Coat. 65, 366–374 (2009).
23. S. Pandey, S.K. Rath and A.B. Samui, Structure–thermomechanical property correlations of highly branched siloxane–urethane networks, Ind. Eng. Chem. Res. 51, 3531–3540 (2012).
24. B.C. Chakraborty, FRP for marine application, in: *Fiber-Reinforced Plastics*, M. Masuelli (Ed.) IntechOpen, London, U.K. (2022). DOI: http://dx.doi.org/10.5772/intechopen.101332
25. D.W. Chalmers, Experience in design and production of FRP marine structures, Marine. Struct. 4, 93-115 (1991).
26. A.P. Mouritz, E. Gellert, P. Burchill and K. Challis, Review of advanced composite structures for naval ships and submarines, Composite Struct. 53, 21-41 (2001).
27. K. Galanis, Hull construction with composite materials for ships over 100 m in length, M.S. dissertation, Department of Ocean Engineering, Massachusetts Institute of Technology (2002).
28. F. Rubino, A. Nisticò. F. Tucci and P. Carlone, Marine application of fiber reinforced composites: A review, J. Marine Sci. Eng. 8, Article no. 26 (2020), doi:10.3390/jmse8010026.
29. Y. Gao, J. Xiong, S. Zhang and Y. Zuo, Aging performance of epoxy adhesive in salt water, Advanced Mater. Res. 11-12, 399-402 (2006).
30. M. Bordes, P. Davies, J.-Y. Cognard, L. Sohier, V. Sauvant-Moynot and J. Galy, Prediction of long term strength of adhesively-bonded steel/epoxy joints in seawater, Int. J. Adhesion. Adhesives. 29, 595–608 (2009).
31. R. Leger, A. Roy and J.C. Grandidier, Non-classical water diffusion in an industrial adhesive, Int. J. Adhesion. Adhesives. 30, 744–753 (2010).
32. M. Lai, J. Botsis, J. Cugnoni and D. Coric An experimental-numerical study of moisture absorption in an epoxy. Composites A, 43, 1053–1060 (2012).
33. C. Alia, M.V. Biezma, P. Pinilla, J.M. Arenas and J.C. Suárez, Degradation in seawater of structural adhesives for hybrid fiber-metal laminated materials, Advances. Mater. Sci. Eng. Article ID 869075, (2013). http://dx.doi.org/10.1155/2013/869075
34. B.P. Blackburn, J. Tatar, E.P. Douglas and H.R. Hamilton, Effect of hydrothermal conditioning on epoxy adhesives used in FRP composites. Constr. Build. Mater. 96, 679–689 (2015).
35. G. Viana, M. Costa, M.D. Banea and L.F.M. da Silva, Water diffusion in double cantilever beam adhesive joints, Latin Amer. J. Solid Struct. 14, 188–201 (2017).

36. P. Fernandes, G. Viana, R.J.C. Carbas, M. Costa, L.F.M. da Silva and M.D. Banea, The influence of water on the fracture envelope of an adhesive joint, Theor. Appl. Fracture Mech. 89, 1–15 (2017).
37. A. Ilioni, Influence of seawater ageing on the behaviour of adhesives: a rapid characterization of the evolution of mechanical properties of bonded joints. Mechanics of materials physics.class-ph Université de Bretagne occidentale - Brest, France (2017). https://tel.archives-ouvertes.fr/tel-01744438. Accessed on May12, 2022.
38. A. Rudawska, The effect of the salt water aging on the mechanical properties of epoxy adhesives compounds. Polymers, 12, Article no. 843 (2020); doi:10.3390/polym12040843.
39. B.C. Chakraborty and D. Ratna, Impact, shock and vibration characteristics of epoxy-based composites for structural adhesive joints, in: *Structural Adhesive Joints: Design, Analysis and Testing,* K.L. Mittal and S.K. Panigrahi (Eds) Ch.9, pp.247-287, Wiley-Scrivener. Beverly, MA (2020).
40. M. Patri, U.G. Suryavanshi and P.C. Deb, A process for preparation of epoxy-resin based structural putty, Indian Patent No. 40/2007. Assigned to The Chief Controller, Research and Development, Ministry of Defence, Govt. of India. New Delhi (2007).
41. Advanced Materials: Araldite AW 106 / Hardener HV 953 U. Structural Adhesives. Technical Data Sheet. Huntsman Advanced Materials GmBH, Basel, Switzerland (2010).
42. D. Ratna, N. R. Manoj, B. C. Chakraborty and P. C. Deb, Shock-resistant ambient temperature curing epoxy adhesive, J. Adhesion Sci. Technol. 17, 623–632 (2003).
43. M. Smith and P. Hutapea, Surface engineering for adhesively-bonded metal composite joints, J. Ship Production. Design. 23, 72–81 (2007).
44. I. Grabovac and D. Whittaker, Application of bonded composites in the repair of ships structures – A 15-year service experience, Composites Part A, 40, 1381–1398 (2009).
45. R. Cruz, L. Correia, A. Dushimimana, S. Cabral-Fonseca and J. Sena-Cruz, Durability of epoxy adhesives and carbon fiber reinforced polymer laminates used in strengthening systems: Accelerated ageing versus natural ageing, Materials 14, 1533 (2021)
46. S. Cabral-Fonseca, J. P. Nunes, M.P. Rodrigues and M.I. Eusébio, Durability of epoxy adhesives used to bond CFRP laminates to concrete structures. DSpace Repository, (2009), http://repositorio.lnec.pt:8080/jspui/handle/123456789/16864. Accessed on May 22, 2022.
47. O. Sapronov, P. Maruschak, V. Sotsenko, N. Buketova, A.B.D.G. De Deus, A. Sapronova and O. Pienkowskis, Development and use of new polymer adhesives for the restoration of marine equipment units, J. Marine. Sci. Eng. 8, Article no. 527 (2020). doi:10.3390/jmse8070527.

48. D. Ratna, N.R. Manoj, L. Chandrasekhar and B.C. Chakraborty. Novel epoxy compositions for vibration damping applications, Polym. Adv. Technol. 15, 583-586 (2004).
49. C. Kerr, N.C. Macdonald and S. Orman, Effect of certain hostile environments on adhesive joints, J. Appl. Chem. 17, 62-65 (1967).
50. M. P. Zanni-Deffarges and M. E. R. Shanahan, Diffusion of water into an epoxy adhesive: comparison between bulk behaviour and adhesive joints, Int. J. Adhesion. Adhesives. 15, 137-142 (1995).
51. P. Blais, D.J. Carlsson, G.W. Csullog and D.M. Wiles, The chromic acid etching of polyolefin surfaces, and adhesive bonding 1, J. Colloid Interface Sci. 47, 1403-1404 (2001).
52. J.W. Carroll and L. Hill, Aluminium etchant compositions, U.S. Patent No. 3,083,165. Assigned to Pennsalt Chemicals Corporation, Philadelphia, PA. (1963).
53. S.K. Rath, U.G. Suryavanshi, M. Patri and B.C. Chakraborty, Adhesion improvement of LDPE by radiation induced oxidation, J. Polymer Materials 24, 91-100 (2007).
54. S. Yamakwa and F. Yamamoto, Surface modification of polyethylene by radiation-induced grafting for adhesive bonding. V. Comparison with other surface treatments, J. Appl. Polym. Sci. 25, 41-49 (1980).
55. V. K. Shiryaev, Utilization of N, N-Dichloro-p-chlorobenzene sulfonamide, Russian J. Appl. Chem. 74, 1403-1404 (2001).
56. P.J. Keenan, Underwater application of adhesives in ship husbandry, Naval. Eng. J. 103, 50-58 (1991).
57. Wessex 4701-Underwater Adhesives: Technical data sheet, Wessex Resins and Adhesives, Southampton, England, (1986). http://www.wessexresins.com/. Accessed on May 23, 2022.
58. M. Patri, U.G. Suryavanshi and P.C. Deb, A structural adhesive, Indian Patent No. 03/2008. Assigned to The Chief Controller, Research and Development, Ministry of Defence, Govt. of India, New Delhi (2008).
59. Michigan Technological University, How to make underwater glue with a biomed engineer. ScienceDaily, (July 26, 2016). www.sciencedaily.com/releases/2016/07/160726124348.htm. Accessed on May 23, 2022.
60. N. Gao and K. Lu, An underwater metamaterial for broadband acoustic absorption at low frequency, Appl. Acoustics. 169, Article No. 107500 (2020). https://doi.org/10.1016/j.apacoust. 2020.107500.
61. H. Zhao, J. Wen, H. Yang,. L. Lv, and X. Wen, Backing effects on the underwater acoustic absorption of a viscoelastic slab with locally resonant scatterers, Appl. Acoustics. 76, 48–51 (2014).
62. S.N. Panigrahi, C.S. Jog and M.L. Munjal, Multi-focus design of underwater noise control linings based on finite element analysis, Appl. Acoustics. 69, 1141–1153 (2008).

63. P. Méresse, C. Audoly, C. Croënne, A. Christine and H. Hennion, Acoustic coatings for maritime systems applications using resonant phenomena. Comptes Rendus Mecanique (2015). http://dx.doi.org/10/1016/j.crme.2015.07.004.
64. K. Shi, G. Jin, R. Liu, T. Ye and Y. Xue, Underwater sound absorption performance of acoustic metamaterials with multilayered locally resonant scatterers, Results Phys. 12, 132-142 (2019).
65. Y. Fu, I.I. Kabir, G. Heng and Z. Peng, A review on polymer-based materials for underwater sound absorption, Polym. Test. 96, 107115 (2021).
66. Vinyl Ester Based Structural Adhesive - Resins Materials : Eternal Materials Co. Ltd. Kaohsiung, Taiwan. https://www.eternal-group.com/Product/Detail/2072?level2=277&lang=en&level1=81. Accessed on May 28, 2022.
67. M. Patri, J.G. Chavan and P.C. Deb, A process for preparation of styreneless polyester, Indian Patent No. 290947. Assigned to The Chief Controller, Research and Development, Ministry of Defence, Govt. of India. New Delhi (2008).
68. K. Sasikumar, N.R. Manoj, R. Ramesh and T. Mukundan, Carbon nanotube-polyurethane nanocomposites for structural vibration damping, Int. J. Nanotechnol. 9, 1061-1071 (2012).
69. M.A. Mendelsohn, F.W. Navish Jr. and K. Dongsik, Characteristics of a series of energy-absorbing polyurethane elastomers, Rubber Chem. Technol. 58, 997-1013 (1985).
70. B.C. Chakraborty and D. Ratna, *Polymers for Vibration Damping Applications*, Ch.5, Elsevier (2020).
71. V.B. Møller, K. Dam-Johansen, S.M.G. Frankær, and S. Kiil, Acid-resistant organic coatings for the chemical industry: a review. J. Coat. Technol. Res. 14, 279-306 (2017).
72. R. Pekelney, *C.F. 'O' Class Submarines - Training Note Book: Electrical* - Ch. 11. pp. 11.1-11.3, San Francisco National Maritime Park Association. (2011). https://maritime.org/doc/oberon/electrical/index.htm. Accessed on May 25, 2022.
73. Technology Developed by DRDO: Press Information Bureau, Govt. of India, Ministry of defence, New Delhi (2022). https://pib.gov.in/Pressreleaseshare.aspx?PRID=1805751. Accessed on May 25, 2022.
74. Rubber Chemical Resistance Chart. MyKin. Inc. South Lion, MI, USA. https://mykin.com/rubber-chemical-resistance-chart. Accessed on May 25, 2022.
75. Z.T. Khodair, A.A. Khadom and H.A. Jasim, Corrosion protection of mild steel in different aqueous media via epoxy/nanomaterial coating: preparation, characterization and mathematical views. J. Mater. Res. Technol. 8, 424-435 (2019).
76. M.A. Osman, V. Mittal, M. Morbidelli and U.W. Suter., Polyurethane adhesive nanocomposites as gas permeation barrier. Macromolecules 36, 9851–9858 (2003).

77. E. Sancaktar and J. Kuznicki. Nanocomposite adhesives: Mechanical behavior with nanoclay, Int. J. Adhesion. Adhesive. 31, 286-300 (2011).
78. S.K. Rath, M. Patri, K. Sudarshan, P.K. Pujari and D.V. Khakhar, Reinforcement mechanism of polyurethane-urea/clay nanocomposites probed by positron annihilation lifetime spectroscopy and dynamic mechanical analysis, AIP Conf. Proc.1313, 301-303 (2010).
79. S.K. Rath, M. Patri, P.U. Sastry, A. K. Patra, and D.V. Khakhar, Structural investigations of polyurethane-urea/clay nanocomposites, AIP Conf. Proc.1349, 212-213 (2011).
80. S.K. Rath, S.S. Edatholath, T.U. Patro, K. Sudarshan, P.U. Sastry, P.K. Pujari and G. Harikrishnan, Solvent transport through hard–soft segmented polymer nanocomposites, Phys. Chem. Chem. Phys. 18, 2682-2689 (2016).
81. COMPRO EP-1 and EP-2. Solvent Free Epoxy Primer and Coating. Compro Chemicals Pvt. Ltd. Gurgaon, India. https://www.conprochemicals.com/epoxy-flooring.html. Accessed on May 25, 2022.
82. S. Burla, Barrier properties of polymer nanocomposites during cyclic sorption-desorption and stress-coupled sorption experiments, Ph.D. Thesis, Department of Engineering and Mineral Resources, West Virginia University, https://researchrepository.wvu.edu/etd/1696 (2006).
83. F. Taheri, Application of nanoparticles in adhesives: current status, in: *Handbook of Adhesive Technology*, 3rd Edition, A. Pizzi and K.L. Mittal (Eds.), pp.95-141, CRC Press, Boca Raton, FL (2018).
84. S. Kenig, H. Dodiuk, G. Otorgust and S. Gomod, Nanocomposite polymer adhesives: A critical review, Rev. Adhesion. Adhesives. 7, 93-168 (2019).
85. H. Yi , T. Wei, H. Lin and J. Zhou, Preparation and properties of hybrid epoxy/hydroxyl-terminated polybutadiene/modified MMT nanocomposites, J. Coat. Technol. Res. 15, 1413-1422 (2018).
86. D. Ratna, N.R. Manoj, R. Varley, R.K. Singh Raman and G.P. Simon, Clay reinforced epoxy nanocomposite, Polym. Int. 52, 1403-1407 (2003).
87. N. Merah, F. Ashraf and M.M. Shaukat, Mechanical and moisture barrier properties of epoxy–nanoclay and hybrid epoxy–nanoclay glass fiber composites: A review, Polymers 14, 1620 (2022).
88. W. Liu, S.V. Hoa and M. Pugh, Fracture toughness and water uptake of high-performance epoxy/nanoclay nanocomposites. Composites. Sci. Technol. 65, 2364–2373 (2005).
89. J.K. Kim, C. Hu, R.S.C. Woo and M.L. Sham, Moisture barrier characteristics of organoclay–epoxy nanocomposites, Composites. Sci. Technol. 65, 805–813 (2005).
90. M. Al-Qadhi, N. Merah and Z.M. Gasem, Mechanical properties and water uptake of epoxy-clay nanocomposites containing different clay loadings. J. Mater. Sci. 48, 3798–3804 (2013).
91. P.N. Patil, K. Sudarshan, S.K. Sharma, P. Maheshwari, S.K. Rath and M. Patri, Investigation of nanoscopic free volume and interfacial interaction in an

epoxy resin/modified clay nanocomposite using positron annihilation spectroscopy. ChemPhysChem. 13, 3916-3922 (2012).

92. P.M. Jacobs and F.R. Jones, Diffusion of moisture into two-phase polymers. Part 1 The development of an analytical model and its application to styrene-ethylene/butylene-styrene block copolymer, J. Mater. Sci. 24, 2331-2336 (1989).
93. P.M. Jacobs and F.R. Jones, Diffusion of moisture into two-phase polymers, Part 2 Styrenated polyester resins, J. Mater. Sci. 24, 2343-2347 (1989).
94. J.D. Garcia-Espinel, D. Castro-Fresno, P.G. Parbole and F. Ballester-Muñoz, Effects of seawater environment on glass fiber reinforced plastic materials used for marine civil engineering constructions, Mater. Design, 66, 46-50 (2015).
95. A. Kootsookos, A.P. Mouritz and N.A. St. John, Seawater durability of glass and carbon-polymer composites. Composites Sci. Technol. 64, 1503-1511 (2004).
96. K.H.G. Ashbee and R. Wyatt, Water damage in glass fiber/resin composites. Proc. Royal Soc. A, 312, 553-564 (1969).

Index

Also of Interest

Check out these published and forthcoming related titles from Scrivener Publishing

Biobased Adhesives
Sources, Characteristics and Applications
Edited by Manfred Dunky and K.L. Mittal
Published 2023. ISBN 978-1-394-17463-8

Structural Adhesives
Properties, Characterization and Applications
Edited by K.L. Mittal and S.K. Panigrahi
Published 2023. ISBN 978-1-394-17472-0

Rubber to Rubber Adhesion
By Dinesh K. Kotnees and Anil K. Bhowmick
Published 2021. ISBN 978-1-119-76889-0

Surface Science and Adhesion in Cosmetics
Edited by K.L. Mittal and H.S. Bui
Published 2021. ISBN 978-1-119-65482-7

Ice Adhesion
Mechanism, Measurement and Mitigation
Edited by K.L. Mittal and Chang-Hwan Choi
Published 2020. ISBN 978-1-119-64037-0

Structural Adhesive Joints
Design, Analysis and Testing
Edited by K.L. Mittal and S. K. Panigrahi
Published 2020. ISBN 978-1-119-73643-1

Progress in Adhesion and Adhesives, Volume 5
Edited by K.L. Mittal
Published 2020. ISBN 978-1-119-74806-9

Bioadhesives in Drug Delivery
Edited by K. L. Mittal, I. S. Bakshi and J.K. Narang
Published 2020. ISBN 978-1-119-64019-6

Advances in Contact Angle, Wettability and Adhesion Volume 4
Edited by K.L. Mittal
Published 2019. ISBN 978-1-119-59254-9

Adhesives for Wood and Lignocellulosic Materials
By R.N. Kumar and A. Pizzi
Published 2019. ISBN 978-1-119-60543-0

Progress in Adhesion and Adhesives, Volume 4
Edited by K.L. Mittal
Published 2019. ISBN 978-1-119-62525-4

Advances in Contact Angle, Wettability and Adhesion Volume 3
Edited by K.L. Mittal
Published 2018. ISBN 978-1-119-45994-1

Laser Technology: Applications in Adhesion and Related Areas
K.L. Mittal and Wei-Sheng Lei
Published 2018. ISBN 978-1-119-18493-5

Progress in Adhesion and Adhesives, Volume 3
Edited by K.L. Mittal
Published 2018. ISBN 978-1-119-52629-2

Textile Finishing: Recent Developments and Future Trends
Edited by K.L. Mittal and Thomas Bahners
Published 2017. ISBN 978-1-119-42676-9

Progress in Adhesion and Adhesives, Volume 2
Edited by K.L. Mittal
Published 2017. ISBN 978-1-119-40638-9

Adhesion in Pharmaceutical, Biomedical and Dental Fields
Edited by K.L. Mittal and F.M. Etzler
Published 2017. ISBN 978-1-119-32350-1

Interface/Interphase in Polymer Nanocomposites
Edited by Anil Netravali and K.L. Mittal
Published 2016. ISBN 978-1-119-18491-1

Progress in Adhesion and Adhesives
Edited by K.L. Mittal
Published 2015. ISBN 978-1-119-16219-3

Advances in Contact Angle, Wettability and Adhesion Volume 2
Edited by K.L. Mittal
Published 2015. ISBN 978-1-119-11698-1

Particle Adhesion and Removal
Edited by K.L. Mittal and Ravi Jaiswal
Published 2014. ISBN 978-1118-83153-3

Laser Surface Modification and Adhesion
Edited by K.L. Mittal and Thomas Bahners
Published 2014. ISBN 978-1-118-83163-2

Adhesion in Microelectronics
Edited by K.L. Mittal and Tanweer Ahsan
Published 2014. ISBN 978-1-118-83133-5

Advances in Contact Angle, Wettability and Adhesion Volume 1
Edited by K.L. Mittal
Published 2013. ISBN 978-1-118-47292-7

Advances in Modeling and Design of Adhesively Bonded Systems
Edited by S. Kumar and K.L. Mittal
Published 2013. ISBN 978-1-118-68637-9

Atmospheric Pressure Plasma Treatment of Polymers
Edited by Michael Thomas and K.L. Mittal
Published 2013. ISBN 978-1-118-59621-0

Atomic Layer Deposition
Principles, Characteristics, and Nanotechnology Applications
By Tommi Kääriäinen, David Cameron, Marja-Leena Kääriäinen and Arthur Sherman
Published 2013. ISBN 978-1-118-06277-7

Atmospheric Pressure Plasma for Surface Modification
By Rory A. Wolf
Published 2012. ISBN 9781118016237